Heinrich Fritsch

The Diseases of Women, a Manual for Physicians and Students

Heinrich Fritsch

The Diseases of Women, a Manual for Physicians and Students

ISBN/EAN: 9783744718226

Printed in Europe, USA, Canada, Australia, Japan

Cover: Foto ©berggeist007 / pixelio.de

More available books at **www.hansebooks.com**

THE
DISEASES OF WOMEN

A MANUAL FOR PHYSICIANS AND STUDENTS

BY

HEINRICH FRITSCH, M.D.,
PROFESSOR OF GYNECOLOGY AND OBSTETRICS AT THE UNIVERSITY OF HALLE

TRANSLATED BY

ISIDOR FURST

WITH 159 WOOD ENGRAVINGS

NEW YORK
WILLIAM WOOD & COMPANY
56 & 58 LAFAYETTE PLACE
1883

CONTENTS.

	PAGE
CHAPTER I.—*Anatomy*	1
A. External Genitals	1
B. Internal Genitals and Pelvic Organs	3
C. Urethra, Bladder, Ureters	8
D. Rectum	9
E. Vessels and Nerves	9
F. Topographical Remarks	11
CHAPTER II.—*Physiology*	15
A. Menstruation and Ovulation	15
B. Fecundation and its Consequences	20
C. Senile Involution	21
CHAPTER III.—*General Diagnosis*	23
A. Anamnesis	23
B. General Remarks on Examinations	25
C. The Touch	26
D. Combined Examination	27
E. Examination from the Rectum	28
F. Examination with the Sound	29
G. Examination with Specula	32
H. Diagnostic Dilatation of the Uterus with Tents	38
I. Mechanical Dilatation of the Uterus	41
CHAPTER IV.—*Gynecological Antispesis*	44
A. Prophylaxis	44
B. After-Treatment	46
C. Treatment of Septic Gynecological Wounds	47
D. Permanent Irrigation	50
E. Antisepsis during Laparotomies	53
F. Concluding Remarks	54
CHAPTER V.—*General Therapeutics*	55
A. General Considerations in Gynecological Operations	55
B. Vaginal Irrigations	56
C. Tamponade	57
D. Vaginal Suppositories	59
E. The Abstraction of Blood	59
F. The Application of Caustics	61
G. The Employment of the Actual Cautery	65
H. The Effect of Remedies on the Internal Surface of the Uterus	67
CHAPTER VI.—*Diseases of the Vulva*	71
A. Malformations	71
B. Inflammation of the Vulva (Vulvitis)	74
Etiology and Pathological Anatomy	74
Symptoms and Course	77
Diagnosis	78
Treatment	79

	PAGE
C. New-Formations of the Vulva.	80
Papilloma	80
Carcinoma	81
Elephantiasis	82
Rarer New Formations on the Vulva	83
D. Lesions of the Vulva (Perineal Ruptures)	84
CHAPTER VII.—*Diseases of the Vagina*	92
A. Inflammation of the Vagina (Colpitis, Vaginitis)	92
Etiology	92
Anatomy	93
Symptoms and Course	96
Diagnosis	96
Prognosis, Treatment	97
B. Cysts of the Vagina	99
C. New-Formations of the Vagina	99
D. Vaginismus	100
E. Recto-vaginal Fistulæ	102
CHAPTER VIII.—*Diseases of the Bladder and Urethra.*	104
A. Fissures and Ectopiæ	104
B. Inflammations of the Bladder	106
Etiology	106
Anatomy	107
Symptoms	108
Diagnosis and Prognosis	110
Treatment	110
C. New-Formations of the Bladder	113
D. Lesions of the Bladder (Fistulæ)	115
Diagnosis and Prognosis	118
Treatment	119
E. Diseases of the Urethra	127
CHAPTER IX.—*Uterine Malformations, Arrests of Development, and the Gynatresiæ*	130
A. Malformations	130
B. Defect of the Uterus	135
C. Infantile Uterus	136
D. Stenoses of the Uterus	136
Stenosis of the External Os	136
Stenosis of the Internal Os	140
E. The Gynatresiæ	142
Symptoms and Results	145
Diagnosis	146
Prognosis and Treatment	147
CHAPTER X.—*Inflammation of the Uterus (Acute Metritis, Chronic Metritis).*	151
A. General Remarks	151
B. Acute Metritis, Etiology, Anatomy	152
Symptoms and Course, Diagnosis and Prognosis, Treatment	153
C. Chronic Metritis	154
Etiology	154
Anatomy, Symptoms, Course	155
Diagnosis and Prognosis	157
Treatment	158

CONTENTS.

	PAGE
CHAPTER XI.—*Diseases of the Endometrium*	161
A. Acute Endometritis	161
B. Chronic Endometritis	161
Etiology and Anatomy	161
Symptoms and Course, Diagnosis and Prognosis	163
Treatment	164
C. Pathological Conditions of the Cervix Uteri	168
Etiology	168
Anatomy	169
Symptoms and Cure, Diagnosis	173
Treatment	175
CHAPTER XII.—*Dislocations of the Uterus*	179
A. Anteflexion	179
Etiology and Anatomy	179
Symptoms and Course	182
Diagnosis and Treatment	184
B. Anteversion	188
Etiology and Anatomy	188
Symptoms and Course	189
Diagnosis and Prognosis, Treatment	190
C. Retroversion	191
Etiology and Anatomy	191
Symptoms and Course	192
Diagnosis, Prognosis, Treatment	192
D. Retroflexion	194
Etiology and Anatomy	194
Symptoms	195
Diagnosis	197
Treatment	200
E. Prolapsus of the Uterus	204
1. Isolated Descensus of the Vaginal Wall	204
2. Descensus of the Vaginal Wall with Descensus of the Uterus	206
3. Primary Descensus of the Uterus with Inversion of the Vagina	207
4. Prolapse of the Uterus due to Increased Pressure from Above, etc.	208
Anatomy	208
Symptoms and Course	212
Diagnosis	212
Treatment	213
F. Inversion	222
Etiology	222
Anatomy, Symptoms, Course	222
Diagnosis and Prognosis	223
Treatment	224
G. The Rarer Displacements of the Uterus	225
CHAPTER XIII.—*New-Formations of the Uterus*	226
A. Myomata	226
Symptoms and Course	230
Diagnosis	233
Treatment	237
B. Carcinoma of the Uterus	240
Anatomy	240

THE DISEASES OF WOMEN.

CHAPTER I.

ANATOMY.

A. External Genitals.

The lower end of the abdomen is limited by the *mons Veneris* (Fig. 1, *n*), situated over the bony symphysis of the pelvis. It is a subcutaneous accumulation of fat peculiar to the female sex. At puberty it becomes covered with hair.

Below, in a sagittal direction, lies the *pudendum muliebre*, the *vulva* (Fig. 1). It is formed, in the first place, by the *labia majora* (Fig. 1, *a*), the analogues of the scrotum. In virgins, the labia majora are in contact; in multiparæ and in nulliparous prostitutes they are gaping. Posteriorly, the labia majora converge and imperceptibly blend with the surroundings; at times they are joined by a cutaneous fold which becomes especially distinct when they are drawn apart. Above the latter is a depression, the fossa navicularis (Fig. 1, *k*). Between the labia majora are the *labia minora* (Fig. 1, *d*), having but half the length of the former.

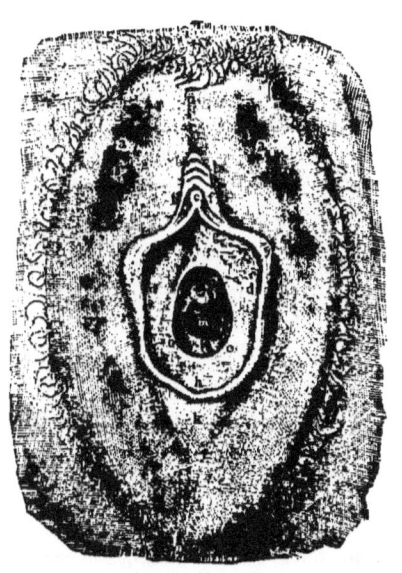

Fig. 1.—External Genitals (schematic). *a a*, Labia majora; *b*, præputium clitoridis; *c*, clitoris; *d d*, labia minora; *e e*, openings of the efferent ducts of Bartholin's glands; *ff*, the upper ends of the labia minora coalescing into the præputium clitoridis; *g g*, the crura of the clitoris, likewise springing from the labia minora; *h*, the frenulum, the posterior commissure of the labia minora; *i*, the opening of the urethra; *k*, fossa navicularis; *l*, hymen; *m*, introitus vaginæ; *n*, mons Veneris.

They are cockscomb-like, not very thick cutaneous lobes of varying length and breadth. In virgins they can be made visible only by separating the labia majora. In multiparæ and in women excessively indulging in sexual intercourse, the labia minora project far outward from between the labia majora, in the shape of triangular, wrinkled, brownish-colored, flabby cutaneous lobes devoid of fat. Posteriorly the labia minora are frequently joined by a membranous fold, the frenulum (Fig. 1, h). This fold is variable, so that some authors ascribe it to the labia majora. After a perineal rupture has destroyed the frenulum or the posterior end of the vulva, the labia minora are seen exposed, and between them the transversely corrugated, descended vaginal wall. Above, the labia minora divide into two pair of crura, the upper of which forms, as it were, a roof above the clitoris (Fig. 1, b), as the præputium clitoridis (Fig. 1, ff). The lower or inner pair joins the clitoris as the crura clitoridis (Fig. 1, gg). Posteriorly at e are the external openings of the efferent ducts of Bartholin's glands.

The *clitoris* (Fig. 1, c) represents a miniature penis. It is formed by two converging cavernous bodies springing from the pubic arch. It is erectile, is amply provided with vessels and nerves, and in its erect condition it doubles its size.

The *urethra* (Fig. 1, i) terminates in the triangle limited by the clitoris, its crura, and the introitus vaginæ. The urethral orifice is often quite smooth, often too it is fringed, irregularly surrounded by processes of the mucous membrane which look like hymenic remnants.

In front of the entrance to the vagina, but back of the fossa navicularis and the vestibule, a duplicature of the mucous membrane, usually concave externally, is expanded—the *hymen* (Fig. 1, l). The thickness and form of this membrane are variable; it may completely occlude (*h. imperforatus*) the introitus (Fig. 1, m), it may be attached all around, or it may have several, usually two adjoining openings (*h. cribriformis*). Coition causes several lacerations of the hymen, but its form in the main is preserved. Parturition, however, destroys the hymen, nothing remaining to mark its place but some quite small membranous prominences termed *carunculæ myrtiformes*. Although this is the usual condition, the presence of larger portions or even of the *entire* hymen does not absolutely exclude the possibility of a previous labor.

Posteriorly at the vulva, near the posterior commissure, is a lobulated gland, that of Bartholin, also termed the vulvo-vaginal or vulvo-vestibular gland. It is often irregular in form, is from fifteen to twenty millimetres long, and has an efferent duct which terminates near the middle of the vestibular wall where the hymen arises (Fig. 1, ee). This gland secretes a liquid even in childhood. In virulent catarrhs the gland often becomes inflamed and suppurates.

B. Internal Genitals and Pelvic Organs.

Above the hymen, and extending up to the uterus, is the *vagina*, a muscular tube which in its superior concavity nearly reproduces the curvature of the posterior pelvic wall. The vagina is from seven to eight centimetres long. If measured with a rule from the posterior commissure, the vagina being stretched as much as possible, we obtain a length of twelve centimetres. The anterior wall is shorter than the posterior, representing a segment of a circle concentric with the latter, but with smaller radius. Above, the vagina is joined to the vaginal portion of the uterus, with which the vaginal mucosa blends. Below, the vagina is connected firmly with the rectum and urethra, more loosely in its upper portion. The vagina is excessively distensible and contains in its wall numerous plexuses of veins. On dividing the lower third of the vagina transversely, we observe that the anterior and posterior walls project between the parallel lateral walls, approximating the shape of an H. Above, the vagina is smoother and its walls thinner than below. But in the immediate neighborhood of the uterus the walls again grow thicker. Below anteriorly and posteriorly are the so-called columnæ rugarum, ridges projecting into the lumen, which become somewhat flattened by coition and childbirth.

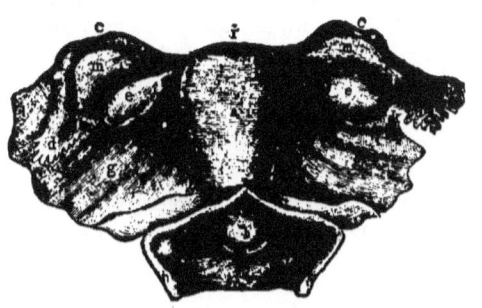

FIG. 2.—Internal Genitals. *a*, Body of the uterus; *b*, vaginal portion with the external os; *c*, tube; *d*, abdominal opening of the tube, fimbriæ; *e*, ovary; *f*, ovarian ligament; *g*, broad ligament; *h*, vagina; *i*, fundus of the uterus; *k*, ovarian fimbria of the tube; *l*, infundibulo-pelvic ligament; *m*, mesosalpinx; *s*, ala vespertilionis, inclosing the parovarium between its two plates.

In descent of the vagina, the columnæ are pendulous and may even become visible as narrow or broad projections. That part of the vagina which lies above the distinctly palpable edges of the levator ani has thinner walls and is not tubular, but has somewhat the form and position of the empty bladder. The wall of the vagina contains both elastic and muscular fibres, the latter of which form a circular accumulation around the introitus—the sphincter cunni. The superficial membrane of the vagina is in reality not a mucous membrane, but a cutis, for it is devoid of glands. Although glands have been found in some vaginæ—a fact, by the way, which has not been firmly established—their number is at all events very small; too small to make the secretion of the glands of any importance. The moisture of the vagina is due rather to the uterine secretion; in part, perhaps, also to that of Bartholin's glands. If inversions and prolapsus have rendered this moistening of the vagina impossible, it becomes dry and assumes the character of the external skin.

The *uterus* (Fig. 2, $\imath\ a\ b$), is exactly pear-shaped and is united to the vagina in such a manner that its lower part projects more or less into the latter; hence the term, the *vaginal portion of the uterus*. The uterus is divided into the *body* and the *cervix*, the demarcation being at the internal os. Externally in front, the limit of the body and cervix is the point of duplicature of the peritoneum; internally, the termination of the palmæ plicatæ, superficial folds of mucous membrane.

The uterus consists of smooth muscular fibres. The uppermost layer, situated like a cap above the fundus, its intimately connected with the peritoneum, so that the latter cannot be divided anatomically. From this upper layer bundles of muscular fibres extend to all the ligaments springing from the uterus, and to the entire pelvic peritoneum. The middle layer, which is the thickest and intimately interlaced, contains the greatest number of vessels. The inner layer forms ring-shaped accumulations of muscular fibres around the termini of the tubes and the internal and external os. The cavity of the uterus ends below at the *external os* (Fig. 2, *b*), called also os tincæ; it is invested with a mucous membrane provided with vibratile epithelium and pierced by numerous tubular glands likewise covered with vibratile epithelium. The mucous membrane blends immediately with the musculature, so that it cannot be separated from it with the knife. Strictly, the entire uterine muscle must be conceived as muscularis mucosæ uteri. The periglandular tissue is uncommonly soft and consists of roundish connective-tissue cells. For this reason, the mucous membrane readily permits an unequal repletion with blood. The uterine mucosa or its glands secrete a glairy mucus which often depends from the external os in the shape of a viscid drop. In the cervical canal the glands have more of an acinous quality. Especially in the puerperium, during pregnancy, and in pathological hypertrophies of the vaginal portion, numerous acinous gland are found in the cervical canal. They occur also on the external surface of the vaginal por-

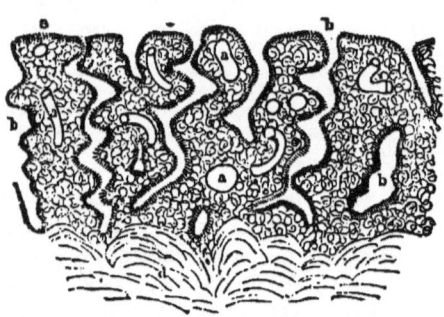

FIG. 3.—Uterine mucosa in a somewhat hyperplastic condition. *a*, Vessels in section; *b*, glands, tubulous, convoluted; *c*, tissues of the mucous membrane; *d*, musculature of the uterus; *e*, surface of the uterine cavity.

tion, where some observers consider them as pathological, others as physiological formations.

The *peritoneum* invests the entire pelvic cavity; the latter is divided into two parts in woman, for the uterus is, as it were, pushed upward from below, forming a fold in the peritoneum. The septum consists of the uterus and the broad ligaments (Fig. 2, *g*). These extend laterally from the uterus to the bony pelvic wall, directed upward and somewhat posteriorly. Between the layers of this ligament the vessels and nerves extend to the uterus. The posterior layer of the broad ligaments extends farther downward, especially in the middle. Here the peritoneum covers even the upper part of the vagina, while in front, at the level of the internal os, it is reflected on to the bladder. Looking behind the uterus, we see in the first place two lateral, triangular flat spaces, limited in front by the broad ligament, laterally and posteriorly by the pelvic wall, medially by a semilunar fold extending from the uterus—the fold of Douglas, plica semilunaris Douglasii. Between the two semilunar folds we penetrate deeply into the fossa of Douglas.

The folds of Douglas spring from the posterior surface of the uterus, at about the level of the internal os, where the uterus is normally flexed forward. At the uterus both folds blend with each other, so that on drawing at the fold we observe a semicircle at the uterus with the concavity backward. Posteriorly the folds likewise converge, thus frequently producing a perfectly circular opening leading into Douglas's fossa. The nucleus of the folds of Douglas is formed by muscular fibres derived from the uterus; they cannot, however, be followed as far as the bones of the pelvis. Therefore, the name utero-sacral ligaments is false, because it is likely to give rise to the erroneous assumption that these folds or ligaments fasten the uterus to the sacrum. The muscular fibres radiate rather into the peritoneum, becoming progressively sparser above. Therefore, on drawing on the uterus at the one point of attachment of the ligaments of Douglas, we draw on the entire dorsal parietal peritoneum, the other point of attachment of the folds of Douglas. In the fossa of Douglas lies the empty large intestine which is often lifted out in postmortem examinations.

The space in front of the uterus is termed the vesico-uterine excavation. However, there is no free space here, for the uterus rests directly upon the bladder. The much described utero-vesical ligaments do not exist.

Three strands spring from the upper lateral margin of the uterus and extend to the broad ligaments:

1. The *round ligament* or *ligamentum teres*, from twelve to fifteen centimetres long, is a band composed of smooth muscular and elastic fibres. It extends first into the broad ligament in an external and anterior direction, penetrating the inguinal canal in company with the epigastric artery. Externally it radiates into the tissue of the labium majus. The round ligament is the former inguinal ligament of the Wolffian body, and is of

particular importance in the exact determination of pathological conditions. For instance, if we find a left uterus unicornis, and, at some other spot in the abdomen a muscular body from which a distinct round ligament extends downward, then that body is the rudimentary right horn of the uterus.

2. The *oviduct*, the *salpinx*, the *tube*, the *Fallopian tube* (Fig. 2, *c*). It extends in a wavy line superiorly and externally, as if its retaining band, the *mesosalpinx, ala vespertilionis* (Fig. 2, *m*), were too short. Three portions are to be distinguished in the tube: the interstitial part blending with the uterine parenchyma, the abdominal part, and the ampulla terminating in the fimbriæ. From within outward the tude widens, and contains cylindrical epithelium vibrating toward the uterus. While in its interstitial part the mucous membrane is merely folded, these folds progressively increase outwardly, so that completely felted, dendritic anastomoses fill the lumen. At the terminus, the interlacings of the mucous membrane proliferate out of the ostium abdominale, giving rise to numerous villi, the fimbriæ. One especially large and long fimbria fastens the abdominal opening to the ovary, whence it has received the name of ovarian fimbria (Fig. 2, *k*). Not uncommonly we find even in advance of the abdominal opening defects in the wall from which the villi of the mucous membrane project: an accessory ostium. At the end of the fimbriæ is the terminal hydatid which was formerly explained as the end of Müller's duct, but is by Waldeyer believed to be an attenuated portion of Müller's canal. The tubal mucosa is very vascular; separated by a circular layer of connective tissue (Fig. 4, *c*), it surrounds a muscularis (Fig. 4, *b*) which again is covered by peritoneum. The power of resistance of the tube against centrifugal pressure does not appear to be considerable, inasmuch as nearly all tubal pregnancies lead to rupture in the first months, and as in all accumulations of blood in the tube (hæmatosalpinx) a break occurs in it with especial facility.

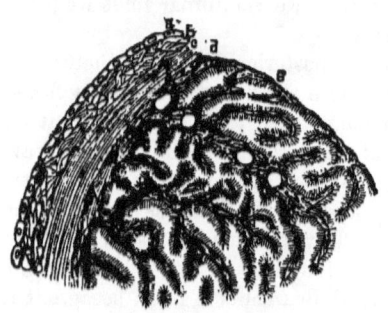

FIG. 4.—A piece of the wall of the tube, ampulla. *a*, Peritoneum; *b*, longitudinal fibrous layer, muscularis; *c*, circular fibrous layer; *d*, mucous membrane; *e*, villi of the mucous membrane, covered with cylindrical epithelium, seen in transverse section to the villous vessels.

For a long time a dispute has been raging as to whether the tube is patulous or not. That may be decided to this effect, that in post-mortem preparations fluid may indeed be injected from the cavity of the uterus through the tube, without particularly great pressure and without tying the cervix around the canula. Likewise, in disease of the tubal mucosa and in dilatation of the tube, indubitable cases of that nature are well known. But in these the necessary force of the stream is probably much greater. But it is very questionable whether fluid penetrates while the mucous membrane is in perfect health and the tube of normal width, *during life*. Especially as the liquids employed in treatment excite the peristalsis of the tube, or coagulate the mucus within it, that danger is certainly not present for the physician who proceeds according to the rules of his art. On the other hand, with rude force everything is obviously possible. If, with a wide uterine ostium of the tube, the canula of a syringe project immediately into the tube and injection be vigorously made, of course every fluid will penetrate through the tube.

3. The *ovarian ligament* (Fig. 2, *f*) extends from the upper uterine canal, most generally in a posterior direction. This ligament fastens the ovary to the uterus. The *ovary* (Fig. 2, *e*) is not covered by peritoneum. In the neighborhood of the hilus the peritoneum ends frequently in a perfectly straight, often in a somewhat jagged but sharp line which may be observed equally well in recent and old preparations. Judging from pathological conditions, the ovary is sometimes more, sometimes less deeply inserted into the peritoneum. At least tumors of the ovary occur which are not at all, others which to but a slight extent are covered with peritoneum, and again others which are intraligamentous almost in their greater portion.

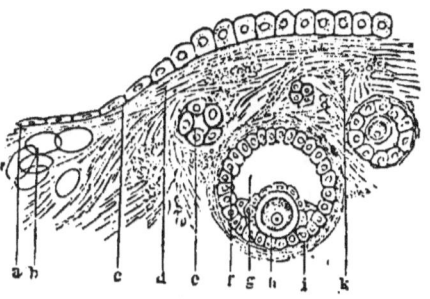

Fig. 5.—Ovary. *a*, Flat epithelium of the peritoneum; *b*, fat cells in the peritoneum; *c*, transition of peritoneal into germinal epithelium of the ovary; *d*, dense connective tissue, the so-called albuginea; *e*, unripe follicle; *f*, epithelium of the Graafian follicle, membrana granulosa; *g*, liquor folliculi; *h*, ovum lying in the cumulus proligerous, an accumulation of epithelium of the Graafian follicle, zona pellucida, yolk, vesicle, and macula germinativa; *i*, theca folliculi; *k*, connective tissue of the ovary, longitudinal and transverse fibres.

The ovary is invested with the so-called germinal epithelium (Fig. 5, *c*) which in the ovarian fimbria continues into the epithelium of the tube.

On section of the ovary, the external layer forms an accumulation of connective-tissue fibres which is so dense that it was formerly considered to be a special albuginea (Fig. 5, *d*). Next follows the parenchymatous layer containing the most essential formations of the ovary—the follicles. These, in their unripe condition (Fig. 5, *e*), are grouped together; when farther developed they represent more isolated small cysts (Fig. 5, *g*) invested with cylindrical epithelium. The smaller or more recent cortical or primordial follicles are close together, in groups near the surface; the larger, riper, more fully developed ones are situated more deeply and are less numerous.

The lesser follicles are completely filled with small cells so that there is no free space. The vicinity is deficient in vessels. Around the larger follicles a vascular network can be demonstrated. Through a separation of the cells in the follicles, there gradually arises a cavity which enlarges by degrees. Those follicles are regarded as fully ripe which have reached a considerable size. Within them is a free cavity containing the liquor folliculi (Fig. 5, *g*). The wall is covered with layers of epithelium, the so-called membrana granulosa (Fig. 5, *f*). In the latter again is found, usually at the side farthest from the surface of the ovary, an accumulation of cells: the cumulus proligerus s. ovigerus (Fig. 5, *h*). It incloses the ripe ovum, to which, even after its expulsion, some cells of the membrana granulosa still adhere. The ovum consists of a pellicle—the membrana pellucida, perhaps formed of the above cells of the granulosa. Within is the vitellus, and inside the latter, likewise situated eccentrically, the vesicula germinativa with the macula germinativa (Fig. 5, *h*).

Surrounded by the parenchymatous layer—the layer of follicles—is the stroma: the hilus, the interstitium of the ovary which, as vascular connective tissue, carries the blood-vessels in between the follicles. It contains no muscular fibres besides those of the vessels; therefore, myomata of the ovary are not likely to occur. Lymphatic vessels, arteries, veins derived from the spermatic, and sympathetic nerves enter the ovary at the hilus.

Between tube and ovary, between the two plates of the ala *vespertilionis* (Fig. 2, *m*), lies the *parovarium*, the *organ of Rosenmüller;* it consists of a number of connective-tissue canals invested with vibratile epithelium. Sometimes one of these canals is especially large, so that it projects from the parovarium, piercing the peritoneum, in the shape of a small pediculated cyst-hydatid. According to Waldeyer, the parovarium extends into the hilus of the ovary. Kölliker makes it appear even probable that the cells of the membrana granulosa originate from the parovarium, so that the ovary would be formed by a sort of intergrowth of the germinal gland with the Wolffian body. This view would have great influence on the doctrine of the growth of ovarian tumors.

C. Urethra, Bladder, Ureters.

The *urethra*, the most inferior part of which is firmly attached to the bone, runs normally in a vertical direction, horizontally if the woman is recumbent. The urethra may become bent, so that the upper part forms with the lower an angle open inferiorly. The *bladder*, situated between the uterus and vagina, does not contract in a spherical form like the puerperal uterus, but its walls during life appose themselves in the same way as an exsected bladder, that is to say, the upper wall rests upon the lower like two plates set one on the other. When the bladder expands, it may assume any shape by pressure. Posteriorly the bladder is united with the uterus and the vagina by a loose connective tissue which may become the seat of inflammations. Farther down, particularly from the level of the internal end of the urethra, the latter is so intimately connected with the vagina that it can no longer be separated from it.

The *ureters* open into the bladder. Familiarity with their course is very important for a great many operations. The ureters proceed downward along the sides of the pelvis, and at the base of the broad ligaments they join the bladder through the parametria. To be sure, the distance between both ureters at their entrance into the bladder is three centimetres, and at the insertion of the uterus into the vagina eight or even ten centimetres, but still the ureter, especially the left which physiologically lies rather more medially, may be drawn close to the uterus by inflammations in the parametrium. Its lesion leads to fistulæ which are very hard to cure.

D. The Rectum.

The *rectum* forms a large space above the anus—the ampulla recti, which extends far forward, so that the finger at the anterior end of the ampulla feels the lower surface of the body of the uterus. The upper wall rests upon the lower, and usually there is but little fecal matter between the soft folds of mucous membrane. At about the level of the internal os commences the tube of the rectum; it ascends irregularly, generally at the left, but often also at the right or in the median line. When the rectum is empty, it lies partly collapsed in Douglas' cul-de-sac. The well-filled rectum pushes the uterus forward. Its continuation, the sigmoid flexure, may, when distended by fæces, reach in the form of a hook as far as the right iliac fossa.

E. Vessels and Nerves.

The afferent and efferent vessels of the female organs of generation are very numerous and are derived from various sources.

In the first place, the two spermatic arteries run through the broad ligaments to the uterus and supply the tubes, ovaries, and fundus uteri with an ample, numerously anastomosing vascular network. Secondly, the uterine artery, derived from the hypogastric, extends laterally through the broad ligament to the point of junction of the uterus and cervix; frequently it is found double on the left side. From the hypogastric are also derived a number of smaller arteries which supply the vagina (vaginal arteries). The vessels of the external genitals come from the common pudic artery. Finally, the epigastric artery sends to the uterus a branch which goes to the upper uterine angle together with the round ligament. These arteries anastomose largely with each other, and alongside of them a great number of valveless veins with multiple intercommunications extend to the points of origin of the arteries; that is to say, the pampiniform plexus upward to the spermatic, the uterine plexus laterally outward to the hypogastric which also receives the vaginal plexus, and a third set with the round ligament outward to the abdominal coverings.

In the neighborhood of the veins and arteries, lymph-vessels likewise course through the broad ligaments.

Hence the uterus is not an organ such as the liver or kidney which, being nourished through a hilus by *one* large arterial trunk, could become diseased alone, without direct implication of the surroundings. A hyperæmia in any *isolated* portion of the uterus is inconceivable; during an arterial congestion toward the sexual organs, the uterus and its annexa must always participate equally. On the other hand, during venous stases in the region of the inferior vena cava, especially in that of the hypogastric vein, the uterus will always be implicated. Particularly at the sides of the uterus the limit of the organ will scarcely be clearly demonstrable. Accordingly, in order to deplete the uterine vessels, the general abdominal plethora must *first* be diminished; and *secondly*, besides direct withdrawal of blood from the uterus, we may, by abstracting blood from the external genital organs, the perineum, or the epigastric—in the inguinal region—act upon the amount of blood in the uterus.

This impossibility of conceiving the uterus as an isolated organ becomes particularly evident also by considering the relations of its muscle. The superficial muscular cap is so intimately connected with the peritoneum that the latter cannot be pulled off. Therefore, an isolated affection of the uterine parenchyma, or of the peritoneum situated above it, is impossible. If we consider the circumstance that the musculature of the uterus is continued into all the ligaments starting from it, it will be evident that the surroundings of the uterus cannot fail to become involved in its inflammatory affections. At any rate, an inflammation having a pronounced progressive character will never find its limits at those of the uterus.

Nor is the uterine mucosa separated by a connective-tissue layer from the musculature, as is the case in that of the stomach or intestine. The mucous membrane here and there sinks with the bases of the glands deeply between the muscular bundles of the uterus, so that in general hyperæmia of the uterus the mucosa cannot remain exempt. For we may observe every day that during menstruation the general congestion rapidly finds expression in the mucous membrane by rupture of vessels.

All these facts form the anatomical explanation of the peculiar circumstance that an inflammation of the uterus hardly ever remains confined to it, or that in simple plethoric conditions the uterus participates so frequently—an injunction on every gynecologist to combine general treatment with the local.

The *nerves of the uterus* and ovary are derived from the sympathetic, namely, its lateral hypogastric plexus which is situated within the broad ligaments by the side of the uterus. The third and fourth sacral nerves also send a few twigs to the tube and the ovary.

Of greater importance is the knowledge of the pelvic nerves which may perhaps be made to suffer by pressure or involvement in inflammations. These are especially the nerves of the lower extremities springing from the sacral plexus. The fourth and fifth lumbar and the sacral nerves issuing from the foramina in the sacrum, and the superior coccygeus, together form the sacral plexus. The upper nerves cross the innominate line in their course over the wing of the sacrum. Here irritation of the nerves and neuralgiæ may be set up by the pressure of an obstetric instrument, the inflammation in the train of a wound, or the pressure of a long persisting exudation.

The parts of the sacral plexus lying in the pelvic cavity may also become irritated, so that not infrequently we observe neuralgiæ and pareses in all regions of the lower extremities in consequence of inflammations of the pelvic connective tissue.

F. Topographical Remarks.

The position of the female genital organs in the living subject can be determined only by the examination of the living woman. It needs no further discussion that no cadaveric sections, with or without previous freezing, are ever able *alone* to give information regarding the position during life. It is B. S. Schultze's great merit to have secured general assent to this principle, as well as to have given a new, and assuredly the right, direction to the doctrine of the position of the uterus.

The only objection that could be raised against examination of the living is that the finger in the vagina, as well as the hand palpating externally, unconsciously displace the genital organs. In the place occupied by the one or two fingers in the vagina, were previously the apposed vaginal walls; it is natural, therefore, that the fingers displace them. But in proportion as the dexterity to be demanded of the gynecologist is greater, this source of error becomes less. Moreover, as the fingers have the same curvature as the vagina, the latter is not at all displaced. At most, a collapsed, somewhat shortened vagina becomes stretched by the fingers. In the recumbent woman, the vagina lies almost horizontally, concave superiorly. Between it and the uterus, overspreading the pelvic contents, lies the plate-shaped empty bladder; immediately upon it, the uterus, so that in the normal position of the uterus there is no free space between uterus and bladder. The posterior surface of the uterus extends upward and backward. The fundus is directed toward the upper margin of the symphysis. The vaginal portion extends backward and downward. It lies about at the point of junction of the sacrum with the coccyx. As there is an angle at the internal os, the vaginal portion points somewhat downward; the entire uterus lies below the pelvic inlet. Behind the uterus, the rectum descends on the left, but often extend beyond the middle line, so that in sections the part above the anus is situated entirely on the right side.

It is obvious that, if the uterus rests upon the empty bladder, the latter must raise the former while it is filling. But if we bear in mind that the bladder has very thin walls; that—as we may ascertain by catheterization during laparotomy—it is easily expansible in all directions, it can be readily understood that the urine may also give way in front, to the right, and to the left. During digital examination it is always possi-

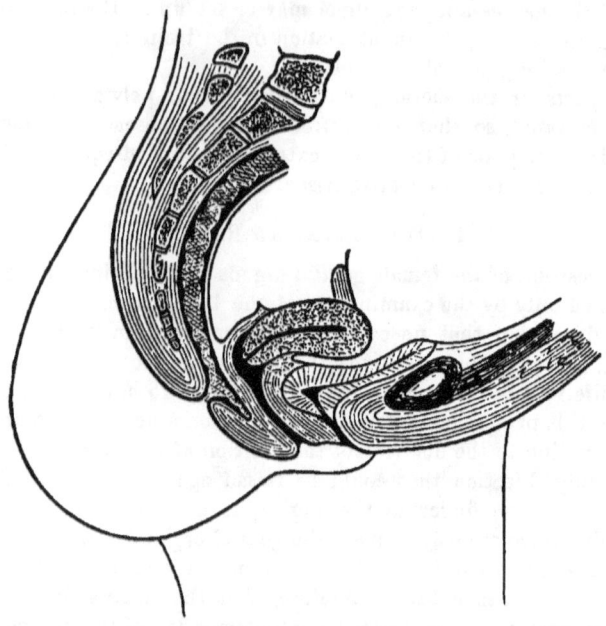

Fig. 6.—Position of the Uterus (schematic), bladder and rectum being empty. Designation by letters has been omitted so as not to disturb the view. The plate-shaped bladder is seen in section; resting upon it, the physiologically moderately anteflected uterus. On the latter, in front and above, the ovary and broad ligament; posteriorly, the point of origin of the fold of Douglas. Posterior to the vagina, the ampulla of the rectum.

ble to displace the urine when the bladder is full and to reach the anterior surface of the uterus. Thus the uterus is not lifted as if, perhaps, a firm ball were pushed under it, but it is elevated only when the filling is coniderable, and even then but slightly. Strong intra-abdominal pressure, as for instance in retention of fæces, will have a disturbing influence on the elevation.

In like manner, by great fæcal accumulation, the uterus is crowded forward a little.

It is of particular importance to remember that, *firstly*, the uterus *in toto* is displaceable in all directions, upward and downward; *secondly*, that it may be moved in such a manner that the upper longer arm of the lever, the body, imparts motion to the lower smaller arm, the cervix—in the opposite direction—and inversely, the cervix to the body; and *thirdly*, that the uterus is mobile in itself, so that there is a cessation of the lever relation of the movements, and that the fundus alone moves in any direction.

All those notions, to a slight degree, occur physiologically.

When standing erect, the uterus sinks downward, even in the virgin, thus shortening the vagina. And during coition the uterus is elevated as much as five or six centimetres without giving pain. Likewise without causing any considerable pain, can the nulliparous uterus be drawn down close to the introitus vaginæ. These changes in position are very evident in the lax pelvic organs of a multipara. When examining such a one when recumbent with the pelvis elevated, the uterus often sinks so far upward that the finger but barely reaches the vaginal portion; and when exploring in the same woman while erect, especially when made to bear down, the vaginal portion comes so low down as to be reached when two joints of the finger are inserted into the vagina.

Therefore, we cannot speak of a "fastening by ligaments," in the ordinary sense, with such considerable physiological mobility. On the contrary, *that condition* is pathological in which the above-described ligaments become really tense. For physiologically all these ligaments are quite relaxed; they resist excesses of motion, but by no means do they hinder the just-related extensive physiological displacements upward and downward.

The uterus is fixed between the abdominal and pelvic organs somewhat like a book is held between two apposed hands. On loosening the lower hand, the book becomes movable and sinks downward: pressing with the upper against the slightly yielding lower hand, the book is likewise displaced downwards. Changing the surfaces of the hands to horizontal without holding the book tight, inclined planes occur which allow the book to glide down. Just so the uterus is interposed between the floor of the pelvis and the intestines. As long as the pressure from above and the resistance from below are physiologically in equilibrium, so long the uterus lies in normal position. The ligaments do not immediately enter into the consideration. But if, for instance, the pelvic floor relaxes, or the pressure becomes unilateral, or generally excessive from above, the uterus will change its position and drag upon one of its ligaments. If the ligament is stretched, it in turn drags upon the organs to which it is fastened. That is the peritoneum. Finally it must not be forgotten that the position of an organ is also dependent on its form. If we place a sphere on an inclined plane it will roll down, but a cube will remain at rest. If the sphere be heavy, its intrinsic weight will favor the speed of its rolling. Thus the form and absolute gravity of the uterus will have some influence upon its position.

While nearly all modern gynecologists accept the above-described position of the uterus as correct, there is *still some dispute about the position of the ovaries and the tubes.* If we again deduce the normal position from the condition found during life, we shall reach the positive conviction that he ovaries are not hidden behind the uterus, but that they are placed laterally at the pelvic brim. But here it is particularly difficult to eliminate the sources of error—the displacement by the examining fingers from within and without. Therefore it is hardly possible to determine with absolute certainty the direction of the axis of the ovary, that is, its longest diameter.

His has chosen a different course: hardening the muscles previous to the anatomical examination by injecting dilute chromic acid into the vessels. In this manner he found that the tube surrounds the ovary like a tie with two ends. The end of the tube nearest the uterus rises upward, while the other again descends. Between the two lies the ovary in such a manner that its longest diameter points vertically. The medial end of the ovary, to which the ovarian ligament is fastened, is directed downward and backward; the lateral end, upward and forward. The free margin of the ovary points backward, the hilus forward. One surface, the lateral, of the ovary adjoins the pelvic wall, while the other touches the uterus below, the tube above. At all events, the fastenings of the ovary permit great mobility: in dislocations of the uterus, the latter, by drawing on the ovarian ligament, must exert some influence on the position of the ovary. Especially during retroflexions the ovary must be considerably dragged downward. Regarding the position of the bladder and rectum, whatever is necessary has been said on page 8. Some anatomical explanations in reference to the pelvic connective tissue are added to the description of the inflammation of these formations.

CHAPTER II.

PHYSIOLOGY.

A. MENSTRUATION AND OVULATION.

THE ancients already sought for an explanation of the surprising phenomena of menstruation. They believed that the body of the woman purified itself of certain substances. Even to-day the [German] people have still retained the expression "monthly purification." These substances, a kind of excess of strength, were said to be destined to build up the child during pregnancy. After birth, the lochial secretion was thought to serve in carrying off matter which should not remain in the body under physiological conditions. The more amply the lacteal secretion was formed, the more freely was a fresh efferent channel opened for the substances in question, and as soon as the lacteal secretion diminished, another excretion of these matters again occurred through menstruation. As soon as the circulation suffered disturbance, the organism must become diseased, or if the body grew sick, disturbance of the conditions described was said to be a symptom. Thus if the lochial or lacteal secretion were to be suddenly suspended, the substances in question vainly sought to escape, and "threw" or "displaced" themselves to an organ to which they did the most serious injury. According to this view, puerperal peritonitis was a metastasis of the lochia or the milk to the peritoneum; the puerperal psychosis, a displacement of milk to the brain, etc. These doctrines prevailed about two thousand years. They had the one great advantage that they were plausible—an advantage often contributing more to the spread of a doctrine than logic and truth.

At the time of the doctrine of crises, it was assumed that menstruation was a crisis. A crisis usually occurred on the seventh day, so menstruation set in after three times seven days.

The greatest step forward was the discovery of the connection between menstruation and ovulation. While formerly the menstrual hemorrhage was regarded as something essential, a symptom, the requisite; the assertion was now made that the principal matter was the ripening of the ovum and its expulsion from the ovary—ovulation.

The following explanation was given of the process: From puberty onward, the follicles in the ovary ripen, that is to say, the follicular fluid gradually increases. During the growth of the follicle, the surrounding tissue becomes displaced, and this displacement irritates the nerve termini in the tissue. Although this irritation is slight, it is continuous, and the sum of these slight irritations finally suffices to effect, in a reflex way, an arterial congestion, a hyperæmia of the internal female genital organs. This congestion has two effects occurring simultaneously: first, the sudden increase of the follicular fluid, leading to rupture and setting the ovum free—*ovulation;* second, the hyperæmia of the uterus which causes the rupture of the delicate vessels of the mucosa, the hemorrhage —*menstruation.*

The proofs for this assumption—the connection of ovulation and menstruation—were obvious.

Menstruation and ovulation occurred simultaneously and disappeared at the same time. Whenever menstruation shows itself, the possibility of impregnation exists, and as soon as menstruation disappears, for instance, in the climacteric period, impregnation no longer occurs. If a woman becomes pregnant, ovulation ceases with menstruation, and the possibility of conception begins only with the recurrence of menstruation. With the removal of both ovaries menstruation ceases.

As all scientific truths become established through attacks and defences, so it was with the doctrine of the connection of menstruation and ovulation.

If it were shown that either one occurred alone, this doctrine would be seriously shaken. Therefore, proof was sought, in the first place, that pregnancy occurred without menstruation, or ovulation without menstruation.

A number of cases were adduced of very early pregnancy in the ninth, tenth, and twelfth year, in which no menstruation had been observed. In like manner it was thought that another counter proof was furnished by the every-day occurrence of pregnancy very soon after delivery, in the physiologically amenorrhœal period of lactation. To this, of course, it may be replied that, on the one hand, the cases were not accurately observed, and on the other, that menstruation was just about to occur when conception took place. Moreover, it was possible that, for instance, the force of the coition or the congestion present during the sexual excitement caused premature rupture of the follicle, thus anticipating the menstrual congestion. The same could occur in a woman impregnated soon after parturition. Here, too, it was possible, just as in chlorotic and anæmic persons, that in a general anæmia there was an absence of the hyperæmia necessary to the rupture of the vessels. For it could be observed every day that phthisical women became amenorrhoic, while both the habitual difficulties indicated the menstruation and recent follicles were found at the post mortem.

Finally as regards the assertion that menstruation had recurred even after extirpation of both ovaries, this objection is invalidated by the fact that, with ovaries in cystic degeneration, it would be very difficult to furnish the proof that every, even if but microscopic, little piece of the ovary has been removed. And the most striking proof was furnished by the results obtained by Hegar, who invariably saw menstruation disappear after castration.

Nor is it any better with the assertion that menstruation occurs without ovulation. Whenever a woman in the climacteric period stated that her menstruation continued, while the power of conception had ceased long ago, the hemorrhages were mostly quite atypical ones due to heart disease, polypi, or carcinoma. In fine, it is obvious that a woman eventually declares every hemorrhage to be menstruation, as well as that here there may be some resemblance to the menstrual type. After a hemorrhage, the woman had to regain a certain degree of strength before she could furnish an additional hemorrhage. When anæmia recurred, the hemorrhage ceased but to be renewed after re-invigoration. The accidental finding of blood in the uterus of a patient dead of cholera or typhoid fever could just as little serve as proof that in this case menstruation without ovulation had been present. For it was exactly the typical element which was lacking here. It was an accidental menorrhagia, not the typical hemorrhage of menstruation. The same explanation holds good for cases in which menstruation is said to have occurred during pregnancy. Early authenticated cases of continued menstruation during pregnancy are lacking. A single return proves nothing, as some blood may be lost during gravidity for many reasons.

Therefore, all attempts to deny the connection of menstruation and ovulation must be considered as having failed. .

Also as regards the manner in which the hemorrhage is brought about the views have altered of late. The investigations bearing on this subject cannot as yet be declared concluded. According to the investigations of Kundrat and Engelmann, Wynn Williams, Leopold, and others, the facts are the following. The uterine mucosa changes continually; there exists here a kind of ebb and flood. To commence with the entrance of menstruation, the mucous membrane is then, as it were, ripened. A considerable congestion has brought to it much blood; the mucosa is swollen, thickened, and now begins to undergo fatty metamorphosis in its uppermost layers. This fatty alteration extends not only to the epithelium, but also to the upper layers of the mucous membrane and to the vessels coursing close underneath the epithelium. These vessels, therefore, undergo fatty metamorphosis, become eroded, opened, and hemorrhage ensues—menstruation. Of course, the bleeding of the mucosa leads to regression of its swelling, it gradually grows thinner again. The epithelium which had suffered fatty change is re-formed, that is to say, this epithelium grows out of the lumen of the glands across the entire surface of the mucous membrane, analogous to the reconstruction of the mucosa in the puerperium. In about two weeks the process has terminated, the mucous membrane is at ebb, that is, quite thin. Then the mucosa begins to swell anew, gradually increases in thickness so as to ripen again for the next menstruation. Accordingly, menstruation is a distorted image of pregnancy. In the latter, the ovum reaches the uterus, the swollen mucosa of which renders a downward migration of the ovum impossible. The ovum is retained and enters into organic connection with the mucosa. The entire process is to the mucous membrane a mighty impulse to further development. It becomes a decidua and gradually undergoes fatty metamorphosis. After two hundred and eighty days, during birth, the uppermost fatty layers are cast off with accidental hemorrhage. During menstruation an ovum likewise reaches the uterus, but it drops out. That terminates the process. The impulse of the fructification is absent. The fatty change occurs at once, and with it the hemorrhage. The great improbability which was inherent to the old theory, that the impregnated ovum was to become implanted in a "wounded" mucosa covered with blood, was thus done away with, for if hemorrhage, menstruation, occurred, pregnancy had not taken place. The process was at an end. But if impregnation had occurred, there was an absence of the hemorrhage—the fatty change —which did not occur until after nine months.

In favor of this theory were also the earliest ova which have been described. They showed conditions which rendered it possible only to date their fructification from the menstruation which had failed to occur, not four weeks previous to the date when menstruation had been present.

Quite recently, serious objections have been raised against these views. Particularly, it is incorrect that the hemorrhage should be the consequence of fatty metamorphosis. It seems that, with perfectly intact epithelial covering, the blood transudes through the vessels by diapedesis. Thereby we again approach the older view which looked upon the hemorrhage as the consequence of congestion of the greatly distended, reploted vessels. Perhaps the arrangement of the vessels in the mucosa is of importance. According to Leopold, there are more vessels leading into than out of the mucous membrane.

However, as has been stated above, the question is not yet fully decided, which is partly owing to the insuperable difficulty in securing sufficient material. And even if present, the subjects are very perishable, hard to examine, and difficult to judge.

Again summarizing what has been stated, menstruation should be explained in this wise: An ovum ripens; this swelling of the Graafian follicle irritates the nerve termini in the ovary. The irritation is propagated to the central organs. Through reflexes, by vaso-motor processes, an arterial congestion of the internal female sexual organs is set up. This in turn increases the liquor folliculi, so that the theca folliculi breaks and permits the ovum to escape—*ovulation;* and, second, the uterine mucosa becomes so hyperæmic that there occurs a bursting of the peripheral vessels, hemorrhage upon the surface of the uterine mucosa—*menstruation*.

Menstruation sets in in the fourteenth to the seventeenth year, but may occur earlier (menstruatio præcox) or later (menstruatio serotina). At first it is often irregular, often very copious (menorrhagia), or absent (amenorrhœa). It is frequently coupled with pain (dysmenorrhœa). Should there be a regular monthly hemorrhage, not from the uterus, but from other organs, for instance, the lungs, nose, etc., it is called xenomenia or vicarious menstruation. As menstruation ceases with age, this time is termed the climacteric period or simply climacteric; the cessation being called the entrance of the menopause.

In hot climates, precocity is physiological. In the tropics, girls are menstruating already in the tenth to the twelfth year.

Education, sedentary mode of life, mental stimulation cause an earlier appearance; therefore, in general, the inhabitants of cities menstruate at an earlier age than those of the country. Brunettes likewise are said to menstruate earlier than blondes. The onset of menstruation designates the puberty of the individual. The beginning of puberty, in the female as in the male sex, is associated with general changes of the mind and body. Menstruation disappears in the fifth decennium. In general, it ceases earlier in virgins than in parous women. Should a childbirth occur at the close of the thirties or after the fortieth year of life, menstruation will continue longer.

The hemorrhage lasts from one to five days; it is exceedingly variable quantitatively, so that one woman may believe an amount of blood to be physiological which in another would be pathological. A quantitative determination, therefore, is impossible. The reflex effect on the general condition alone will decide whether the amount of blood is physiological or pathological. The interval between two menstruations lasts from twenty-one to twenty-eight days; here, too, there are many individual variations. If the menstruation is quite regular and painless, the women themselves mostly are ignorant of the type.

The popular designations for the menstruation are: the monthly period, the monthlies, the courses, the period, being unwell, etc.

Still less clear than the mechanical process of menstruation is that of *ovulation*. According to the laws of diffusion, it is impossible that fluid should enter the follicle while its contents are under greater pressure than its environment. Therefore, it is physically inconceivable to make the rupture depend solely upon the increase of the liquor folliculi, upon the centrifugal pressure. Hence there must be peculiar processes in the follicle itself which lead to rupture. These processes can consist only in an activity of the cells of the membrana granulosa; whether these cells swell, disintegrate, or secrete fluid has not yet been determined. Rindfleisch supposes that a colloid substance perhaps forms physiologically, that this rapidly augments the contents and thus ruptures the follicle.

The very frequent pathological occurrence of colloid matter points to the possibility of physiological conditions underlying the pathological process. Or it would be necessary to assume that only those follicles rupture which are situated close to the surface—so close that the follicles could expand toward the abdominal cavity, as it were into the free air, without finding any material counter-pressure. For it is certain that a great number of ova derived from the deepest layers of the ovary perish abortively. Nevertheless, a large number of ova are expelled from the ovary. The enormous quantity of the ova on the one hand, and that of the spermatozooids on the other, is an offset, so to speak, for the fact that the fructifying apparatus otherwise functionates rather imperfectly and is subject to many accidents.

After the ovum has been actually expelled, the cells of the membrana granulosa proliferate, and in between them grow the vessels of the theca folliculi, from which at the same time a large number of white blood-corpuscles emigrate. Very soon the circulation of blood ceases in the proliferated capillaries; fatty metamorphosis and retrogression occur. To this, and not, perhaps, to an altogether irregular extravasation of blood into the follicles, is due the yellow color—the corpus luteum.

Should pregnancy occur, the process becomes greatly intensified, owing to the largely increased blood-supply. The corpus luteum, in such a case termed "verum," is larger, continues growing until the third or fourth month of gestation, and thenceforth atrophies. But this atrophy proceeds so slowly that at the autopsy of a puerpera the corpus luteum verum is always still recognizable.

B. Fecundation and its Consequences.

Among the physiological processes in the female sex belongs pre-eminently *fecundation*. The first requisite—the expulsion of ova from the ovary—has been described above.

The fimbriæ of the tube, covered with vibratile epithelium, incite a kind of vortex in the peritoneal fluid which carries toward and into the tube everything lying near it. In animals, this process has been demonstrated experimentally, by the introduction into the peritoneal cavity of coloring matters which were subsequently found in the tube and the uterus. In man, likewise, the distant effect produced by the vibratile motion cannot be inconsiderable; because, for instance, in occlusion of the left tube, the right may attract to it, engulf, and propel onward the ovum derived from the left ovary. This process, too, the so-called *transmigration*, has been experimentally demonstrated in animals, and in man has been rendered an almost certain probability by pathological cases. After the ovum has entered the tube, it is propelled toward the uterus by the vibratile motion, and, in the narrower part of the tube, by peristalsis.

Should no fecundation occur, the ovum soon perishes, but it seems possible that it may reach the uterus, as has been demonstrated.

The other necessary element, the seminal fluid, the spermatozooids, are poured into the vagina during coition. Without any co-operation of the female genital organs, the spermatozooids migrate in every direction, hence also upward into the os uteri. Only few, perhaps, get into the uterus, but of these few a single one suffices for the fecundation of the ovum. The female genital organs, therefore, act merely a passive part in copulation. Even in chloroform narcosis, and in a countless number of cases without a trace of sexual excitement, conception has occurred. If the os be patulous, the form of the vaginal portion or of the mouth of the womb is almost immaterial. *There is no form of the vaginal portion which has been described as the cause of sterility, in which conception does not occur spontaneously.*

The most weighty objections have been raised against all theories which presuppose the reception of the spermatozooids into the uterus through causes other than the movements peculiar to the spermatozooids. These, often quite visionary, views may therefore be shelved.

The question still remains to be discussed, where, how, and when the ovum becomes fructified. As in abdominal, ovarian, and tubal pregnancy the fecundation has decidedly not taken place within the uterus, it is assumed that physiologically the spermatozooids also migrate into and through the tubes. Perhaps the external, dilated part of the tube, the so-called ampulla, is the place where ovum and spermatozooid meet.

At all events, the spermatozooids retain their vitality for several days. If they are effused at any time into the vagina, they migrate as far as the tube, even to the ovary. The ovum is fructified and consumes perhaps seven or eight days in getting into the uterus. In the latter it is intercepted, probably mechanically, by the meantime proliferated mucous membrane: pregnancy is established. After its termination by labor, the former relations gradually become re-established in the puerperium. But ordinarily the vascular system of the female genital organs remains wider, vagina and uterus being rather more lax, both as regards the organs themselves and their attachments. The mucous membrane, on the other hand, re-forms exactly in its previous condition, and in the cervical canal the glands often remain larger than before and retain a certain tendency to independent proliferation.

C. Senile Involution.

After the above-mentioned climacteric period, the female sexual organs begin to undergo senile involution. The vulva becomes smaller and less adipose; the vagina narrower, smoother, and especially less yielding and expansible. The vaginal portion of the uterus gradually disappears completely, the vagina above terminating smoothly, at its end being an opening, the os uteri. The uterus diminishes greatly in size, shrinking to a length of five centimetres. It becomes quite lax and thin-walled. The mucous membrane may so change its character that no glands remain discoverable and that the mucosa with its investing epithelium is supplanted by a kind of granulation tissue. Where the latter is in close juxtaposition, the walls become adherent, especially at the internal os. Then the secretion may stagnate above, leading to senile pyometra with atresia of the internal os. The ovaries likewise atrophy; but in them we still find follicles, small cysts which have originated from follicles, and accumulations of connective tissue which can be traced to corpora lutea. At the surface, the old cicatrices may almost always still be found.

A similar atrophy occurs in the entire pelvic peritoneum. the uterine ligaments also becoming shorter.

CHAPTER III.

GENERAL DIAGNOSIS.

THE eminent progress of recent years, in the recognition of all diseases, is to be attributed to the gradually more and more developed methods of examination. But while the physically-diagnostic methods for other diseases—as, for instance, those of the lungs—have been perfected for decades with the greatest practical results, there was hitherto a lack of "method" in the examination of the female sexual organs. Gynecological diagnosis is an acquisition of the most recent time, is as yet by no means a settled doctrine, and is annually more and more developed and elucidated. It is vastly more difficult to describe critically the permanently useful and practically important, than to enumerate all propositions. But as the science of gynecology *in general* is indebted to improved diagnosis for its progress, so *the individual* will have to make diagnosis the first and most important part of his study.

A. ANAMNESIS.

Of the means for making the diagnosis we have first the anamnesis. Let us commence with the entrance of the patient. Even from the behavior, the experienced gynecologist will often know whether a virgin or married woman is entering. The latter will obviously constitute the great majority. Mostly, we shall have to deal with persons of somewhat anæmic aspect. Considerable anæmia points to past or present hemorrhages. The age, again, may indicate the cause of these hemorrhages. Cautious, slow sitting down, points to pain in the pelvis—pelvic peritonitis.

The anamnesis must be taken down with especial accuracy, but the systematic method, permissible at a man's bedside, contains much that is offensive to an educated lady. Therefore, the physician may at first enter upon the sufferings of the patient and listen patiently. Withal, the conversation is to be adroitly directed to the important points, always bearing in mind that the necessity of an examination must, as it were, spontaneously occur to the patient from the tenor of the conversation.

Of particular interest are the questions, whether the patient is married; if she has borne children, with facility or difficulty, how many, whether in rapid succession; how long since the last child. Have there been abortions; have diseases or prolonged illness succeeded the birth or abortions?

Then we inquire after the *menstruation*. When did it first occur; has it continued in the same way from the beginning to the present; or has the menstruation changed quantitatively or qualitatively after marriage, after eventual diseases, labors, abortions, or causelessly? Are there any pains previous to, during, or after menstruation; are they confined to a certain spot, or not; have they a fixed character? Have the pains been made better or worse by any remedies thus far employed? Is there any discharge of liquid or clotted, bright or dark blood, with or without mucus? Does the hemorrhage continue uniformly, or does it intermit and return; does anything, such as rest or motion, affect the hemorrhage and other symptoms? Is the bleeding copious or not; is it followed by a sense of weakness or not? Does the menstruation continue long; does it return quite regularly and always produce the same group of symptoms? For all these questions such expressions are to be employed as are adapted to the degree of education of the patient.

The questions regarding the hemorrhages are naturally succeeded by those relative to some other *discharge* or *flow*. What is its nature; glairy, purulent, bloody, mixed with shreds? What stains does it leave in the linen; greenish, yellowish, bloody, or none at all? Is the discharge offensive; does it excoriate the external genitals? Is the flow continuous; is it occasionally more profuse, and when?

We then inquire after the *pains*. Are there real pains; can they be accurately located and characterized? Do they consist more of an indefinable sense of pressure; or are they real lancinating, boring, pulsating, cramp-like, or labor-like pains? Are the pains always present, or only during certain causative acts—difficult defecation, urination, sitting upon hard-bottomed chairs, walking upon uneven pavement, rapid walking, ascending stairs, lifting or carrying heavy weights, stooping, riding, dancing, coition—are the clothes worn tightly or loosely? As a criterion of the violence and the persistence of the pain, we ask whether the patient must lie down during their continuance; whether they render work or social intercourse impossible; whether they disturb sleep; whether they make the temper irritable, morose, or changeable.

After these questions, referring to consensual symptoms, we inquire after pains in distant organs, especially the breasts, the head, and the stomach. Also the appetite, digestion, the quality of the stools, and diuresis are of importance.

If, after all this, we fail to form an exact idea of the affection of the patient, we ask in conclusion, What is *the trouble* which caused the patient to seek medical aid? If it be something definite, as, for instance, a tumor, we inquire after special anamnestic data. Frequently the patients expect the physician to suspect the cause of the visit without having given it distinct utterance. Thus, ladies will undergo cross-examination for a quarter of an hour, and still the wish will not escape them "that they desire to have children."

On the other hand, in dealing with a hysterical patient, we must be careful not to direct to her every question above indicated. Any and all symptoms ever described will be assented to by many a hysterical woman, and, with a supreme self-satisfaction, the patient will assert that all these remarkable phenomena are present in her case. Thus, we may examine *into* some patients anything we please.

The physician must proceed with appropriate tact; for instance, not inquire after children first, and then if the patient be married, nor, perhaps, begin with questions referring to eventual pain during coition. Still less shall we gain the patient's confidence, by making sport or laughing at the often quite ludicrous complaints, narrations, and descriptions of abnormal sensations. We must ever be mindful to preserve professional dignity, and, equally far from the fixed, complaisant amiability of the routine practitioner, and the brutal severity of the inconsiderate, exact investigator, be guided in our manner solely by true humanity.

B. GENERAL REMARKS ON EXAMINATIONS.

The medical examination of the female generative organs begins with *the touch;* the internal exploration. To this the external mode is to be added, forming the *combined examination.* In the case of abdominal tumors, the external examination, palpation with both hands, is likewise to be practised. The more detailed description of the latter will be reserved for the differential diagnosis of abdominal tumors. Then follows *inspection,* during which the vagina must be dilated and held apart by instruments—*specula*—so as to make the vaginal portion of the uterus visible. Moreover, in order to determine the relations of the cavity of the uterus, it requires to be sounded. In some cases, the examination with the sound does not suffice; it then becomes necessary to dilate the uterine cavity so as to permit the finger to palpate it.

During the medical examination, everything must be so arranged as to make the procedure as convenient as possible for both patient and physician. Therefore, the patient must be placed comfortably. The position varies in different countries, both in reference to the couch and to the patient. Just as—perhaps decenniums ago—the first great teacher in gynecology examined, so do his pupils even to-day. Habit makes the method appear handy, correct, and practical. It is difficult to shake it. Thus in America and England the lateral decubitus, in Germany, usually, the dorsal position is preferred. In Germany, where the power of authority is divided in many parts, what is of foreign origin is most readily accepted if it be good.

An *examination in the standing posture* is always insufficient; it should be practised for two reasons only. First, to learn how a descent of the uterus or vagina *differs* from its extent in the recumbent position; second, to feel whether a pessary lies as well, or as firmly, in the erect as in the recumbent posture.

The *ventral and knee-elbow positions* offer no advantages, and are not employed. Only a few operators have used the knee-elbow position for operations on the anterior vaginal wall.

The *lateral position* has the disadvantage that the combined examination is impossible; but it is preferable in the employment of specula and for minor gynecological operations, where we shall recur to it. The position most convenient for patient and physician is the *dorsal decubitus*, because it is not necessary to change this position for any method of examination.

A large number of cheap and expensive, simple and complicated, practical and impractical examining tables and chairs have been constructed. Here it is to be borne in mind that all chairs or tables having the form of an operating-table, frighten the patient. The large tables which the patients have to mount on steps like a scaffold detain many from the examination. Without trying to ignore that it is purely an individual matter what every one would believe to be good and convenient, I should recommend a *chaise longue* or a simple sofa without back. It should easily move on casters that it may be brought near the window and that, if necessary, examination can be made from either side. The sofa must be very firmly upholstered, or a hard pillow must be placed under the pelvis of the patient. The sofa should not be too high, so that the patient may comfortably lie upon it and be able to descend. The cover should be of wax cloth so that it will be possible to wash it off. The physician sits upon the edge of the sofa or on a chair. The patient should lie perfectly straight, with legs moderately spread and drawn up, the head placed firmly on the head-rest.

C. THE TOUCH.

The *touch* is practised by introducing the finger coated with carbolized oil or carbolated vaseline (1:10) from behind into the vagina. The following is to be noted. Is the *frenulum* still present? Is the *vulva* wide, owing to a *ruptured perineum?* Is there a tumor at the vulva, for instance at a labium majus; is the introduction painful or not? Is the finger met by the anterior or posterior vaginal wall, or perhaps even by the uterus (prolapsus or descensus)? Is pressure on any part of the vagina painful? Is the uterus in its proper position; is the vaginal portion soft or hard, open or closed? Is the uterus movable; that is, can the vaginal portion be dislocated forward, backward, to the right and left with equal readiness and painlessness? Is there any resistance, any tumor beside the uterus? Is this tumor movable; what is its consistence; is it tender; can it be distinctly circumscribed and its form described, or is it connected

GENERAL DIAGNOSIS. 27

anywhere with the neighborhood? What is the relation of the uterus to the tumor; is it displaced, merged with it; can tumor and uterus be limited, moved separately, or are both firmly adherent?.

I am constrained to add right here that it is necessary for every exact gynecological diagnosis to introduce two fingers into the vagina. The second (middle) finger is always inserted behind the first; even in nulliparæ the expert causes little pain, for the penis is thicker than two fingers. Inside, the tips of the two fingers are spread and the margins of the uterus felt. In this way the tip of one finger fixes one margin of the uterus, while the other moves to and fro on the other margin. Two fingers are necessary especially to feel the ovaries, small parametric tumors, resistances, possible indurations in the uterine ligaments, and pre-eminently for the fixation of the uterus during the combined examination. Artificial displacements of the uterus, such as drawing the vaginal portion forward in anteversion, reposition of the fundus in retroflexion, can be performed only with two fingers. But aside from all other advantages, we can obviously reach much higher with two fingers than with one.

D. THE COMBINED EXAMINATION.

Then the other hand is placed upon the abdomen. With gentle pressure, during which a look is cast on the patient's face to observe whether we give pain, the hand is approximated as far as possible to the

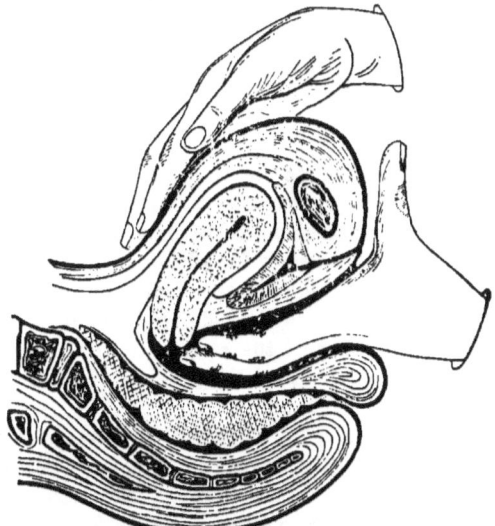

FIG.—7. Position of the hands during combined examination.

vertebral column. Then the tips of the fingers are crowded downward, and while the internal fingers press upward, both hands are sought to be approximated. Everything between the two hands is to be palpated to gain an idea of the size, consistence, and connection with the pelvic organs. By the combined examination we determine the position, size, form, and mobility of the uterus. Is there a tumor anywhere in the pelvis, and what is its nature? Is this tumor connected with any other organ, or is perhaps a tumor detected on abdominal palpation connected with one felt previously in the vagina? Is fluctuation continued from one into the other? Are both movable together or separately? Judging from its locality, from which organ did the tumor spring or originate? By negative results we determine the absence of any tumor, or even of the entire internal generative organs.

Besides this combined examination from the abdominal coverings and the vagina, we can also, from the rectum and the vagina, examine tumors located *between* rectum and vagina or in Douglas's fossa. This examination is best performed with the thumb in the vagina and the index finger in the rectum. In the absence of the vagina or with intact hymen, we can also perform combined examination from the rectum and the abdominal coverings.

From the rectum and bladder the examination is also necessary, especially in malformations when the vagina is absent. Then the urethra may be dilated either with the finger or with the specula devised by Simon.

In tumors of the anterior wall of the uterus which were difficult to diagnose, a finger has been brought to the tumor from both the bladder and the vagina, or the rectum and the bladder, while an assistant at the same time pulled on the vaginal portion held with tenaculum forceps, and moved it to and fro.

E. Examination from the Rectum.

The examination from the rectum, as has been stated, is to be practised in the absence of the vagina or with intact hymen. Only in the most urgent cases is it permissible to destroy a hymen. Moreover, the examination per rectum is necessary for all tumors of the fossa of Douglas, even to corroborate the results of the vaginal exploration. Besides, palpation from the rectum alone will decide whether a hard tumor springs from the bone or lies in front of the rectum in the fossa of Douglas. Here it is not necessary to go beyond the internal sphincter which is felt by the finger in the shape of a funnel.

Examination per rectum is of particular importance if there be in the vagina a pessary holding the retroflexed uterus in position. If the vagina be narrow, we must ascertain through the rectum the relation of the vaginal portion to the pessary, and if it performs its part. Rectal examination is best made in Sims's lateral position (see below). On entering, the finger first touches the anterior lower surface of the uterus, and must be bent backwards in the form of a hook to feel the concavity of the sacrum.

GENERAL DIAGNOSIS.

The exploration by the introduction of the whole hand into the rectum, first practised by Simon, has been again abandoned. Nearly all gynecologists who have tried this method believe its results to be inadequate to its difficulty and danger. The hand cannot move freely in the rectum, and hence cannot feel with delicacy. I myself was present when Simon erroneously diagnosed a small secondary cyst of an ovarian tumor to be the uterus. This uncertainty is coupled with danger. There are some cases in which the rectum ruptured longitudinally, and the patient died of peritonitis. If one finger should not suffice, it may be advantageous to introduce two fingers into the rectum, so as to reach somewhat higher.

F. EXAMINATION WITH THE SOUND.

The *examination with the sound* is not necessary in all cases, but, in certain circumstances, it alone can furnish positiveness in diagnosis.

The instrument is best made of copper with wooden handle. It must be flexible, but not too thin or too yielding, lest it bend on slight pressure. In such a case we should never know where the point may be. The length is thirty to thirty-five centimetres. The end terminates in a knob. Six or seven centimetres below the latter is a projection indicating the normal length of the uterus. Several gynecologists have had the whole sound divided into centimetres, or have had special measuring contrivances attached. The knobs of the sound are of variable thickness, three to eight or more millimetres. Three sizes suffice, but of course a greater number may be made, at pleasure.

Fig. 8 *a* represents a copper sound with wooden handle; Fig. 8 *b*, that devised by B. S. Schultze, made entirely of copper. Of late, Schultze has had a number of marks added to the sound, so that the length of the uterus may be ascertained by a single finger passed along the sound into the vagina.

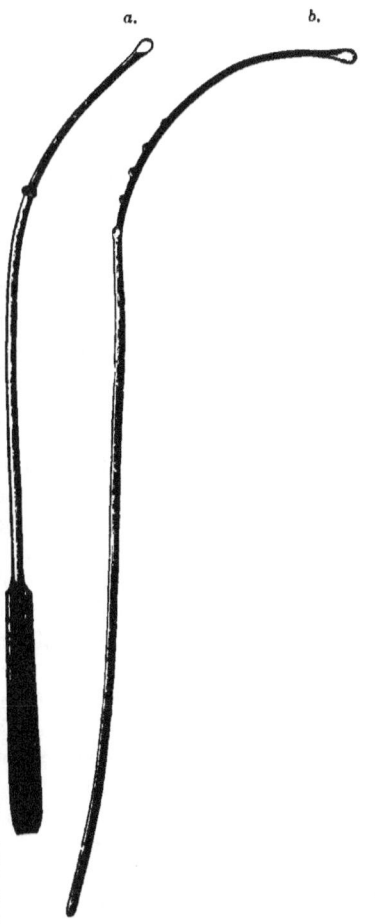

FIG. 8 *a.*—Uterine Sound.   FIG. 8 *b.*—Uterine Sound of B. S. Schultze.

Each sound, even if it has been cleaned every time after use, must be dipped into three to five per cent carbolic solution before being employed. The sounds are best preserved in a bottle filled with carbolic solution, into which the sound is immersed for two-thirds of its length. The statement that a sound, if not warmed to the heat of the body, will, by reason of the difference of its temperature, excite the uterus to contraction and resistance against the sounding, is quite erroneous.

Before sounding, we must know exactly how the uterus is situated. But should we intend to ascertain the position of the uterus by the sounding, for instance, in irregular tumors, the greatest possible care must be exercised, and the procedure should be temporarily interrupted should there be pain or difficulty.

Having bent the sound corresponding to the position of the uterus —slightly curved in retroflexions, more strongly in anteflexions—the knob of the sound, pressed against the intravaginal finger, is passed into the vagina. Close to the tip of the finger fixing the os uteri, the sound glides into the cervical canal and is guided so that the tip of the sound advances in the presumed direction of the uterine cavity. Meeting with resistance, force must never be used; it is better to abstain from all sounding than to force it. By withdrawing, advancing in various directions, different curvatures of the point, cautious lever motions of the handle, the attempt is renewed. Should this fail, sounding in the speculum (see below) or in the lateral position may be tried. Especially in Sims's lateral position, if the anterior lip of the os be drawn straight and forward with a tenaculum, the sound often glides with surprising facility into the uterus, while previously it was impossible to advance. Thick sounds will encounter some resistance at the internal os; should this resistance be considerable, the diagnostic sounding should be done with a thinner sound. After removal of the sound, it should be closely examined to see if there be any blood upon it.

The following can be and is to be ascertained with the sound:

1st. How long is the uterine cavity? To this end the tip of the index finger is placed on the sound at the internal os. Then finger and sound are removed together, and the measurement is taken, or the distance from the prominence marking the normal length to the tip of the finger is estimated.

It cannot be denied that the simultaneous removal of the finger and sound, in strong flexions of the uterus, may injure the mucosa of the uterus; therefore Schultze had a number of marks applied to his sound. While in position, the prominent marks are counted by the tip of the finger gliding over them, and thus we know how much of the sound is within the uterus or how long the latter is.

2d. Is the uterine cavity wide or narrow? This is learned, in the first place, by the readiness or difficulty with which the sound is advanced; in the next place, by attempts to move the sound laterally.

3d. Which is the direction of the uterine cavity? For instance, in uterus unicornis or if a large tumor be present in one wall, the point of the sound will deviate laterally.

4th. Is a uterine cavity or a uterus present, or is this cavity anywhere occluded or stenosed?

5th. Where is the uterus situated? In tumors upon or around the uterus, it is often very important to know where the uterus lies. If it be impossible to gain this information from the form of the tumor, a careful sounding is permissible. Then the direction of the cavity will indicate the position of the uterus.

6th. Is the uterus movable or not? Of course, in this instance, too, sounding will only then be permissible if the knowledge of the mobility of the uterus *cannot be otherwise obtained and yet is urgently needed for diagnosis and treatment.*

The greatest care in sounding is necessary because several dangers of this procedure are known. In the first place, as for instance in catheterization of the male urethra, rigors or other nervous symptoms have followed sounding. Again, the pain on passing the internal os may be so great as to excite labor-like pains and uterine colic, and even collapse and syncope. In tumors, or hemorrhagic tendency of the mucosa, violent hemorrhages may be provoked which can only be stilled by the tamponade. The same, with succeeding abortion, occurs when sounding a pregnant woman, though this is not an invariable consequence.

Moreover, should the uterine mucous membrane be injured by unclean instruments, infection, septic metritis may be set up. This may extend to the peritoneum, as is demonstrated by cases, though rare, where fatal peritonitis ensued after sounding.

Two noteworthy circumstances that may occur during sounding remain to be mentioned. *First, the sounding of the tubes.* If the uterine opening of the tube be wide, and the uterus be placed somewhat obliquely so that the knob of the sound is directed to the tubal ostium, the sound may glide into the tube and pass through it. Although these cases are rare, they are still possible and authenticated.

Second, it is much more frequent to have the sound pierce the entire uterus and penetrate between the intestines until it can be felt close under the abdominal coverings. These accidental uterine perforations are not at all of rare occurrence and usually they terminate without hemorrhage or inflammatory phenomena. This circumstance demands the utmost caution. Puerperal uteri and those of sick and weakly women appear to be particularly easily perforated. This accident would have to be feared in an organ predisposed thereto, if a change of position were sought to be effected by lever motions of the handle of the sound, although the uterus is fixed.

In hysterical women with a tendency to syncope, the first sounding should not be performed during office hours, as the patient may be so affected by the procedure as to render transportation to her home impossible for hours. After difficult soundings which required force, having more of the character of dilatations, the patient should be kept in bed for one hour as a rule, and be made to use disinfecting vaginal irrigations, for one or two days. Occasionally colicky pains occur some time after sounding. For these, as well as for violent pains after sounding generally, a subcutaneous injection of morphine is the simplest remedy. However, in general it is not at all objectionable to make even a first examination with the sound at the office. I have never met with such severe accidents. But violence should be guarded against, lest a careful diagnostic sounding be perhaps transformed into a forcible dilatation.

G. Examination with Specula.

The vulva and the vestibule, the relations of the hymen, of the orifice of the urethra, and of the posterior commissure are inspected without auxiliaries. But if the vagina or the vaginal portion of the uterus is to be examined with the eye, instruments must be employed which have long been known under the name of specula or vaginal specula.

There are, in the first place, cylindrical specula, formed of glass, clear or opaque; metal, rubber, or wood. Their external end is expanded in funnel shape, so that they can be held without obstructing their lumen. Their internal end is bevelled off. The specula of opaque white glass are the cheapest and best; they do not require the strong illumination necessary for the dark rubber tubes. The various chemicals employed do not destroy the glass. Cleansing is very easily done. Of like form are the so-called Fergusson's specula. They are made of silvered glass protected by a rubber covering. These specula will be found requisite only if the patients cannot be examined near the window and by daylight. Even with moderate illumination the relations of the vaginal portion are readily observed, and if any portion is to be very closely inspected, a Fergusson's speculum will be required after the use of the opaque glass speculum. The main fault of the instrument is its high price with slight durability. Similar instruments have also been formed of thin German silver. They are very useful at first, but the polished surface is rapidly dimmed. The rubber specula have the advantage of infrangibility, the the disadvantage of being applicable only in very bright light. The wooden specula are employed for the actual cautery of the vaginal portion, because the heat cracks the glass, melts the rubber, and renders the metal unbearably warm.

If a speculum is to be introduced, we examine first with the finger, to estimate the capacity of the vagina and select the proper size of speculum.

Besides we learn by the touch where the vaginal portion is, and on introducing the speculum can direct its point to the appropriate place. With two fingers of one hand we first separate the vulva in such a man-

GENERAL DIAGNOSIS. 33

ner as to draw the labia minora well downward at the same time. Then the point of the tube is placed into the vagina above the posterior commissure, and the speculum is passed along, rotating, in the ascertained axis of the vagina. Corresponding to the upward concavity of the vaginal axis, we first direct the point, the *inner* end of the speculum, downward, and then, passing in farther, depress the *external* end.

The urethral prominence is very sensitive; should it perhaps be just in front of the margin of the speculum and be rubbed and pressed upon by the latter, the patient will feel pain and unconsciously bear against it, so that the speculum will have to be withdrawn and the prominence avoided. Every beginner will learn to overcome this little difficulty within a short time.

Fig. 9.—Opaque glass speculum.   Fig. 10.—Cusco's Speculum.

While the instrument is pushed up, we observe through its lumen the unfolding vaginal walls. If a peculiar position of the vaginal portion has been previously ascertained, it must be borne in mind during the passing. The edge of the tube scrapes off the vaginal mucus and carried it upward, so that the observer gets also some idea of the secretions. With rotatory motions, withdrawals, and re-introductions the vaginal portion is caught in the opening of the tube. Hereby the os uteri must be rendered plainly visible. Difficulties exist only in very fat persons where the tube is too short.

Moreover, the speculum may be introduced in such a manner as to press the opening upward. The point must extend beyond the urethral prominence before we twist or push. With this introduction, owing to the tenser anterior vaginal wall, we have the advantage of catching the vaginal portion more quickly in the lumen. If the couch be soft, so that the pelvis of the patient is much depressed, the last-described introduction is advantageous.

Another, still older kind of speculum is that composed of two or more valves. The best of this class, and the only one still in use, is that of Cusco.

The chief advantages of this speculum are that, in case the introitus is painful, it is more easily introduced. Owing to its conical shape, the introitus is but gradually dilated. Moreover, the speculum is self-retaining. Should we desire to perform any therapeutical manipulation, it will always be annoying to have one hand occupied with the speculum. But if the glass speculum be not held, it is often extruded by the abdo-

FIG. 11.—The patient in Sims's lateral position, upon Dr. Chadwick's gynecological examining table.

minal pressure, and if the patient hold the instrument, the vaginal portion will often slip out, and we again have the difficulty of replacing it. But if, at first only for diagnostic purposes, the Cusco speculum has been introduced, or the vaginal portion engaged, the speculum can be allowed to remain while we select and fetch remedies or instruments for therapeutic purposes. Hence for the practitioner the Cusco speculum is excellent and indispensable.

The Cusco speculum is introduced so that the width of the upper valve points at first vertically, then it is turned around, advanced toward the vaginal portion, and opened by turning the screw. By lever motions the vaginal portion is caught in the lumen, but should they fail, the portion is drawn into the lumen with the tenaculum, or pressure with the sound. Bright illumination is requisite. Previous to removal, the vaginal portion is to be pushed out of the speculum while making traction. If this be not done, we might with the speculum draw down the vaginal portion or the uterus and thus often cause quite severe pain.

Formed altogether differently and based on other principles are the grooved specula. The idea of these instruments originated with Marion Sims, who may be called the father of the whole of modern gynecology. Although not everything which he asserted is correct, he still has pointed out new paths to investigation. He it was who first made gynecology surgical *in principle*, and it is the persistence in this direction which has secured to gynecology its modern triumphs. Sims made the peculiar observation that in the knee-elbow position the abdominal coverings and all the intestines sink downward. The uterus and vagina take the same direction; but if the posterior wall be held fast in the sacral excavation, only the uterus and the anterior vaginal wall follow that course; at the same time the vulva opens; the air rushes, often with audible noise, into the vagina which becomes inflated, and we see the vaginal portion and the anterior vaginal wall. Instead of the knee-elbow position, Sims soon chose another more convenient to the patient—the one called after him Sims's lateral position.

In that position the woman lies near the edge of a table, on her left side. The left arm being extended backward, the breast and abdominal surface are directed nearly downward. If now the whole body be so placed that the pelvis is rather near the edge, the trunk and thighs slightly bent forward, the vulva becomes very plainly visible. Then an assistant, standing behind the patient with his face toward her lower extremities, separates the buttocks, Sims's grooved speculum is introduced, the posterior vaginal wall is drawn strongly toward the sacrum, and the speculum given to the assistant to hold. Although it is more convenient for the physician to have the woman placed upon the table, still any sofa will do.

Fig. 12.—Sims's Speculum.

If the examination is to be prolonged, the physician sits down upon a chair opposite the vulva. If the parts be not sufficiently visible, for instance, if the anterior vaginal wall be too long, this wall must be pressed against the anterior pelvic wall with an instrument (sound, Sims's depressor), or the uterus be drawn downward with a tenaculum hook inserted into the anterior lip of the os. The illustrations (Figs. 13 and 14) show a single and a double tenaculum hook with long stem, for traction on the vaginal portion. If greater force is to be used, this may be effected with Simon's forceps depicted in Fig. 15. The latter represents two double tenaculum hooks joined like scissor blades which can be fixed by dentated plates (crémaillère). Such forceps with different curvatures are known by various names. Usually they are called Muzeux's forceps.

The disadvantages of Sims's position are the following. *First*, the combined examination cannot be performed in this position. *Second*, we cannot see the patient's face which is of great value to us in manipulations; for instance, during sounding, to observe whether we cause pain or not, that is to say, whether the patient unconsciously contorts the face or not. *Third*, an assistant is required. This is a deciding point why, for instance, this position is unfit for the office hour of the practitioner. However, Sims's lateral position has many important advantages in another direction.

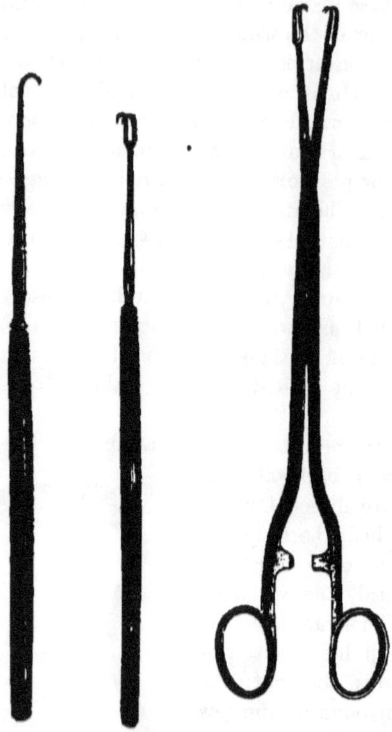

In the first place, it is undeniable that the vaginal portion and its neighborhood are presented to us only in Sims's speculum without distortion or alteration by pressure. For instance, should we desire to form an opinion about the shape and extent of a lateral laceration, or the conditions of the os uteri, the gaping of the lips, etc., the best method is the examination in Sims's speculum.

Fig. 13. Fig. 14. Fig 15.
Fig. 13.—Sharp hook,
Fig. 14.—Double tenaculum hook, } employed in various manipulations in the vagina and at the vaginal portion of the uterus.
Fig. 15.—Muzeux's forceps, after Simon.

Moreover, the customary minor manipulations, the scraping of the vaginal portion, the introduction into the uterus of instruments such as sounds, tents, caustics, brushes, and injections of the uterus, can be performed with particular facility in the lateral position. The excursions of the lower part of any instrument are not in the least hindered in it. In version of the uterus, especially forward, it is often uncommonly easy to get into the uterus, while in the dorsal decubitus we often strike the posterior cervical wall with the point and cannot get any farther.

To sum up, Sims's position is excellent for the execution of minor operations in *a case with which we are familiar*, but the dorsal position should be chosen for the formation of a diagnosis when first meeting the patient.

A very material improvement of Sims's speculum has been effected by Simon. That speculum can also be used in the dorsal decubitus, and Simon has named the position in which he usually operated, the ano-dorsal position. The blades are formed exactly after those of Sims, but are

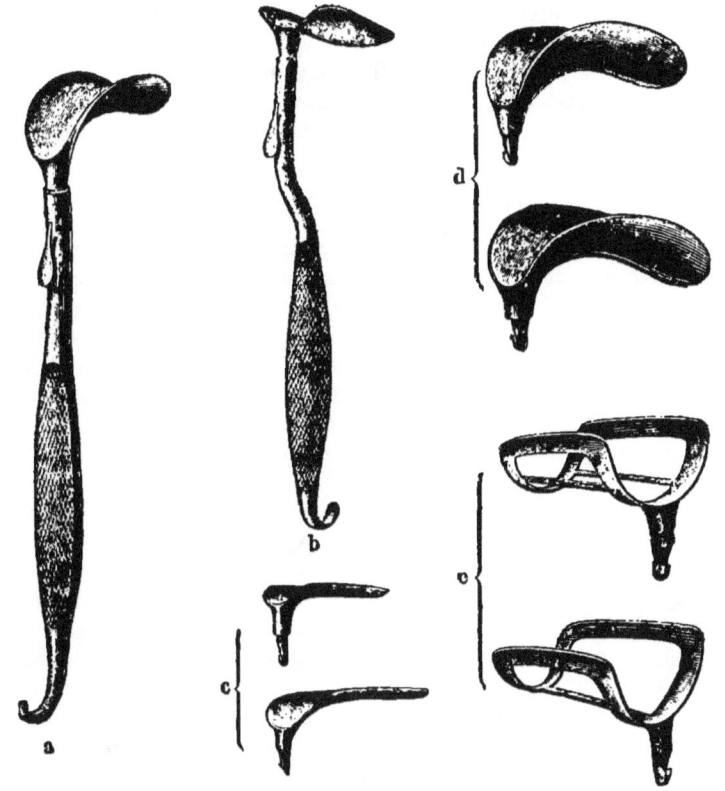

FIG. 16.—Simon's specula. *a*, lower part; *b*, upper part; *c, d*, smaller and larger, upper and lower attachments; *e*, fenestrated perineal specula devised by Simon for colpoperineorrhaphy.

movable on a handle, so that they may be changed according to the size of the genitals. As in the dorsal position the anterior vaginal wall naturally remained down, Simon devised a particular holder for this wall (Fig. 16, *b*). By drawing on both handles, the vaginal walls were separated and the vaginal portion became visible. In order to remove also the soft parts on both sides, special lateral holders were constructed for use in operations on the vagina. The latter are unnecessary if we employ my specula, to be described hereafter. With this speculum the vaginal portion can be made very freely accessible. Held by deft hands, the speculum does not hide the field of operation; by oblique position, withdrawal, or further introduction of either half, various parts can be so placed as to be easily accessible for operation. This, too, is the main purpose of the instrument. It is less used for diagnosis. But all the advantages which were claimed for Sims's speculum are likewise present in that of Simon. During operations, Simon's modification of Sims's speculum is in universal use.

Besides the specula named, there is any quantity of others. Especially is the attempt made over and over again to construct specula combining the advantages of Simon's division with self-retention. Of course, all these instruments have their advocates without having come in general use. Their price alone, of some very high, is a bar to their employment by many physicians.

H. The Diagnostic Dilatation of the Uterus with Tents.

With vaginal specula we can inspect the lower uterine surface, at times even a part of the cervical canal, but a view into the uterus is impossible. All instruments—uterine specula—which have been constructed for this purpose have proved worthless. However, if we desist from inspection of the uterine cavity, there are methods which at least render it accessible to the palpating finger and to larger instruments than the sound. These are the methods of dilatation.

For a long time already has the uterus been dilated, by tents, by forcible stretching, and by incisions.

Tents are made of three substances: of fine-grained sponge (sponge tents), of sea tangle (laminaria tents), of Nyssa aquatica (tupelo tents).

The oldest, the sponge tent, has even to-day yet many advocates, although, owing to the great difficulty of their disinfection, the majority of gynecologists inclines more toward laminaria and tupelo tents.

The sponge tent acts not alone mechanically, but also irritatingly. Its leads to congestion and softening of the tissue in its neighborhood. When expanding, it, as it were, grows into the mucous membrane, so that on withdrawal a part of the mucous membrane is removed with it. Despite all disinfection, it soon becomes offensive; decomposition occurs in the sponge which gives rise to infection. As there are lesions, as well as sanious secretions, a strict carrying out of antisepsis will be a matter of impossibility with sponge tents. *But if we conditionally demand of every method of treatment the possibility of keeping clean the wounds made, we are bound to discard the sponge tent, in view of our inability to carry out this principle when employing it.* The dangers of infection are not counterbalanced by the favorable irritation of the tissues, particularly as this irritation is not even always desirable. And the wounding of the mucous membrane can be done with far less danger by the curette or caustics. Besides, a large number of cases have been published of sickness and even death due to sponge tents. Even the carbolized sponge tents, obtainable in the market for about five years past, become soon offensive and offer no guaranty against infection. *For this reason I urgently advise to abandon the sponge tent altogether.*

The old method of its application I shall describe in a few words.

Sponge tents must be made of one piece. We first select a very thin one, no thicker than the knob of the sound which can pass into the uterus. This tent, grasped in a dressing forceps, is *rapidly* pushed into the uterus. The introduction must be done through the speculum. Sims's lateral position is especially appropriate for it. Previous to every introduction, the uterine cavity must again be sounded. For if the direction is well known, the introduction will be rapid; if this fail, the point swells and further advance is impossible. Moreover, the sponge must not be too thick, or it will pass the internal os with difficulty. After the sponge tent has been brought in position, it is watched for one or two minutes. After that, it has expanded somewhat and will retain itself. This expansion is hastened still more by a warm disinfecting injection. A sponge tent can be removed again after five or six hours. This must be done carefully, lest the string or a piece of the sponge break. Especially out of the palmæ plicatæ the sponge must be almost enucleated by the finger. If the dilatation be insufficient, another sponge is introduced after irrigation of the uterus. It should hardly ever be necessary to use many sponge tents. The mechanical effect of dilatation is less than that of fluxion, so that the finger may enter the softened uterus with some degree of force. Should it fail, then a new sponge would not be likely to mechanically dilate the uterus still more, but there would rather be the danger of a rapidly spreading inflammation in the swollen turgid tissue. The removal of the sponge tent is immediately succeeded by the therapeutic measure eventually required—irrigation of the uterus, injection, cauterization, curetting, etc.

The *laminaria tent* dilates much more slowly, not reaching its greatest volume in less than thirty-six hours. If laminaria tents removed from the uterus are laid in water, they usually continue to swell a little. Particularly old laminaria tents dilate much more slowly than fresh ones. There are solid and perforated laminaria tents; the latter dilate more, but cannot be produced in very thin and long shape, so that the solid tents are more often needed. The danger of infection may be avoided in the employment of laminaria. Aseptic tents I produce in the following manner. The tents are tied to a cord so as to make the latter pierce the entire cavity and a transverse hole near the lower end, and to surround the entire tent in a groove. The tents are then thrown into a reagent glass containing boiling alcoholic solution of salicylic acid. After boiling one minute, the tent is withdrawn by the cord and dipped into another reagent glass containing salicylic acid dissolved in hot fluid wax. Even the cord is dipped to the end in the boiling liquid, the tent is rapidly again withdrawn and allowed to cool, the unwaxed piece of cord is cut off, and the tent is wrapped up and preserved. Before use, the tent must be dipped in boiling salicylic or carbolic acid solution to render it flexible. This solution dissolves the thin coating of wax and the tent is ready for insertion.

If the uterus is to be dilated with laminaria tents, the vaginal portion must be exposed by Sims's speculum in the lateral position, or by Simon's specula in the dorsal decubitus. Then the vaginal portion and the cervical canal are again cleansed with moist carbolated cotton. Many gynecologists also insert the tents in the dorsal position of the patient. This is much more difficult. We require complicated instruments to hold or pierce the tent, and the point must necessarily be pressed into the mucous membrane at the angle where the uterus bends. Whoever has once practised the introduction in the lateral position will always choose this method again. Any person is qualified to hold Sims's speculum. The exposed vaginal portion is seized with a tenaculum at the anterior lip which is drawn taut. Thereby the bend of the uterus is obliterated or at least made less acute. Now the sound is passed once more, so as to thoroughly familiarize ourselves with the course the tent is to take. We then choose the tent the circumference of which exactly corresponds to that of the readily passed knob of the sound, and cautiously push it up, while the tenaculum fixes the uterus.

The end of the tent must project somewhat out of the external os uteri, lest the tent dilate within the uterus, leaving the os undilated. For the removal of the thick, soft, and friable tents from the narrow and hard os uteri may be very difficult: incisions have to be made, and if the cord break or cut its way through, or if on traction the lower end of the tent separate, the removal may last for hours and cause the patient much suffering.

The tent being placed, a tampon is laid over it. Then the speculum is withdrawn and we wait eighteen or twenty-four hours. Then by traction on the cord, tent and tampon may be removed together. If that be not easy, the tampon may be removed first, then the tent. Not rarely the tent is unable to dilate the internal os. While the tent expands in the uterus and cervical canal, it is firmly surrounded by the internal os. Only with very vigorous traction, often accompanied by a loud cry of pain by the patient, the tent yields, and the deep groove in its middle is, as it were, an impression of the internal os. If the desire had been to render the os patulous, the object has not been reached in these cases. After removal of the tent, we first irrigate the uterus with my own (see below) or Bozeman's catheter.

We then attempt to introduce the finger or the necessary instrument. Regarding expansibility, the tupelo tents stand between sponge and laminaria. Unfortunately the tupelo tents are generally too short so that they can barely be used.

I. The Mechanical Dilatation of the Uterus.

All tents are open to the objection that their employment is slow, troublesome, and not free from danger. Even without the occurrence of infection and resorptive fever, there may be recrudescences of a chronic perimetritis. Not rarely, however, the tent is insufficient to dilate the

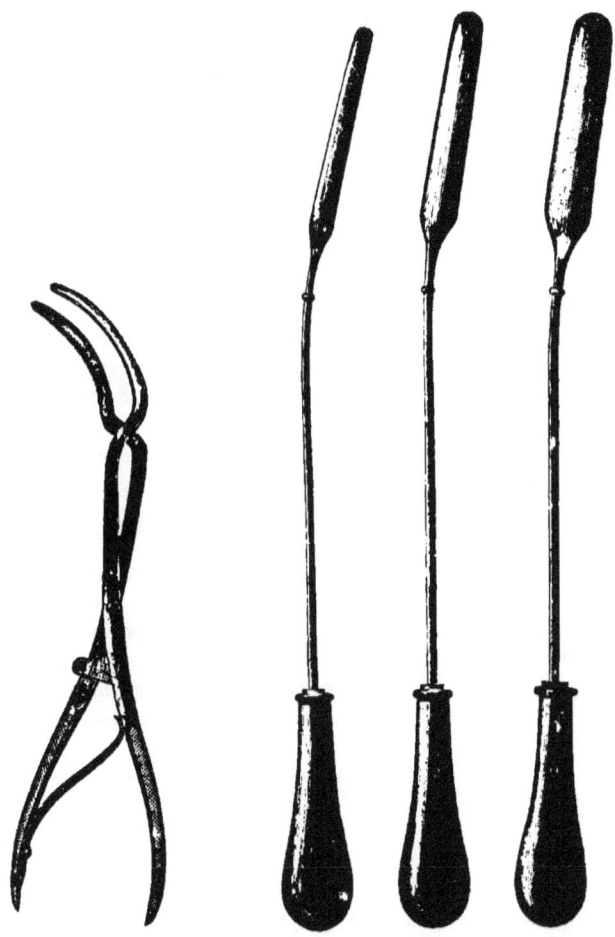

Fig. 17.—Schultze's dilator.   Fig. 18.—Fritsch's uterine dilators.

internal os, and we have to resort to instruments to overcome this resistance. For this purpose Schultze constructed the forceps depicted in Fig. 17.

This instrument is to be employed if the dilatation secured by laminaria should prove insufficient. Great stress is laid on the softening of the tissues. This is a consequence of the employment of laminaria.

For some time past, a reaction has set in against all kinds of tents. A number of authors have proposed mechanical dilators, and have held the method of rapid, forcible dilatation to be less dangerous and more practical.

In the first place, it should not here be forgotten that, in a large number of cases, the tissues have been favorably prepared for access by the very disease which is to be diagnosticated by entering or palpating the uterine cavity. Moreover, we have more frequently to deal with multiparæ than with primiparæ, and in the former the uterus is wider, laxer, and more accessible. In these cases, a *preparatory* dilatation of any kind would be superfluous. We proceed in the manner directed by Schroeder, dividing the cervix bilaterally, fixing the os with Muzeux's forceps, and, using the finger as dilator, working upward. In many cases, even the division of the cervix is unnecessary, as for instance in the retention of remnants of abortion, in which we can gradually insinuate the finger to the fundus for diagnostic purposes. The same procedure is not rarely possible in tumefactions and hemorrhages of the mucous membrane. However, should the division have been done, the lateral incisions may be sewed up again unless we prefer to leave them open for the future performance of similar manipulations.

Schroeder's method is particularly advantageous also in cases where the diagnosis of intra-uterine polypoid tumor is to be made. In these, every other method is inferior and more dangerous, and the frequently very thin cervix permits the incisions very easily. Besides, as they are any way required for the ready removal of the tumor, we might as well incise the cervix for diagnostic purposes. In disease of the internal surface of the cervix, Schroeder's method of dilatation is likewise to be recommended. By drawing down and inverting the two halves of the vaginal portion forward and backward, the internal surface of the cervix is exposed to view, and the diagnosis of the affection in question is easily made.

Other authors have employed instruments. There exists a whole series of dilators so constructed that two upper arms situated within the uterus become separated by screwing or pressing. Undoubtedly many cases in which the inventor succeeded "easily and rapidly" the uteri would not have presented any great difficulty to the finger. Of these numerous instruments, the one of Schultze, given above (Fig. 17), is the best. This, of course, is not intended to mean that other gynecologists could not be satisfied with the instrument they have used heretofore. Based on a similar principle are the instruments intended not only to press the uterus apart, but also to incise it. These hysterotomes have more of a therapeutic character, and therefore will be considered later.

Of late, I have devised instruments which correspond about to Simon's urethral specula. Peaslee and Hegar, too, had before constructed similar instruments. Given the surgical task to stretch a narrow canal in a resistant tissue, the surgeon will not be in doubt that this dilatation is best performed *quickly*. For this purpose my dilators will answer. If the abdominal coverings are very lax and the uterus movable, they are introduced in the dorsal position, one after the other, into the uterus, the latter being fixed from the abdomen.

In nulliparæ, the uterus can seldom be so controlled from without as to be quite securely held. Then we employ Sims's lateral position (see Fig. 11), sound first with a small instrument; then, having introduced a Sims speculum (see Fig. 12), the vaginal portion is seized with Muzeux's forceps, one number of the dilators after the other is gradually and slowly pressed in. During the dilatation, vagina and uterus are irrigated with carbolic-acid solution. According to the purpose of the dilatation, either only the smaller numbers, or all of them up to the largest, are introduced. All manipulations of this class are dangerous and not to be employed unless the indication is quite clear. It would be quite erroneous to believe that the danger of infection is the only one in forcible dilatation.

CHAPTER IV.

GYNECOLOGICAL ANTISEPSIS.

A. Prophylaxis.

A LARGE number of gynecological surgical procedures have hitherto been objected to as too dangerous. It was not to be wondered at that the whole body of practitioners occupied a negative position toward a practice which would remove minor ailments by very dangerous manipulations. A condition, in itself not dangerous to life nor even materially affecting general health, demanded a treatment which might be followed by fatal results. However, the observation of spontaneous recoveries, even if delayed to the climacteric, induced many a patient physician to comfort his still more patient clients with hopes for the future. The almost annually recurring reports of exudations following. incisions, or peritonitis and even death succeeding intra-uterine treatment, positively forbade the practitioner to admit such methods, were they ever so rational, into the legitimate means of cure.

The greater portion of these dangers, however, were the consequence of infection. Since antisepsis has found systematic employment also in gynecology, the dangers of many methods have entirely disappeared.

The gynecologist is bound to proceed according to the following, partly self-evident rules:

Every instrument is to be most scrupulously cleansed after use. Sounds and knives, tenacula, scissors, etc., previous to their employment must be dipped into five-per-cent carbolic-acid solution, be rinsed in it, and washed off with it. Whoever does much sounding had better preserve the sounds in a solution of carbolic acid. The specula are to be greased with a disinfecting mass. I use for this purpose carbolated vaseline (1 : 10), but note that, though rarely, cases occur in which this mass causes great burning. Therefore borated vaseline (acid. borac., 2 gms.; vaseline, 8 gms.) may be likewise employed. For the mopping up of vaginal or cervical mucus, always pure dressing cotton, never old linen or lint, is to be used. Should there be any lesion, benzoated or salicylated cotton must be employed, and the same is necessary for those tampons which are to remain for any length of time. Previous to any minor surgical manipulation, such as scraping of the vaginal portion, scarifications, etc., the vaginal portion and the accessible part of the cervix are carefully wiped clean with a pledget of cotton dipped in three-per-cent carbolic acid solution. This likewise is to be done especially if an instrument is to enter the uterus. If there be no leucorrhœa, and the vagina be narrow, sounding may be done without preceding irrigation and cleansing of the vagina; but in the presence of leucorrhœa, the vagina must always be cleansed through the speculum previous to sounding, lest virulent or infectious secretions be carried into the uterine cavity.

One or two days before every major surgical procedure, such as amputations of the vaginal portion or discissions, if at all possible, the vagina should be irrigated with disinfectants in the recumbent posture, at least three or four times in the twenty-four hours. It is obviously impossible to completely disinfect the genital passage by a single irrigation, even if combined with wiping of the cervix. But if certain quantities of fluid remain in the vagina, they must also spread upward at least as far as the internal os. Hence if the irrigation is repeated from three to six times, the guaranty against infection is greater. The repeated moistening of the external genitals with disinfecting fluid, besides, has the advantage of preventing, during surgical procedures, the transportation by the speculum or finger of any infectious masses from the external to the internal genitals. Therefore, in the case of an uncleanly patient, a sitz-bath or washing of the genitals in the *bidet* belongs to the preparations. Especially sanguinolent remnants must be carefully removed from the hair. Since I have had these preparatory irrigations and cleansings made, I have never seen the least inflammations after minor operations.

If carbolic acid is to be employed, the two-per-cent solution should be prescribed in large quantities or be made *personally* in the house of the patient. Despite the most careful directions, mistakes will occur: too strong solutions are used or such in which an undissolved portion of the acid has settled at the bottom. Then we have some bad burns which are exceedingly painful. Therefore a non-caustic disinfectant is preferable. I usually prescribe a powder of three grammes of salicylic acid which is to be dissolved in some Cologne water or alcohol, and added to the irrigation or to one litre of warm water. In this way, we obtain a three-per-cent salicylic acid solution. Potassium permanganate solutions may also be employed; their only drawbacks are the dark color and easy decomposition, so that portions remaining in the genital canal have no lasting effect.

During the actual operation the wound is kept aseptic by frequently syringing with disinfectants. Schroeder has introduced a method of permanent irrigation during operations in the vagina. Any one who ever employed this method will rapidly become an advocate of it. The irrigation can be most easily done with my gigantic speculum. This is so contrived that a flat tube is soldered to the upper wall situated beneath the symphysis and opens at the upper margin of the speculum; externally it ends in a knob beneath which a faucet is attached. If the knob is connected with an irrigator by means of a rubber tube, and the faucet opened, the disinfecting solution flows from within, from the vaginal vault, across the wound. Of course, if the vaginal portion is drawn far down, another stream must be directed against the raw surface possibly situated in front of the vulva.

B. After-Treatment.

Immediately after any therapeutic procedure, all blood should be removed if possible. But as arrest of parenchymatous oozing does not always succeed, a tampon is to be inserted or pressed against the walls. In such a case the tampon is best saturated with three-per-cent carbolic solution, but squeezed dry. The removal of every tampon is to be followed by an irrigation. If there has been only a cauterization or abstraction of blood, warm water will suffice which will remove the crusts, coagula, and possibly retained remnants of the cotton. After operations, disinfecting irrigations are necessary.

In general, the following may be said respecting the antiseptic treatment of gynecological wounds.

We must be convinced that the chief requisites for an antiseptic dressing of wounds of the vagina or uterus are furnished already by the natural conditions. The wound lies deeply, as it were subcutaneously. As protective silk guards the wound against every influence, so the opposite wall is apposed and exerts some degree of pressure which is also thought desirable in the Lister dressing. A long canal—the vagina—fulfils about the functions of the drainage tube and of the layers of gauze. Therefore, if the vagina has been irrigated, and if the dangerous air of a hospital is not present in front of the genitals, a uterine or vaginal wound will be under the most favorable conditions for healing. Experience, too, has taught us on innumerable occasions that wounds of these parts do not become infected even in inferior hospitals, and heal promptly. I would but call to mind the vesico-vaginal fistulæ, the prolapsus operations, the countless discissions and operations on the vaginal portion which some clinicians have performed. Assuredly, therefore, we have far more favorable conditions in wounds of the genitals than in those of any other part of the body. If a wound has not been infected during the operation, a subsequent infection is hardly possible unless copious secretions or dead particles of tissue remain in the vagina.

Less favorable, however, are the conditions if the lesion or the secreting wound is not at the vaginal portion, but in the uterus. Our practice teaches us that "absolute antisepsis" unfortunately is impossible. If a dressing remains too long, the secretions putrefy underneath and act infectiously, despite all antisepsis. In like manner, secretions stagnating in the uterus will become septic and act infectiously. For this condition, too, nature has natural remedies. The process of the immediately adjoining inflammation softens the uterus so that its portals become readily pervious. According to the law of gravity, foreign bodies in the uterus sink into the vagina, or the uterus is excited to contractions and expels the dangerous contents. The capacity of every organism to exclude a certain quantity of infectious material by processes of demarcation, or to absorb them and render them innocuous, overwhelm them, so to speak, makes a favorable result possible even without any therapeutic measures.

C. Treatment of Septic Gynecological Wounds. Uterine Catheter.

But it is not always permissible to leave nature unaided. We must try to avert the danger. This is done by rendering the discharge possible and disinfecting it according to general surgical principles. As the uterus is at an angle to the vagina, it is necessary to irrigate the uterine cavity, like a deeply situated, irregular abscess cavity, the efferent channel of which does not run in a straight line. For such irrigations we employ the uterine catheter. A large number of these instruments have been constructed. I myself have devised various uterine catheters for purposes of irrigation which I have had in use for nearly ten years.

Fig. 19 represents one of the thinner numbers of my uterine catheter. It is a plain tube open above and can be employed for all disinfections of the uterine cavity.

Fig. 20 likewise is a very useful instrument having three openings on each side. The conical smooth extremity renders this catheter easy of introduction into the uterus. The catheter with open extremity is more liable to injure the mucous membrane despite the greatest care.

If the cavity be very wide, Fig. 21 may also be employed. This catheter, originally constructed by me for irrigating the puerperal uterus, allows a large quantity of fluid to pass. The latter issues from all openings, as may be easily ascertained, so that the uterine cavity is thoroughly cleansed.

During all irrigations of the uterus some fear was justly felt lest the fluid penetrate into the tubes. If the internal os was narrow, or the uterus contracted in consequence of the irritation of the inflowing disinfectant, the os being firmly applied around the catheter prevented the fluid from returning. Great pressure, therefore, could force the fluid into the tube. For this reason instruments are preferred which permit the return of the fluid. The prototype of these instruments is the old catheter *à double courant*, Fig. 22. There are two openings at its extremity. One tube of the double-current catheter opens into one of them, while the fluid flows back into the second and returns by the other tube. This instrument is not well adapted for disinfecting the uterus, because the

48 THE DISEASES OF WOMEN.

efferent tube does not answer the purpose, being too narrow and easily plugged by mucus or coagula. For other manipulations, such as irrigating the bladder, this catheter is used very extensively.

Of the many uterine catheters intended to permit the return of the fluid despite a contracted os, the best instrument is that devised by

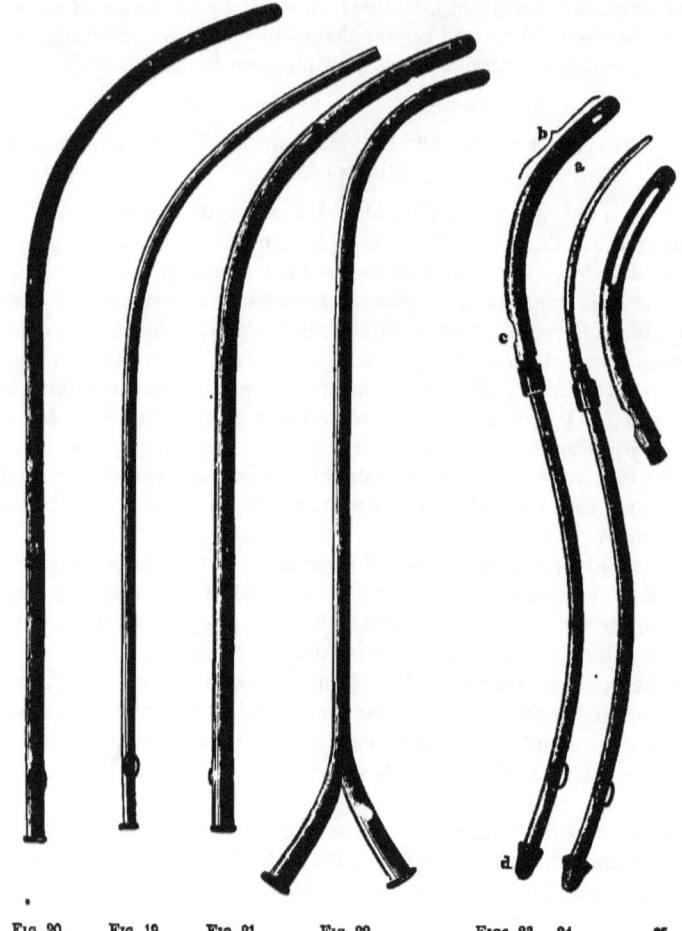

Fig. 20.   Fig. 19.   Fig. 21.   Fig. 22.   Figs. 23, 24,   25.

Figs. 19, 20, 21.—Various uterine catheters (Fritsch).
Fig. 22.—Catheter à double courant.
Figs. 23, 24, 25.—Bozeman's uterine catheter modified by Fritsch.

Bozeman, Figs. 23 to 25. I had this catheter made separable and effected some modifications which certainly tend to materially improve it.

Bozeman's catheter is so constructed that a thin tube is surrounded by a thicker, uterine tube (Fig. 23, *a b c*). The fluid, entering at *d* from the irrigator, issues from the thin tube at *a* and spreads in the uterus. Even if the os or the cervix closely surrounds the catheter, the fluid can re-enter through the large lateral fenestræ *b*, and be discharged through *c; c* lies in the vagina. By using the catheter through a speculum, we can convince ourselves of this fact. The great lumen of the ensheathing cylinder permits even larger, tenacious masses of mucus to pass through the catheter.

The original Bozeman catheter has three disadvantages. *First*, the distance from *a*, the discharging point, to the tip of the catheter is too great. If the uterine wall hugs it closely, unless we use great water pressure which is always dangerous, the neighborhood of the *upper* extremity of the fundus remains non-disinfected. The fluid returns below. *Second*, the extremity is made hollow, so that here, in spite of every care, particles of tissue adhere which are hard to remove. *Third*, the double canula can be cleansed only with difficulty if it should become plugged, which happens often. I therefore modified the catheter as follows: The point is solid and is as short as possible; the discharge opening is high up, close under the point. Besides, the thin tube is fastened to and in the thicker one by a peculiar screw which permits removal of the thinner tube for the purpose of cleansing the thicker part. The whole instrument is somewhat larger and therefore handier. If it is to be cleansed, after use, of the mucus remaining in the thicker tube, it is unscrewed and slipped off the thinner tube. The mucus then drops out or is removed by agitation in a vessel of water. In multiparæ it is almost always possible to insert this catheter into the uterus without preceding dilatation.

This catheter has first of all the advantage of being very handy on account of its curvature. Besides, there is no possibility that, in close application of the os or cervix, the fluid will penetrate into the tube. While the catheter is in position, the passage from the uterus into the vagina is moreover so slightly curved that the fluid and what it is intended to remove will readily pass from the uterus into the vagina. Flexible copper catheters have also been constructed to permit entry into a fixed and flexed uterus.

But if ichorous masses have remained for some length of time in the uterus, it can be assumed that the tissues adjoining the surface have become penetrated by septic germs so that a simple irrigation will not suffice. Just as noxious matters are absorbed by the lymph-vessels, so undoubtedly are those substances which render them innocuous. The therapeutic paradigm for putrid infection is carbolic acid intoxication. We may hope, therefore, that the disinfecting solutions remaining behind after irrigations become absorbed, and as it were follow the septic matters in their passage into the depths and render them harmless. Hence in many cases, after preceding careful removal of all decomposed secretions, repeated irrigation of the uterus suffices. But this result is not always so easily reached. To keep up the parallel with surgical cases, in these the surgeon would cauterize. By cauterization we destroy the uppermost layers containing *most* of the germs, and protect the wound by another appropriate dressing against renewed infection. It is not always possible to conquer infection in this manner, for the cauterization does not always reach as far as the invasion of bacteria. But the body assists the physician in natural manner. If there be no further introduction, the body will overwhelm the old ones. The fever moderates; the septic germs have been definitely destroyed or excreted by the organism, assisted perhaps by appropriate drugs, such as quinine or sodium salicylate.

If we apply these general principles to the internal female genital organs, we have the disadvantage of being unable to see the wounds. The important guiding points gained by the surgeon from the quality of the granulations, the covering of the wound, etc., are usually lacking to the gynecologist. Therefore, besides the results of palpation and the touch, there are but two things remaining to indicate energetic procedures—the general condition, and the secretion of the wound.

Far more carefully than in any other specialty must the temperature be measured and the pulse observed after gynecological operations! Should we conclude therefrom that fever is present, local examination, especially that of the secretions of the wound, will show whether the fever is referable to the wound. In that case the wounds have to be attacked.

The locality makes it impossible to employ the energetic caustics employed by surgeons. *The surfaces to be irrigated are not such as will have to heal together in future; therefore, strongly destructive measures are excluded eo ipso.* But an irrigation of the uterus with five-per-cent carbolic solution is to be permitted. The burning sensation in the vagina is to be very simply avoided by directing the tube of a second irrigator filled with water into the vagina and using it simultaneously.

In numberless cases this treatment will be followed by obvious improvement or cure. Then the greatest care in observation is pre-eminently necessary. Even in private practice this is possible—by self-registering thermometers which are to be shown several times daily, or with more intelligent people by the construction of a temperature chart, the further course is to be carefully observed.

D. PERMANENT IRRIGATION.

Should the fever persist in spite of these measures, there remains to us one energetic procedure—*permanent irrigation.* It alone enables us to employ strict antisepsis in the depth of the genital tract. Both in cases in which the above-described measures fail and in those in which the danger of infection would be especially grave, it is necessary to use permanent irrigation, or, as I have termed it, instillation. For this purpose we require an apparatus which I have called "the hydrostatic disinfecting machine" (Fig. 26).

This apparatus consists of an irrigator filled with disinfecting fluid. A Schücking dropping tube, depicted in the adjoining Fig. 27 in full size, is interposed in the efferent tube. This ingenious little apparatus regulates the flow. At the opening, provided with a faucet, the fluid enters

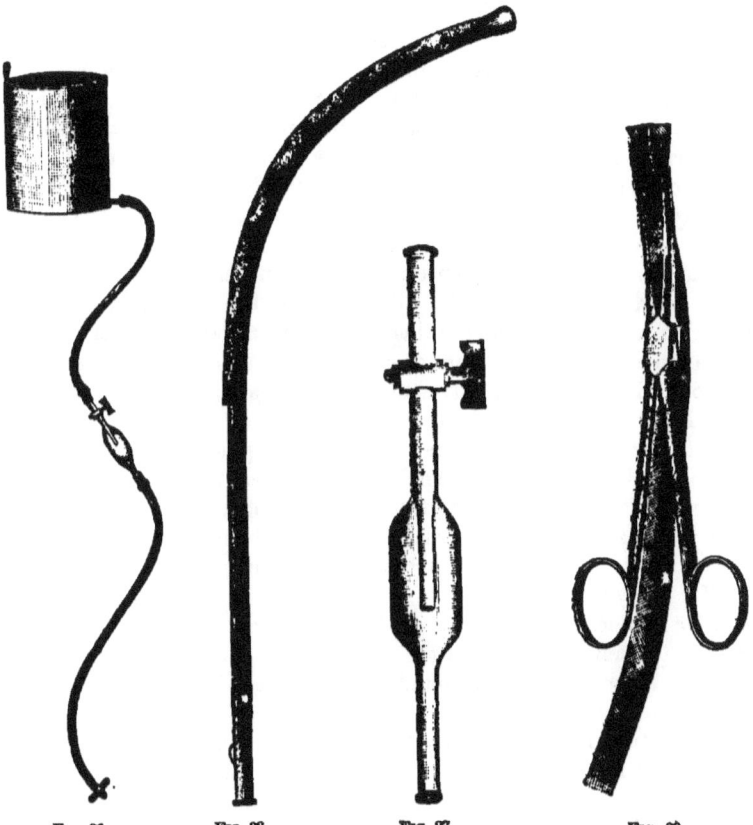

FIG. 26.   FIG. 28.   FIG. 27.   FIG. 29.

FIG. 26.—Hydrostatic disinfecting machine for the permanent irrigation or instillation of the internal female genital organs.
FIG. 27.—Schücking's dropping tube.
FIG. 28.—Schücking's catheter.
FIG. 29.—Drainage tube with transverse tube, ready for introduction into the uterus.

into a hollow cylinder through the continuation of the thin afferent tube, and thence continues its flow in the rubber tube. By the faucet we can regulate the flow from the full stream to a rapid or slow succession of drops. The layman is always able to convince himself, by observing the dropping-tube, whether the apparatus is acting. Whenever this is not the case, that is, whenever there is obstruction, the hollow cylinder fills and the apparatus must be looked after or set agoing again.

Schücking originally constructed this apparatus for the treatment of diseased parturients, and devised a catheter (Fig. 28) to remain permanently in the uterus. To an ordinary Fritsch's uterine catheter a metallic drainage-tube was soldered which was intended to serve for leading the fluid out of the uterus. This catheter is too heavy and large; it keeps the uterus in an unnatural position and hence may easily give rise to injury during movements. For this reason it is better to choose a quite thick-walled rubber tube as the intra-uterine part of the apparatus.

At the lower end of the apparatus shown in Fig. 26, in the afferent rubber tube, there is a small transverse tube intended to keep the tube in the uterus. It is passed through two holes in the larger tube; these holes should be of about half the diameter of the smaller tube, then the latter is firmly held. Below the crossing point two or three openings are cut to permit the disinfecting fluid to escape.

The introduction into the uterus may be done in various ways; with the finger if the uterus be wide, with the sound if narrow, just as drainage tubes are passed deeply in other places. I have found it very handy to use the nasal polypus forceps or curved dressing forceps for the introduction. The short ends of the tube are bent *upward;* then the apparatus is inserted, assisted by the other hand placed externally.

If the ends of the transverse tube are bent upward, the upper extremity of the larger tube must tend upward; when the elastic transverse tube rights itself, the drainage tube will rise higher. Were the transverse tube inserted in the opposite way, the tube would sink after the forceps are removed, when the short tube straightens.

Not rarely it is possible to lay the transverse tube into the wound itself, for instance after enucleation of large tumors. Often it is sufficient to place the transverse tube in the vagina in front of the os. This will be the case especially if the wounds to be disinfected are situated in the vaginal vault or the cervix as far as the internal os.

The patient had best be placed on a frame with a hole beneath which is a bed-pan for the reception of the fluid. But if in private practice such a frame cannot be had, the upper half of the couch must be raised. A bed-pan is placed in front of the genitals to catch the return fluid. The bed is kept clean by waterproof underlays. As the whole apparatus is movable, the patient may make some slight changes of position. If the bedding is to be changed or cleansed, the lower end of the tube is detached from the dropping tube, bound to the thigh, and replaced afterward.

After the apparatus has been started, the full stream is first turned on until the returning fluid is almost clear. Then the faucet is turned to permit the fluid to pass by drops.

Every two or three days the apparatus is to be removed for the purpose of cleaning the tube. Even if the patient be free from fever, the tube must remain for the time being to permit of instant re-irrigation should fever perhaps set in again. If the temperature has remained normal for two days, the discharge has ceased or has become simply purulent, the tube may be removed, and the vagina should be irrigated immediately afterward. Even if we cause some small wounds by the removal—for some blood flows nearly always—there are now no longer any infecting masses.

In cases of great flexion of the uterus, I have also tied or sewed the tube to a German-silver catheter. With Bozeman's catheter alone, we can also effect permanent irrigation. But the instrument is likely to slip out; therefore a rubber transverse tube should be sewed to the fenestra above.

It would be claiming too much to assert that in this manner we can always control the dangers of infection. But in large, ragged, infected wounds of the female genitals, permanent irrigation is assuredly the proceeding offering the greatest security.

E. Antisepsis During Laparotomies.

Besides antisepsis in operations *within* the female genital organs, antisepsis during *laparotomies* remains yet to be discussed. Here antisepsis is of such paramount importance that it must be managed with the utmost care. A laparotomy *without* infection is a trifling procedure void of any danger, a laparotomy *with* infection is almost always followed by fatal general sepsis. For this reason, here a painfully careful, I might say fanatical, prophylactic antisepsis is justified.

Previous to every laparotomy, the patient should be made to take at least two general baths. Previous to each general bath, the abdomen, the vulva, and the anus are well washed with soap. In the bath, the vagina is irrigated with three or four litres of three-per-cent carbolic solution. After the first bath, the patient is put into a freshly covered bed; the same after the operation. I hold this procedure, in which the uppermost layers of the epidermis are fully removed, to be of great importance. The most scrupulous disinfection of all instruments is necessary. I keep every instrument to be employed in laparotomy wrapped in white writing-paper. This paper, of course, is renewed after every cleansing of the instruments. Before the operation, the customary disinfecting soaping, brushing, and shaving of the operating surface are performed.

It is also very important to remove all old fæces from the bowels previous to the operation. The observation has been made so frequently that collections of pus close to the bowel become decomposed and develop septic gases, that we must be spurred on to seek prophylaxis. For instance, if we are to separate large flat adhesions, if there be in the abdomen an extensive raw surface to which the adjoining bowel becomes adherent, it is to be feared that the adhering, inflamed, infiltrated portion of the intestine would be likely to permit the passage of intestinal gases and elements of decomposition. If there be old fecal masses in the bowel, they may incite the decomposition of an exudation, even peritonitis and sepsis. I have thought myself justified in giving, besides evacuants, also disinfecting internal remedies. The possibility of a favorable effect is clear. For instance, in infantile diarrhœas we must disinfect the contents of the bowel and can cause the putrid-smelling evacuations to lose their offensive odor. In like manner we have the power also in laparotomies to obviate a danger *in a harmless way*. I give from five to ten grammes of potassium chlorate or two to five grammes of sodium salicylate twenty-four hours before the operation.

Regarding the spray, it is certainly unnecessary during laparotomies in private practice or good sanitary conditions in hospitals. And as at all events it is not immaterial to have larger quantities of carbolic-acid solution absorbed by the peritoneum, the spray may be omitted in most cases.

If it is possible, as in nearly all cases, to close the abdominal wound exactly by sutures, it becomes completely agglutinated within a few hours. Absorption or sepsis from that channel is not possible, hence a strict carrying out of the Lister dressing after laparotomies is barely serviceable. On the other hand, if a part of the abdominal cavity is to be left open in tumors which cannot be fully removed, or if but a portion of an open wound extends outward, the strictest exclusion of air or Lister dressing is of course absolutely essential.

Concluding Remarks.

Finally it may be remarked that the belief that through the teachings of antisepsis we have about reached the point of perfection would in itself be a step backward! Every uniformation of science is a misfortune! We can never stand still, but every advance opens up only additional, formerly unknown domains of further investigation! The enormous progress for which we are indebted to Lister is but the beginning of a struggle the consequences of which are far from being exhausted! Preliminarily, however, the principles of the Lister method of treating wounds have been generally recognized as correct. The next object must be, to test and to study, *while guarding those principles*, the antiseptic method appropriate to every single case or every group of peculiar cases!

CHAPTER V.

GENERAL THERAPEUTICS.

A. GENERAL CONSIDERATIONS IN GYNECOLOGICAL OPERATIONS.

ALL therapeutic measures are to be joined to certain periods. During menstruation, and shortly before or after it, the female genital organs must be left as quiet as possible. Smaller manipulations, cauterizations, etc., should not be performed later than four or five days before the expected menstruation, otherwise the irritation may bring on the menstruation prematurely and especially copiously. In the same way, local treatment should not be commenced sooner than one week after cessation of the catamenia, lest the return of menstruation be caused. Travelling to and from the physician likewise should not be undertaken in the time immediately adjoining the period. Any larger operation is reluctantly performed at the time near the period. On the other hand, the dread of every manipulation during menstruation is carried too far. I have very often injected liquor ferri into the uterus in menorrhagia with good result. Above all, however, cleansing irrigations and careful washings of the pudenda are not contra-indicated by the menstruation. Explorations during menstruation may possess an important diagnostic value. Thus, in inexplicable hemorrhages, we may occasionally detect a polypus palpable only during menstruation.

The manifold minor gynecological manipulations should be done with consideration and quiet. There should be no unnecessary noise. The exceedingly anxious patients become frightened if the physician rattles the instruments, drops them, or searches too long for the appropriate one. For this reason, everything should be as handy as possible, lest there be too much walking about, appearing excited, and frightening of the patient. Especially knives and other instruments should not be shown unnecessarily. Even the blood should be removed lest the patient see it. In all manipulations we should strive to proceed with the greatest delicacy. A rude uncovering of the patient, a rough touch, is forgiven far less readily than an erroneous diagnosis! Many an older gynecologist owes the greater part of his practice to his tact and behavior rather than to his knowledge and results!

Besides, the physician should proceed as tidily as possible. Drugs which diffuse a disagreeable odor, such as pyroligneous acid, iodoform, etc., should not be employed unnecessarily. Medicaments which stain and destroy the clothing should be kept away from it. The patient's attention should be called in advance to the possibility of soiling the linen.

If some unfavorable occurrence be feared after any manipulation, the patient should be carefully prepared for it; for instance, secondary hemorrhage after large incisions, also the separation of crusts, scabs, shreds after extensive cauterizations, colicky pains after uterine injections, discharge of water after placing glycerin tampons, etc. If this precaution be omitted, many a patient becomes much alarmed, excited, and annoys the busy physician by continual inquiries and summons.

If possible, the purpose, the importance, and the manner of the treatment should be to some extent explained to the patient. Never should too much be promised, lest the confidence in the physician's dictum be shaken after failures. If the case be hopeless, the unfavorable prognosis should be reported, not to the patient, but to the relatives.

B. Vaginal Irrigations.

Injections into the vagina are made both for cleansing purposes and for securing contact of the vaginal mucosa with medicamentous substances. The irrigator, Fig. 26, page 51, is to be chosen as the instrument for conveying fluid into the vagina. It is more practical, cheaper, more durable, less dangerous, and more easily applied than all instruments hitherto in use.

For irrigating the vagina the instrument should be provided with a very long tube measuring two or three metres, to enable the patients to insert the vaginal tube into the vagina beneath the clothing without being hindered by the latter. The vaginal tube is provided with a faucet which is opened as soon as that tube has reached the vagina.

For cleansing purposes, there should be added to the water a teaspoonful of table salt, soda (sodium carbonate), a tablespoonful of common or medicated alcohol.

The temperature of the water should be agreeable to the patient— from 25° to 29° R. (88° to 97° F.).

Astringents are most frequently added to the irrigation.

The ordinary agents of that class are: alum, sulphate of copper, tannin, acetate of lead, sulphate of zinc, sulphate of iron, acetate of zinc, nitrate of silver in weak solution. Also flaxseed tea, weak decoctions of starch, of oak bark, and of cinchona bark, with or without the addition of opium, were formerly employed.

For disinfecting purposes, carbolic or salicylic acid, chlorine water, hypermanganate of potassium, etc., is added to the water. The two may also be combined; for instance, by ordering a mixture of alum (5 : 100) and carbolic acid (2 : 100) for injections. This would be a *strongly* astringent vaginal injection.

For particular therapeutic purposes, injections are also made of hot 40°–41° C. (104°–106° F.) or iced water. If larger quantities of water are to pass into the vagina of the recumbent patient and remain as long as possible, the return water is caught in a bed-pan, the efferent tube of which is closed with a perforated cork. A thick glass tube extends through the hole in the stopper to the bottom of the bed-pan. To the glass tube is attached a rubber tube leading to a chamber vessel or a pail. If the bed-pan be warmed by being filled with hot water, the latter will flow off when the pan is raised. If then the patient be placed on the bed-pan, the vaginal tube inserted, and the stream kept uninterrupted, all the fluid coming from above will be siphoned off. As it is often essential to have large quantities of hot water brought in contact with the vaginal vault, this apparatus, which renders the frequent troublesome emptying of the bed-pan unnecessary, will be found very useful.

Tubular specula have also been constructed which are closed in front by a tube provided with two faucets. Into one the fluid enters and returns through the other. In this manner local baths of the vaginal portion are given.

Of course, as the stream of water does not penetrate into the cervix, the irrigation of the vagina cannot cure leucorrhœa. On the other hand, vaginal irrigations are necessary during the treatment of uterine diseases, to remove blood, secretions, or remnants of the cautery from the vagina.

The simplest method of irrigation for cleansing purposes is the following: Two chairs are brought together, the seats being fifteen centimetres apart. A pail is set underneath. The irrigator is placed near on a stove, bureau, or hung on a nail. The vaginal tube is within easy reach. The patient lifts up *all* her clothes and sits with bare buttocks on both chairs, the vulva being in the centre above the pail. The vaginal tube is then introduced and the faucet opened.

In this way the patient avoids wetting the room and her clothing. Quite minute directions are very agreeable to the patients who usually dread to ask any one's assistance. It is incredible how awkward many women are about it. Not rarely the patients report in despair that all their clothes had got wet, the room soiled, but not a drop had got into the vagina.

Just as liquid remedies are introduced into the vagina, so also has the vagina, dilated with speculum, been insufflated with powdered drugs, such as alum and starch, smeared with salves, painted with solutions, etc.

C. TAMPONADE.

Another therapeutic method acting from the vagina is the tamponade. The vagina is packed with pledgets of cotton. As to the technique, it is best to carry the tampons upward through a tubular speculum. They must not be too large, so that they may be easily pushed out of the upper opening. Each tampon must be inserted singly and pushed out, otherwise all the compressed tampons would be withdrawn with the speculum.

Of course, the tamponade may be done equally well through Sims's or Simon's speculum. We can tampon even without a speculum. Two fingers of one hand crowd the posterior vaginal wall somewhat backward toward the anus. In the groove between these two spreading fingers the tampons are pushed up one by one. Without speculum it is best to tampon in Sims's lateral position. Two fingers are hooked into the vagina, drawing the posterior vaginal wall down, with the other hand the tampons are packed, avoiding the plainly visible urethral eminence.

In order to enable the patient to remove the tampons, every one had best be tied with a strong thread. If this be omitted, the physician must introduce a speculum, and the tampon presenting above be seized with forceps and removed.

Should the patient herself have to apply medicated tampons, she will require an instrument—a tampon-carrier. Many such instruments have been constructed. The one with which women succeed best is represented in Fig. 30.

The capsule into which the tampons are laid I have had made five centimetres long and two centimetres in diameter, so as to permit it to enter narrow vaginæ. The tampon, perhaps saturated with glycerin, with dependent thread, having been placed into the capsule, the instrument is brought as high up into the vagina as possible. The capsule contains a piston which is brought to its upper level by pressing on the top of the handle. This pressure empties the capsule of its contents—the tampon. While the handle is being depressed, the patient must hold the instrument with the other hand. The vaginal walls keep the tampon in place when the instrument is withdrawn.

Fig. 30. Tampon-carrier.

In the first place, tamponing is often done for the purpose of arresting uterine hemorrhages. If the *momentary* danger during uterine hemorrhages is too great, there is nothing left but to plug the vagina with cotton so as to occlude the passage of the flow. In this case many small tampons are taken, dipped in water or carbolic solution, and tightly packed into the vagina. If strong astringents were chosen to saturate the tampons, the vagina would be cauterized, while the bleeding spot would not be touched by them. But the tampons are moistened with astringents if the application can be made to the bleeding spot directly, for instance, to the vaginal portion. Moistening the tampons is necessary, because dry tampons shrink after being saturated with blood, and thus cannot occlude the vagina absolutely.

Besides, a dry tampon is often placed in front of the vaginal portion, to protect wounds against infection from below, or the vagina against caustics employed at the vaginal portion.

Again, as a kind of dressing in ulcers of the vaginal portion, tampons are inserted to absorb the secretions. Formerly, the position of the uterus was even sought to be corrected by the insertion of tampons.

Tampons are very frequently employed where it is intended to have remedies act upon the vaginal portion or the vaginal walls or both. Therefore the tampons are saturated with glycerin with which other drugs are also mixed, such as tannin, iodine, potassium iodide.

Tamponing is also done for the purpose of exerting pressure on the vagina and uterus. Dry, specially elastic, woollen wadding is used in order to compress the vaginal walls centrifugally.

Quite recently, Schultze has employed a "test tampon." A tampon saturated in tannin-glycerin is placed exactly in front of the external os for the purpose of showing quite accurately the secretion of the uterus. After removal, the uterine secretion is seen at the spot corresponding to the os externum.

D. Vaginal Suppositories.

The application of suppositories is a method related to the tamponade.

The ancients already had "medicated pessaries," that is to say, wax globes containing fragrant herbs, which were laid into the vagina. Also nowadays we employ suppositories of cacao butter, glycerin and gumarabic, etc., which melt in the vagina and liberate the medicament, such as morphine, potassium iodide, alum, to act upon the part.

In general this method is not to be recommended. The suppositories drop out of wide vaginæ during walking. Should they melt, however, then the liquid mass flows out of the vagina and soils the vulva and rima pudendi.

But should the patient be very awkward, or if for any reason no irrigations can be made or tampons applied, vaginal suppositories form a substitute for irrigations and tamponade. For this reason, small medicated suppositories are advantageously employed in virgins or in persons with abnormally sensitive introitus.

E. The Abstraction of Blood.

One of the most frequent therapeutic measures is the abstraction of blood from the internal and external female genitals. If we cannot get into the vagina (in virgins), leeches are applied to the perineum, the mons Veneris, or the inner surface of the thigh. The perineum is, under these circumstances, the best point of application. Owing to the many communications of the vulvo-vaginal and uterine veins, we may hope in this manner to deplete also the uterus. As an anastomotic branch extends from the epigastric through the inguinal canal with the round ligament to the fundus of the uterus, the domain of the spermatic artery becomes depleted also by abstraction of blood from the neighborhood of the inguinal canal. If we succeed in some way in inserting a speculum, even if but a small one, the vaginal portion is engaged and blood withdrawn from the uterus directly. In uterine disease, this is far more effective than abstraction of blood from neighboring regions. Formerly leeches were largely used. The whole manipulation consumes much time, and but few midwives have sufficient knowledge to undertake the application of leeches to the vaginal portion. The sudden nervous disturbances, the fact that a leech may crawl into the uterus and there cause colic or may fasten itself to the vagina, the frequently great difficulty of positively arresting the bleeding, and the carelessness of the midwives to whom this duty is usually intrusted, speak against the employment of leeches. With leeches, the after-hemorrhage is the main point, but it is difficult to control, so that we never know how much blood has been abstracted. Much simpler and better are the scarifications of the vaginal portion recommended by all recent authors. The vaginal portion is engaged in a tubular speculum, which is pressed firmly into the fornix so that no blood can escape between speculum and vagina, then the vaginal portion is pricked with a long-handled knife. Quite a number of uterine scarificators have been described. The instrument represented in Fig. 31, resembling a large lance-needle, is to be well recommended. It is also suitable for opening the ovula Nabothi. But the point must be very frequently sharpened, for only a very sharp instrument will easily pierce a soft vaginal portion. If sufficient blood escapes from one opening, then no more are required. However, a single stab rarely suffices. The several openings need but penetrate the mucosa. Incised scarifications may also be made; the entire vaginal portion may be covered with a network of incisions. This method has no advantage over the more easily performed stabbing. If blood is to be abstracted from a higher point, the knife is inserted into the cervical canal, and shallow radial incisions are made. Should the abstraction of blood be intended as a preliminary treatment in catarrh of the cervical mucosa, it is more correct to take the blood at once from the latter. This is most simply effected by scraping it with a sharp spoon. If but few shallow scarifications have been made, the pressure of the vaginal walls against the vaginal portion suffices to arrest the hemorrhage. Nevertheless I always put a tampon of salicylated or benzoated cotton against the vaginal portion, thus securing disinfection with the prompt

FIG. 31.—Two-edged knife for the scarification of the vaginal portion.

arrest of the hemorrhage. Before I had commenced doing so, I saw several times crater-shaped covered ulcers result from the scarifications which healed only after five or six days, after having been cauterized. The prompt arrest of the bleeding has also a psychic value. Many a patient is greatly alarmed by the sudden appearance of blood. The tampon should be removed by the patient at night when in bed, because, if hemorrhage should still recur, it will cease spontaneously in the recumbent position.

F. The Application of Caustics.

Previous to the application of any caustic, the vaginal portion must be cleansed of the adhering secretions. For this purpose, as well as for the application of caustics to the cavity of the cervix and uterus, a large number of partly very expensive and very unpractical instruments have been devised. Right here, in beginning the description of the application of caustics, I shall recommend an instrument which serves excellently, and which I am using for years for all cleansings and cauterizations on and in the uterus. This instrument renders all forceps and caustic-holders superfluous, and will impress every gynecologist who uses it with its simplicity and cheapness. This instrument is Playfair's sound, but modified by me in a manner which renders it a universal instrument.

Playfair's sound consists of a German-silver or aluminium end-piece on a wooden rod; it differs from the uterine sound in having its uterine end-piece roughened to facilitate the wrapping around or fastening of cotton. Thus armed, the instrument was dipped into the caustic fluid and introduced into the uterus. Here we have the disadvantage that the wet cotton is with difficulty removed from the instrument. But if the surface is to be carefully cleansed, we must enter ten or more times and wipe. Nothing could be more tedious and distasteful than after every wiping to remove the soiled cotton from Playfair's sound with the knife and scissors. But if the sound is made conical, taping gradually to the extremity, the cotton is readily pushed off with strong pressure. Soon sufficient dexterity is acquired for twisting large or small brushes, or for covering the entire sound in order to swab out the uterus. The lower layers of cotton must always be wrapped tightly; the thinner the first layers, the more firmly the cotton adheres. If all of the cotton is soiled, a pledget is taken between the fingers, and the wadding is pushed off the sound.

If we desire to cauterize with liquid caustics, but very little cotton is wrapped around (Fig. 32, *a*). It is better to cauterize only a spot of three millimetres in diameter at a time, and slowly and gradually to touch the entire surface, than to have drops of the caustic fluid spread in the surroundings. If but little cotton is wrapped on, but little caustic is retained, and spreading in the vicinity is almost impossible. But if the cervix or uterus is to be cauterized all over, a small cauterizer is much better than a large one. The small instrument penetrates more readily into all the nooks and crannies of the cervix.

Whoever has not tried my uterine rod will raise two objections: first, that the cotton is easily stripped off, and remains in the cervix of the uterus. Some authors even intend to leave a pledget of cotton saturated with the drug in the uterus. It drops out spontaneously amid copious suppuration. Hence the remaining behind would be no great misfortune. But it has never happened to me in many years. The smaller the quantity of cotton wrapped around, the more firmly it is held. Should large pieces remain *in the vagina*, they are easily removed with the rod, that is, scraped out through the speculum. Another objection is that the rod is soon worn smooth, when, of course, the cotton can no longer adhere. To be sure, this objection is just. I have the rods roughened over every three or four months. This has reached its limit after five or six times, so that the rod must be thrown away. Considering that such a rod is worth at most half a dollar, the loss is not very great.

Although I advocate the use of liquid caustics only, I shall briefly enumerate the old methods and agents.

A priori, every one will admit that the object of cauterization can be attained by a number of remedies—silver nitrate, potassium hydrate, chromic acid, liquor Bellostii, zinc chloride, carbolic acid, carbon bisulphide, salicylic and acetic acids, tincture of sesquichloride of iron, nitric acid—all such as destroy the superficial tissue so as to cause a loss of substance which heals by granulation. Moreover, there will be differ-

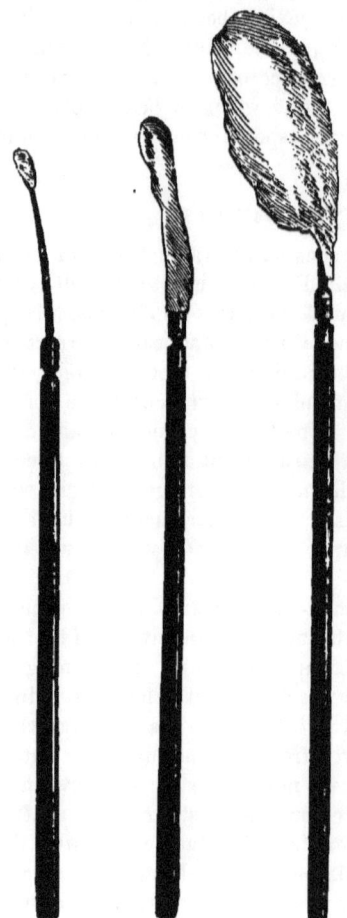

FIG. 32.—Uterine rods.

a, Armed for cautery; only sufficient cotton is wrapped around to permit a drop of liquid to adhere; *b*, armed for cleansing the uterine cavity, also for the introduction of liquid remedies, such as tincture of iodine; *c*, provided with a large tuft of cotton for cleansing the vaginal portion, absorbing blood, removing secretions, crusts, excess of caustic fluid, etc.

ences in the degree of destruction: one agent, for instance, liquor ferri, barely attacks the uppermost layers of the mucous membrane; another, such as chromic acid or potassium hydrate, corrodes deeply. In weak solutions all these drugs will have an astringent and disinfecting effect, while, on the other hand, astringents in substance or very strong solution can be used as caustics, as copper sulphate, zinc sulphate, alum, etc. No gynecologist can test all remedies on hundreds of cases. Every one will use that drug from which he sees good results; will praise it, trust in it, and will not feel called upon to try ten others. On the other hand, it would be wrong to believe that one remedy acts specifically, that with it alone success can be secured.

It is only confusing to lay down special indications for a whole series of caustics. Different indications of course exist in that we sometimes wish to cauterize deeply, as in carcinoma, or at other times desire to produce a more tonic than a caustic effect on the mucous membrane. However, all those caustics or methods respecting which but a *single* fatal case or undesirable after-effect has been reported, are to be at once abandoned as, for instance, the insertion of the stick of silver nitrate into the uterus or the employment of chromic acid.

Probably, the agent most frequently employed is silver nitrate; this remedy is used unspeakably often, without judgment, we might almost say without understanding. Of course it is capable of destroying diseased tissues, so that the normal ones develop anew, and thus performs everything demanded of caustics. Does not "lunar caustic" belong into every instrument case, for "cauterizing granulations?" Silver nitrate in substance was used for erosion of the vaginal portion, and was also introduced into the cervix. In this way the stick becomes at once covered with albuminate of silver, which protects the cervical cavity against its effects. Frequently, only the external os was cauterized and a stricture thus produced which made the entire condition worse.

The stick has been introduced into the uterine cavity, and even left in it. The silver nitrate can also be rendered fluid and fused to a sound, so that in sounding with the silver-covered instrument every spot touched by it is cauterized.

A number of caustic-holders have also been constructed for the silver nitrate stick. I shall neither describe nor figure these instruments, as they are quite superfluous. Should we desire to use caustic sticks, a long dressing forceps, a long quill, or any of the ordinary caustic-holders will suffice.

Other highly corrosive agents are bromine, caustic potassa, and zinc chloride. While in surgery caustics have long been abandoned, and what is removable is cut off, tumors that cannot be operated on being rightly left alone, the otherwise modern gynecologists have, until the most recent times, cauterized and burned carcinomata. I shall recur hereafter to the mistake of instituting such methods of treatment, and therefore desist entirely from describing these methods.

Nitric acid has acquired a special reputation of late. It was first recommended in the shape of the fuming acid, but its employment has many inconveniences for the practitioner. I have therefore ceased using it, and obtain the very same result with the pure acid. This caustic has the advantage of not forming a firm albuminate, but a tenacious brownish fluid which again acts caustically in its spread.

I have never observed any inflammatory symptoms after such cauterizations. I have performed countless cauterizations, both without and with preceding scraping of the mucous membrane.

It is best to take an opaque glass or a Fergusson's speculum, because they are not attacked by the acid. But if the vaginal portion cannot be well engaged, a Cusco speculum (of German-silver) may also be employed. The latter draws the vaginal portion slightly apart, so that it is quite easy to get into the cervical canal. Where assistance can be had, it is of course preferable to cauterize through the Sims speculum.

If a cauterization is to be done, the vaginal portion is first carefully cleansed. I use the word "carefully" advisedly, lest a hemorrhage disturb the accuracy of the procedure. Then we take my uterine rod wrapped with a very little cotton, dip it into some nitric acid, which had better be poured into a porcelain medicine spoon or a watch glass, and by degrees touch the entire ulcerating surface. The cauterized part covers at once with a white crust. This very much facilitates accuracy, and the restriction of the caustic to the required spot, while the healthy tissues are spared.

If the cervical canal is to be cauterized, we likewise take very little acid, advance rapidly to the internal os, and press the small instrument against the walls in every direction. The momentary irritation is very slight, the patients only rarely feel pain which, however, disappears within a few minutes. Contraction does not occur during the cauterizations.

All caustics are open to the objection which is made again and again, namely, that cauterizations easily cause strictures, even atresiæ of the cervix. It is quite obvious that, after complete destruction of the mucous membrane, a granulating surface with a tendency to cicatricial formation and adhesion will take its place. For this reason the introduction of large quantities of caustic fluid into the uterus is dangerous. The surface must only be touched, then the glandular basis remains behind, and from it forms a new mucosa, as in the puerperium. Also from the uncauterized islands always remaining, even after careful cauterization, new epithelium overspreads the denuded portions.

However, if the mucous membrane be given no time for regeneration; if the caustic be introduced in insufficient intervals, about every twenty-four hours, the weakest caustic may lead to atresiæ. *Therefore adhe-*

sions occur with especial facility after rapidly repeated cauterizations. I have seen an atresia occur after the daily application of tincture of sesquichloride of iron. In like manner, frequent superficial cauterizations of the vaginal portion with silver nitrate lead to contractions of the external os. But after nitric acid, I have never seen any contraction. However, it cannot be denied that here, too, the mucosa is destroyed too extensively *if the cauterizations follow each other too soon*. Thereby, of course, the possibility of adhesion is furnished.

If the precaution be taken always to touch with but little acid, and to cauterize at most once a week, a constriction will scarcely ever occur. Obviously, different procedures are required in varying indications; for instance, the cauterization, on the contrary, will be done quite frequently, where an adenoid erosion of the vaginal portion is to be thoroughly destroyed.

Never should so much caustic fluid be taken that it drips down. Should this still happen, every drop must be wiped off. After the cautery, a cotton tampon is laid in front of the vaginal portion to protect the vagina against the effects of remnants of the caustic fluid. The tampon is removed after twelve to at most twenty-four hours. The after-treatment consists of lukewarm vaginal injections, which are intended to carry off the crusts, coagula, and blood from the vagina.

G. The Employment of the Actual Cautery.

Among the caustics belongs also the actual cautery. It is employed in three different ways. The simplest way is to use red-hot irons with pointed or broad extremity. They may be heated in a stove or over a gas or alcohol lamp. A very practical apparatus is the æolipile.

It consists of a tin vessel with double wall (Fig. 33). Through the opening (b) some alcohol (a) is poured in, and the cork closed tightly.

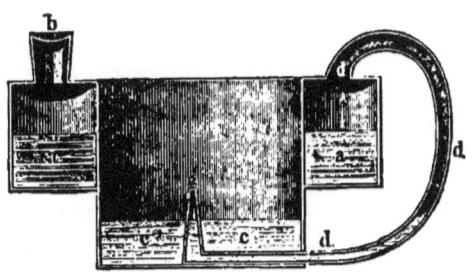

Fig. 33.—The Æolipile in Section.

Then some alcohol is also poured into the central space (c) and ignited. Thereby the alcohol (a) becomes heated and evaporates, escaping through the tube (d) and issuing at e. Here the alcoholic vapor is ignited by the burning alcohol c, so that the inflamed vapor rushes upward with loud blowing noise. The cautery iron is held in the flame or fastened in such a way as to retain the handles between two books placed one on the other. This apparatus has also been provided with a holder for the cautery irons (Fürst). The apparatus is very practical on account of the rapidity with which the heat is obtained. The noise alone terrifies non-anæsthetized patients.

If the actual cautery is to be performed in the depth of the passage, a wooden speculum is required, for those of glass crack and the metal ones become heated. Should metallic instruments still be used, such as Simon's specula, a stream of cold water must frequently be turned on, so as to cool the specula in this manner.

FIG. 34.—Amputation of the vaginal portion with the galvano-caustic knife. The vaginal portion is engaged in Dr. Byrne's speculum, and drawn forward with Byrne's tenaculum.

The galvano-cautery is likewise gaining more and more adherents, as the apparatus have become cheaper and more compact. For the purpose of cauterization, the galvano-caustic apparatus is superfluous; at most in deep fistulous tracks, or in the uterine cavity, the galvano-cautery might find application. But it is a truly indispensable instrument in the ablation of profusely bleeding tumors. Here the galvano-caustic operation may proceed without danger, while the hemorrhage with cutting instruments would be almost uncontrollable. It is even possible to put a wire snare around the base of a tumor which covers the access so completely that we can approach it with no other instrument.

Quite recently a third apparatus has been introduced—Paquelin's thermo-cautery. This apparatus has the great advantage that the red-hot point remains continuously heated. Nevertheless the apparatus has not proved itself in practice indispensable to the physician. Particularly the glowing point often becomes surrounded with crusts, which render a further effect of the heat impossible. Both from the wire of the galvano-caustic apparatus and from the point of Paquelin's thermo-cautery, various knives and even scissor-shaped instruments have been constructed for cutting operations. Thus Fig. 34 shows the amputation of the vaginal portion with a galvano-caustic knife.

To the practitioner, the ordinary hot iron would be most commendable for caustic purposes. The actual cautery finds its application especially in the arrest of hemorrhage.

H. The Effect of Remedies on the Internal Surface of the Uterus.

If remedies are required to act on the internal surface of the uterus, that is to say, of the body of the uterus, they may be introduced in solid or liquid form.

The method of introducing sticks of silver nitrate has been almost universally abandoned, because the effects were not sufficiently under control. Also the cauterization with a stick held in the porte caustique, or with silver nitrate fused on or in, is very little practised. Nor has the employment of bacilli become popular. These were small rods formed of althæa root with zinc chloride or other medicaments which were inserted into the uterus. Nowadays liquid remedies are almost exclusively employed. They may be introduced in the first place with my uterine rod (Fig. 32, *a b*). An internal os which permits the passage of a sound can also be penetrated by a pledget of cotton around a rod, if it be wrapped firmly. For the uterus frequently reacts on caustics by contracting, and a loosely wound tuft of cotton would be stripped off during a difficult withdrawal. The objection has been raised against this method that the caustic fluid remains in the cervical canal, but as the latter is generally diseased, this is especially desirable. The smaller the tuft of cotton, the more readily we can reach every part of the uterus. The uterus may also be previously dilated so as to reach the uterine surface the more carefully and perfectly. The course must always have been explored in advance with the sound, so as to get to the fundus as rapidly as possible. Tincture of iodine or nitric acid are to be especially recommended for painting the cavity of the uterus.

An instrument by which larger quantities of fluids are more readily introduced into the uterus is Braun's uterine syringe, which is much in use by all gynecologists. Braun's syringe holds about ten drops of fluid. Most frequently it is employed for injecting styptic remedies, such as tinct. ferri sesquichloridi. If the os is readily penetrated by a uterine sound, a preceding dilatation is altogether unnecessary, and even increases the danger of the manipulation needlessly.

The following is the procedure: The patient lies in bed; the vagina receives a disinfecting injection. The speculum is introduced, either that of Mayer, or that of Cusco, or best that of Sims in the lateral position. We then ascertain with the sound the precise direction of the uterine cavity and introduce Braun's syringe, filled with tinct. ferri, up to the fundus. The syringe is then drawn back a very slight distance and about two drops are allowed to escape. The patient is asked if she feels any pain. In case it is very violent, which is of quite rare occurrence, we wait some time, palpate externally both sides of the fundus and thereby ascertain its lack of sensibility. Should tenderness be demonstrable anywhere, or should the pains persist, the manipulation is interrupted and we wait. Otherwise the syringe is withdrawn about one-half centimetre and a few more drops are discharged. This is continued until the last drops leave the syringe at the internal os. The syringe is then allowed to remain for some time while the os is observed in the speculum. Should drops of tinct. ferri or black coagula escape from the uterus beside the syringe or immediately after its removal, there is no further danger. But if—what is most likely to happen in hemorrhages after abortion with wide cavity or in strong flexions—the entire fluid remain in the uterus, sounding should be repeated to make sure of the patency of the internal os.

Very rarely, immediately after the injection or a little later, violent labor-like pains occur—uterine colic. They have no significance if, on combined examination, the external surface of the uterus and the parametria are free from pain. An injection of morphine soon removes the pains. In general it may be said that the object of the injection—diminution of the uterus—occurs sooner if uterine contractions, that is pains, are present. But they may become so violent that they must be moderated by morphine. Should there be peritoneal tenderness on pressure, or should fever set in, we must expect beginning peritonitis and treat it antiphlogistically.

The injections of tinct. ferri are so free from danger that I often, in suspected remnants of abortion, as it were tentatively, first injected tinct. ferri and observed permanent cure. In the hemorrhages so frequently occurring in gynecological practice, the application of Braun's syringe is one of the most common minor operations.

Some changes have been made in the instrument. Thus the part of the syringe which lies in the uterus has been roughened externally and perforated with numerous openings. The syringe wrapped with cotton is then introduced, and if it is in good position it is emptied; the fluid must saturate the cotton and thus come in contact with the entire uterine cavity. This modification is most noteworthy; it has the advantage that it renders the entry of fluid into the tubes impossible.

Braun's syringe has a knob at the extremity which is perforated laterally. I have had this knob pierced above, so as to bring the fluid as much as possible in contact with the mucosa all around.

FIG. 35.—Braun's syringe for injecting fluids into the uterine cavity.

Many cases are recorded in literature in which a peritonitis was observed in consequence of intra-uterine injection. It was assumed that the fluid escaped through the tubes. It cannot be denied that a single such case would cause the practitioner to desist from this beneficent treatment. Even if it be asserted that penetration of fluid through healthy tubes is impossible, it must not be forgotten that only pathological cases are the objects of medical treatment. But it is unquestionable that there is absolutely no danger in the method described by me. We have seen above that the tubes may be sounded. Therefore, should in one of these extremely rare cases the syringe enter the tube and injection be practised with great force, fluid could of course be driven into and through the tube. Danger would even then be possible if the injected fluid were to displace a pathological tubal secretion and crowd it into the peritoneal cavity. But as we said above, with care this is impossible.

Altogether, as regards the permeability of the tube for fluids, it must not be forgotten that the tubal epithelium vibrates toward the uterus; therefore that in irritating fluids peristalsis and the vibratile current oppose great obstacles to the penetration. In the dead subject it will be easy to force fluid through. These cases can by no means govern the conditions in the living.

However, that it requires no excessive force of injection to overcome the resistance of the tube is proved among others by the case of Dr. Späth, in which the fluid passed through the tube during a vaginal injection and caused fatal peritonitis. The patient was crouching on the floor, had inserted the uterine tube into the vaginal portion, and suddenly injected a large quantity of water into the uterus with the clysopomp.

Quite recently another method of acting on the internal surface of the uterus has been much employed in the shape of curetting or *evidement*, the scraping of the uterus with the spoon. Simon had very good results with the long sharp spoon devised by himself. With these spoons he scraped off and removed soft tumors. The healthy, hard tissue resisted the spoon, so that even without accurate diagnosis of the extent of a soft tumor it was almost automatically removed by the sharp spoon. In surgery, too, the sharp spoon had been largely used before, for the removal of stubborn exanthemata, unhealthy granulations, pyogenic membranes in fistulæ and sinuous ulcers, or diseased mucous membranes. The sharp spoon replaced knife and scissors in places which could not be reached by cutting instruments, and besides it was more easily controlled in its effects than strong caustics formerly employed, for the destruction of tissues. Hence the idea readily occurred to utilize the sharp spoons also in gynecology. They (Figs. 36 and 37) are made in six or eight different sizes. The largest serve for the removal of large soft tumors; with the smallest spoons it is possible to enter the cervix and even the uterus without any preparatory dilatation.

Besides these instruments, there are also uterine curettes with flexible stem (Fig. 38). On the copper stem fastened to the handle, there is above an oval, not very sharp steel loop with which the inner surface of the uterus can be scraped just as with the sharp spoon. Fig. 39 represents a dull curette provided with a wire instead of the steel loop. It is self-evident that it is possible to construct many other instruments of the

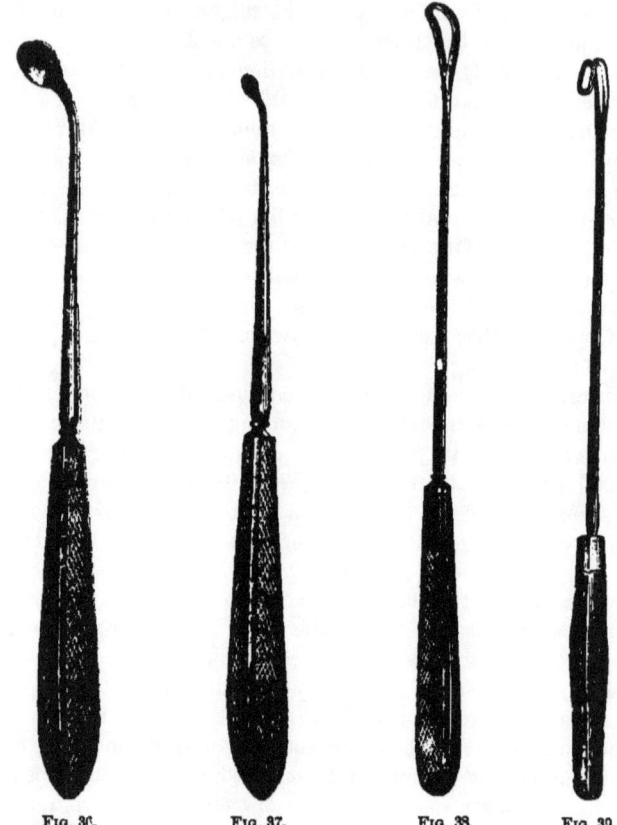

FIG. 36.   FIG. 37.   FIG. 38.   FIG. 39.

FIGS. 36 and 37.—Simon's sharp spoons. The sizes in use range from 3 to 15 mm. in transverse diameter.
FIG. 38.—Sharp curette. The stem is made of copper so as to render the instrument flexible.
FIG. 39.—Dull curette for the removal of masses lying loosely in the uterus.

same type which answer the same purpose. The surgical instrument catalogues contain a number of them.

CHAPTER VI.

DISEASES OF THE VULVA.

A. Malformations.

The defects of development of the vulva may be best understood in connection with the history of development. In referring to what is stated below in regard to malformations of the uterus and the gynatresiæ, I append some schematic drawings for a better understanding.

In the eighth week of embryonal life, there forms below the genital eminence (clitoris or penis) the genital furrow which, gradually deepening, perforates inward at the point where the bladder (allantois) separates from the rectum (sinus urogenitalis). On both sides of this point of perforation arise the labia minora, while the labia majora have to be traced back to the genital protuberances situated on both sides of the genital eminence. The separation of the allantois from the rectum is brought about by the portion of tissue between the rectum and Müller's ducts proliferating downward, forming the perineum and thus dividing the anus from the vulva.

It is possible, in the first place, that neither the peritoneum proliferates downward nor the genital furrow perforates upward. Then the fœtal condition (Fig. 40) remains—atresia ani et vulvæ completa. Or else, the genital furrow fails to perforate upward, but the perineum grows and separates the bladder from the rectum. Then the secretions may accumulate in the bladder and rectum. Such malformations have hitherto been found only in non-viable monsters, combined with other arrests of development.

Not rarely, however, the perineum fails to grow downward, and hence the rectum opens into the urogenital sinus (Fig. 43). The posterior limitation of the vulva is the posterior wall of the rectum, not the perineum unperforated by the anus. These cases are usually designated as atresia ani vaginalis. But even on superficial examination it appears striking that there is an absence of the hymen, also that the lower part is not the vagina, but the urogenital sinus. Therefore, the following designation is more correct: atresia ani vestibularis, or anus præternaturalis vestibularis.

The condition here is something altogether different from atresia ani, where there is often an entire absence of a large part of the rectum which may terminate at the level of the promontory (Fig. 44). In the so-called atresia ani vaginalis, the rectum is present entire.

Such infants being viable, they also come under medical treatment. *Here then the first care must be to provide passages amply sufficient for the evacuation of fæces.* In atresia ani vestibularis, there is often found a considerable fecal accumulation. The fæces cannot readily escape through

FIG. 40.—Primary condition. Bladder *b* and rectum *r* still communicate. Müller's ducts *v* terminate in the allantois *b*. At *c* the genital eminence (clitoris) is forming; posterior to it the genital furrow *h*, and behind that the anus *a*.

FIG. 41.—Bladder *b* and rectum *r* have separated; the perineum *p* has extended between them; in front of the latter is the urogenital sinus in which terminate above the urethra and Müller's ducts 'vagina).

FIG. 42.—Developed genitals. *b*, bladder; *v*, vagina, separated from the urogenital sinus *su* by the hymen *h*; *p*, perineum; *a*, anus; *r*, rectum; *c*, clitoris; *x*, urachus.

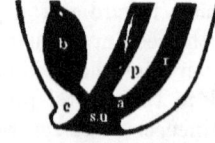

FIG. 43.—So-called atresia ani vaginalis, properly vestibularis. *su*, the long sinus urogenitalis; above the vagina *v*; *p*, rudimentary perineum: *r*, rectum, terminating at *a* in the urogenital sinus.

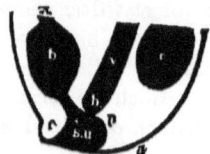

FIG. 44.—Atresia ani; all relations are normal, only the rectum *r* ends in a blind pouch at a great disance from the imperforate anus *a*. The other references as in Fig. 42.

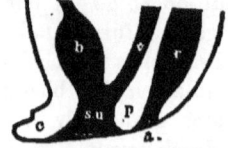

FIG. 45.—Female hypospadias; the bladder *b* terminates directly, without urethra, in the urogenital sinus *su*. The vagina *v* is narrow. *p*, perineum, and *r*, rectum, normal. *c*, hypertrophic clitoris.

the opening in the wall, particularly on account of its frequently great narrowness. It is best, therefore, to freely divide in a sagittal direction the mass of tissue representing the perineum, so as to afford a straight, easy passage to the fæces. As the sphincter ani is absent, it is questionable whether a later restoration of the normal conditions is at all possible. In the new-born the parts are too small to permit of a plastic operation; we have to be content with having at least secured complete evacuation of the fæces, and thereby the possibility of living.

In one case, I tried everything to create an artificial anus, but the opening always contracted again, so that not an anus, but a fecal fistula remained. After this result, of course, I did not venture to close the urogenital sino-rectal fistula.

The prognosis of the operation for simple *atresia ani* is likewise bad. Here, with the lower portion of the rectum, the sphincter is also absent. I have operated three times on male subjects. We wait four or five days, because the end of the rectum is then distended and more easily found. We then dissect preparatorily downward and seek for the rectum. It is often very difficult to find. If it has been reached by accident, as discovered by the escape of meconium, the lower part is drawn down and, if possible, stitched around the artificial anus.

Unfortunately, the threads tear out very easily. Then the artificial anus may be kept open with laminaria. But even that will prolong life for only a few weeks. In one case, the child lived to the age of fifteen months; a fistula, six centimetres in length, led to the cyst formed by the end of the rectum distended by fæces.

Hypospadias, in the female sex, has different degrees. The whole urethra may be absent, so that the bladder merges directly with the urogenital sinus, the urine flows off spontaneously, and in the adult subject the finger easily enters the bladder. Or else the urogenital sinus is very long; at its upper end is the rudimentary urethra, and from behind eccentrically the narrow vagina terminates in the contracted urogenital sinus. In both cases, the clitoris is often hypertrophic, resembling the penis.

Epispadias is rare; it may occur with hiatus of the symphysis, ectopia of the bladder (see below), prolapse of the uterus and rectum, in which case the posterior, respectively inferior, half of the urethra is present. But even with present symphysis, the clitoris may be fissured, and this groove may open directly into the bladder, somewhat like in male epispadias. If there be but a slight degree of hypospadias or epispadias, relief may be afforded by a plastic operation. The method will depend on the nature of the case.

Hypertrophy of the clitoris with accompanying very narrow or rudimentary vagina may cause the external genitals to resemble those of the male. These cases, therefore, have often been considered to be hermaphroditism. *Hermaphroditism*, in the mythological sense, enabling one individual to perform the genital functions of both sexes, has not occurred thus far. In every case, even if the presence of both kinds of genital glands has been demonstrated anatomically or at least microscopically, one of the sexes predominated either anatomically or functionally. But few cases are absolutely certain. Theoretically there are distinguished:

1. *Transverse hermaphroditism,* in which the external genitals are male and the internal female or the reverse. These are generally cases of pseudo-hermaphroditism, *i. e.,* cases of hypospadias with hypertrophied clitoris.

2. *Bilateral hermaphroditism,* in which ovaries and testicles are found on both sides. In the only decided case, there were no vasa deferentia, and no absolutely positive proof was furnished that the bodies called testicles were really such.

FIG. 46.—External genitals of "Katharina Homann." *a,* Entrance to the "vagina" and bladder; *b,* groove leading from the rudimentary penis or hypertrophic clitoris to the entrance into the vagina (hypospadiac groove); *c,* prepuce of the penis or clitoris; *e,* small left labium or scrotum with cryptorchism; *d,* right large labium or scrotum, containing the right testicle.

3. *Lateral hermaphroditism,* in which, for instance, there is said to be a testicle on the right, an ovary on the left side.

4. *Unilateral hermaphroditism,* for instance, a testicle on the right, an ovary and a testicle on the left side.

These cases likewise are by no means as clear as might be thought from the descriptions.

Well known and notorious is especially Katharina Homann, probably a hermaphrodite of male sex.

B. INFLAMMATONS OF THE VULVA (VULVITIS).

Etiology and Pathological Anatomy.

Inflammation of the vulva is not rarely due to trauma. Thus lacerations often occur during defloration. Through uncleanliness, the surroundings of the laceration become infiltrated and the entire vulva inflames. Especially after attempted rape of children, the judiciary physician often finds a quite acute vulvitis, more particularly if infection with gonorrhœal virus is added. Also other traumata, accidental injuries, may cause vulvitis.

Besides, an inflammation of the vulva also arises through the influence of fluids coming from above. Even if they are by themselves of a harmless nature, as for instance menstrual blood, yet the fluids come in contact with dust and air, they decompose, acquire a very offensive odor, and irritate, especially in summer. Through the irritation the numerous sebaceous glands secrete fat, fatty acids are formed which contribute to the decomposition and provoke erythematous processes on the skin. If the discharges by themselves are irritating, as in gonorrhoic vaginitis, profuse flow of pus in consequence of pessaries, carcinomatous ichor, inflammation will occur with especial facility. In urinary fistulæ the vulva is also often inflamed as far as the anus.

In gravidæ an inflammation is particularly favored by the hyperæmia.

Inasmuch as fluid is secreted even physiologically, and dust and dirt easily reach the genitals if the clothing be loose and insufficient, pregnant women of the lower classes are frequently seen with considerable vulvitis. Small or large, few or many furuncles may be the consequence of the cutaneous irritation.

The *syphilitic affections* of the vulva are the most frequent venereal local diseases of women. The primary ulcers are more frequently situated behind than in front. Broad condylomata are often seen. They appear to arise first at the point where the labia majora turn inward. Should a case be left for any length of time without treatment, the whole internal surface of the labia majora as far as the anus changes into a confluent mass which should rather be termed *plaques muqueuses*. The characteristic element of the broad condylomata—the circumscribed elevation—disappears almost completely, only at the borders of the entire surface there is mostly a hard high margin. Bullous syphilides and ecthyma may so fully coalesce with these condylomatous ulcers that a large raw surface extends as far as the inguinal bend. Special forms can hardly be recognized. In the midst of these masses is found now and then a vegetation of pointed condylomata. More rarely the broad condylomata form larger ulcers.

In a later stage of syphilis, indurations are not rarely observed between rectum and vagina. The tissue becomes callous, anæmic, tensely œdematous, very vulnerable, without tendency to healing. In front of the hymen, fistulous, sinuous communications are formed between the vestibule and the rectum. The passing fecal matter prevents the healing of these fistulæ. I never saw such fistulæ in vigorous persons; they occurred generally in infirm prostitutes. But I encountered three such cases among seventy public women. Schroeder, who first recognized and described these fistulæ, ascribes them to syphilis. Cure, even operative, is difficult; but, according to my observations, these fistulæ cause so few symptoms that their cure is not even desired.

Besides, there are found on the vulva now and then singularly lax ulcers which show pale granulations devoid of any macroscopic or microscopic peculiarities. Their limitation is sharply defined; the irregular ulcers lie below the level of the skin. In these cases tertiary syphilis will always have to be thought of, although there is absolutely no induration of the bottom of the ulcer nor any copper-colored surroundings. Several times other serpiginous ulcers and osseous affections were found by which their connection with syphilis was established.

After the healing of syphilitic vulvar affections, cicatricial contractions may form which may obstruct parturition.

In *diabetes mellitus*, a peculiar form of vulvitis occurs together with symptoms of *pruritus*. Quite recently, the discovery has been made that the affection is a mycosis. But pruritus vulvæ is much more frequently an affection of advancing age, of spontaneous occurrence.

But by far the most frequent cause of vulvitis is infection with gonorrhœal virus. This infection may concur with other etiological factors. In gonorrhœal infection, the glands of Bartholin often participate in the inflammation in three ways. *First,* the entire gland may rapidly suppurate. Within three or four days a tumor forms in the labium majus which may attain the size of a child's fist or even greater dimensions. The labium is of phlegmonous hardness and erythematous redness. If not incised, fluctuation appears on the summit of the tumor. A circumscribed portion of the skin may even become gangrenous, so that after perforation at this place a large abscess may occur. As in furuncle, sanguineous pus is evacuated. Or else, the process runs a slower course, and the inner surface of the labium becomes more prominent. In this case, the secretion of the vulvitis sealed the efferent channel. The secretion of the gland cannot escape from the duct, expands it, and lies at the inner surface of the posterior third of the labium majus in the shape of a cyst the size of a hazel-nut or larger. The secretion can often be forcibly expressed. In this latter case, cysts also form which persist for years without incommoding the patients. These cysts contain yellow or dark-brownish fluid and do not refill if *freely opened.*

The third kind of gonorrhoic Bartholinitis is the most pernicious in its consequences; the form probably leading most frequently to the infection of the male. A moderate quantity of pus continually flows from the efferent duct. This pus covers the vulva in a thin layer. It is very probable that precisely the act of coition, owing to compression of the gland by the sphincter cunni or to accidental pressure of the tip of the penis upon the gland, leads to a more copious production of the infectious secretion of the diseased gland. It is natural, too, that in this form of inflammation recurrence easily takes place, particularly through the irritation of too frequently indulged coitus. In prostitutes, according to my observation, the pus often spontaneously disappeared after the patients were a few days in the hospital. Scarcely, however, had the woman resumed her ordinary occupation, when pus was again found at the next examination. Pointed condylomata likewise then recur afresh. *There is no chronic gonorrhoic vulvitis. This chronic Bartholinitis must be placed parallel to the chronic gonorrhœa of the male. It plays a far more important part than colpitis or endocervicitis.*

Phagedænic ulcers—soft chancres—are also found on the labia majora. If badly treated, these ulcers may destroy an entire labium, and leave broad radiating cicatrices. As in the male, they are often complicated with buboes, suppurations of inguinal lymphatic glands.

True *phlegmons* have also been observed on the labium majus. *Gangrene* of the vulva occurs in *typhoid fever, scarlatina,* and *measles* in childhood, and often, moreover, so entirely without symptoms that quite accidentally in the adult atresiæ of the vulva or vagina are found, the origin of which could be explained only in the way indicated. Peculiar to children are cases of *noma* of the vulva running an equally pernicious course as that of the cheek.

Moreover, *diphtheria* of the vulva occurs, probably as a direct infection from diphtheritic affections of the mouth.

A peculiar vulvitis is observed in *scrofulous* children, often of spontaneous occurrence, and hard to cure. It is nearly always associated with discharge from the vagina. Whether some of the secretions are also derived from the uterus cannot be demonstrated on account of the infantile conditions.

In many cases, we find Oxyures vermiculares, small thread worms, which pass from the rectum into the vagina, where they cause continuous itching.

True *eczema* is seen pre-eminently in pregnant women, also *herpes* and *acne* (*follicular vulvitis*) have been described, and, as above stated, multiple *furuncles*, the latter mostly in consequence of acute vulvitis. *Erysipelas* is frequently observed in the puerperium; rarely outside of that state. On the other hand, the vulva occasionally forms the starting-point of an erysipelas migrans neonatorum which ultimately terminates fatally.

Vulvitis runs its course with very pronounced reddening and swelling of the vulva. Muco-purulent, and in acute more violent cases purely purulent or sanguino-purulent, secretions are discharged. On separating the parts, we often see in the vivid-red surroundings countless white, swollen sebaceous glands which may suppurate.

In chronic vulvitis, such as is especially found in pruritis vulvæ, the integument becomes peculiarly stiff, dry, firm, hard, œdematous, leathery, brownish, painful.

Among the affections of the vulva belongs also the so-called *thrombus vulvæ, i. e.*, an extravasation of blood into the parenchyma of the labium majus. Such a tumor may arise from trauma, as a kick or push against the vulva, but more frequently it occurs spontaneously. It is most common in the puerperium, where, however, it by no means ensues immediately in connection with the labor. It increases for a few days, and then usually disappears spontaneously. Despite the livid, even black coloration of the tumor the size of the fist, sloughing does not ensue if infection be kept from it. Therefore—taking it in advance—the treatment is purely expectative: rest and the avoidance of everything injurious. Should the thrombus become ichorous, it must be treated according to surgical principles.

Symptoms and Course.

The symptoms are itching, pain, redness, swelling, secretion. The pains at times are so intense that the patients are unable to walk, and forced to lie with legs spread. Particularly during the development of a Bartholinitis, the pains are quite considerable, especially when the parts are touched. In carcinomata, cases occur where there is only sensitiveness, as in vaginismus. Should the vulvitis be the consequence of itching and scratching, the patients are much tormented by the tenderness provoked by every fresh touch.

In consequence both of the flow of urine over the parts and of the extension of the catarrh to the urethral mucous membrane, urination is painful. Should there be spontaneous complaints of tenesmus of the bladder, and should pain be caused if the urethra is pressed against the symphysis, there is probably gonorrhœal inflammation.

The course is abbreviated by adequate treatment, which is usually demanded rather early, on account of the tormenting symptoms. In many cases, of course, the vulvitis remains in a chronic form, if the etiological factor should fail to be removed.

A peculiar symptom is *pruritus vulvæ*, often described as an independent disease. It consists in itching and burning of the external genitals, and often of the inner surface of the thigh, necessitating perpetual scratching. In many cases, the symptoms make the impression of a hystero-neurosis, at least the symptoms depend rather upon psychic influences than on inflammation of the vulva. The torment caused by the pruritus is quite extraordinary. Patients come under observation who declare, crying with shame and excitement, that they cannot control themselves. In such cases, the orgasm occurs frequently. Real psychoses may be conjoined. Both during pregnancy, in which the pruritus is to be explained through the hyperæmia, and at the time of the menopause, pruritus occurs *temporarily*. Women often have pruritus for some months, and then become free from it without any treatment, while before they had tried all kinds of cure. Moral women of advanced age are suffering from pruritus still more frequently than old prostitutes. Anything leading to congestion, as constipation, the act of defecation, the warmth of the bed, even presence in a heated room, mental excitement, spirituous liquors, carriage riding, etc., reproduces the troublesome itching.

Another symptom of vulvitis is the so-called vaginismus, which will be considered with the diseases of the vagina.

Diagnosis.

The diagnosis is clear from what has been said. At most, the differential diagnosis of a forming tumor in the labium majus might cause some difficulty. Whether there is a phlegmon or a Bartholinitis will be decided by the position, form, and size of the swelling, which in Bartholinitis, of course, is confined to the region of the gland. The differentiation from hernia is important. In hernia of the labium majus, a loop of intestine is present, having descended through the inguinal canal, hence corresponding to scrotal hernia in the male. The likelihood of demonstrating air in the intestine, the entire consistence, the possibility of reposition, feeling of the hernial opening, painlessness on pressure, or, at least, long existence without pain, as well as the history in general, will always permit the differential diagnosis.

Treatment.

The treatment of vulvitis consists pre-eminently in cleanliness. By simple soaping and washing of the vulva, cataplasms with two-per-cent carbolic solution, and sitz-baths, the pain and swelling are soon removed. As the secretion is always mixed with a considerable quantity of fat, a piece of ordinary washing soda (sodium carbonate), the size of a walnut, should be dissolved in the cleansing fluid. In moderate cases, this is followed by painting the vulva with boracic acid ointment (Acidi boracici, 2 gms.; Vaselini, 10 gms.) or carbolized oil (1:10). After twelve hours, the above-described washing with soda is repeated, and again followed by the ointment.

Where there is much reddening, greater extent, and prolonged existence of the affection, this simple treatment will not suffice.

Then also washing with soda is first employed, and the skin thus *freed from grease* is painted with solution of silver nitrate (1:30). If the fat be not removed, the effect is much diminished, the silver nitrate solution does not adhere to the skin. Then the vulva is covered with a carbolized compress, and a carbolized piece of linen is likewise placed between the labia. In this manner, cure is rapidly effected.

At times the secretion is kept up by small depressions of the mucous membrane near the hymenic remnants. In such cases it is best to cauterize every crypt with a pointed stick of silver nitrate.

In *broad condylomata,* excellent effects are obtained from the treatment with salt-water and calomel. After cleansing, the condylomata are first moistened with solution of common salt. Then the calomel is dusted on and rubbed to a paste upon the condylomata with a brush. If the surface is not too large, or there are only isolated condylomata, they will certainly disappear after five or six applications. The procedure is performed once a day. The burning sensation which ensues is never very great. I have continued this method of treatment even during menstruation without untoward effects. If the patient cannot be seen daily, corrosive sublimate ointment (Hydrarg. bichlor. corros., 0.1 gm. [1½ grains]; Vaselini, 30 gm. [1 oz.]) will serve a good purpose. Even this proportion often causes burning, hence it is not advisable to increase its strength. Of course, the local must be supplemented by constitutional treatment. Especially where the ulcerated spots have reached large dimensions, a systematic course of mercurial inunction should not be long delayed. Strong cauterization and scraping with the sharp spoon are of little effect without general treatment. Should the condylomatous masses, however, form a tumor, their surgical removal is indicated.

Those cases are particularly rebellious in which the vulvitis is combined with *pruritus vulvæ*. Here the cause must first be carefully inquired into. A cervical catarrh, quite insignificant by itself, may maintain the pruritus even in women in a state of senile amenorrhœa. Therefore the catarrh should be removed, or the secretion retained by a daily tamponade continued for some time. In mycosis due to diabetes, the treatment will have to be directed against the latter. Should the pruritus not be removed by the treatment above described, a stronger silver nitrate or carbolic solution (1:10) must be applied. Either one, however, only after removal of the fat. I never saw any effect from Peruvian balsam, etc. In a desperate case I succeeded with applications of ice-water. Besides astringents—alum, tannin, decoction of oak-bark—preparations of tar have also been recommended.

During pregnancy, which interdicts all energetic treatment, a continuous application of moist carbolized compresses and borated vaseline ointment is always crowned with success. I have repeatedly observed, particularly in the latter part of pregnancy, a furunculosis of the inner surface of the labia.

Of internal remedies, such as potassium bromide, but little effect may be expected. The cessation of sexual activity is succeeded by no improvement whatever; on the contrary, the itching often increases again in advancing age.

In *Bartholinitis*, it is necessary to open the abscess. Where the incision is to be made is not governed by anatomical conditions, but by the individuality of the case. We simply incise the highest part of the abscess or of the retention cyst. The chronic Bartholinitis described above as the third form I have been unable to cure. The cause probably lay in the non-reliability of the patients. I should advise extirpation of the gland in such cases. The operation would be very easy and free from danger. At all events a definite cure in any simpler manner seems to be impossible.

In *phlegmons*, a purely surgical treatment is justified. *In phagedænic affections, the actual cautery acts most rapidly.* In gangrene, strong carbolic solutions, dusting with salicylic acid, etc., are in place.

Herpes, eczema, etc., are to be treated according to general rules, especially by cleanliness.

The vulvar ulcers, described on page 75, require an energetic treatment. Ointments, irritating and astringent applications are useless. I have dissected out the entire ulcer under an anesthetic and sutured the wound. Where the skin is freely movable this is easily done. I have also secured healing by cauterization with the hot iron.

C. New Formations of the Vulva.

Papilloma.

Papillomata, pointed condylomata (condylomata acuminata), are of very frequent occurrence in the vestibule at the labia minora and majora, as far as the neighborhood of the anus. They are probably always referable to specific infection with gonorrhœal virus. This does not mean that the same form of tumor may not occur elsewhere after other irritations. Thus I often found in children under one year pointed condylomata at the anus, where gonorrhœal infection was not to be thought of.

Isolated pointed condylomata form a white, papillomato-villous, round tumor, ranging in size to that of an apple; although not pediculated, it may be clearly demarcated in its situation on one labium. In other more frequent cases, more or less numerous small villi, two to three millimetres thick and up to one centimetre in length, proliferate singly from the skin all around the vulva. But few condylomata are found just above the frenulum in the fossa navicularis. The efferent duct of the gland of Bartholin may be surrounded with condylomata at its terminus, and some excrescences may also be situated within its lumen. In the same way as the condylomata extend downward, especially in a posterior direction, they may also be found above in the vagina, even on the vaginal portion. This is a rather rare occurrence.

In the examination of prostitutes, which I conducted for some time when police surgeon, I found in about three-fourths of the entire number pointed condylomata, though often very sparse. If few, condylomata are not at all troublesome, and cause no symptoms whatever. Still they should be removed, although it is more correct to say that the virulent catarrh keeps up the condylomata, rather than that the condylomata maintain the catarrh. The diagnosis is very simple, only the isolated larger masses may be mistaken for elephantiasis or other new-formations.

The *treatment* consists in their ablation; *it is not justifiable to waste time with ointments, caustics, etc.*

Previous to the operation, the patient must take a sitz-bath; the entire surface is also soaped off with carbolic solution, cleansed, and disinfected. The patient is chloroformed. Then every single condyloma is lifted up with the forceps, and cut off with curved scissors, together with a piece of skin the size of a lentil. If the excrescences are numerous, this procedure is very tedious, but it prevents relapses and removes all the growths at once and without any pain. Should a large tumor be removed, the wound must be stitched, if but to arrest the bleeding. Otherwise a dry dressing with borated lint, salicylated, carbolated, or benzoated cotton suffices.

Carcinoma.

Carcinoma of the vulva is rarer than that of the uterus, but more frequent than that of the vagina. It forms a node which soon breaks open, exposing a tumor resembling granulations and secreting pus. Carcinoma usually has a tendency to spread superficially, but it also extends into the depth, interspersing cancerous nodes through the entire glutei, and involving the pelvic bones in the carcinomatous degeneration.

The treatment consists in removal; this promises success only *if the carcinoma is movable*. In tumors which have already attacked the bones, we confine ourselves to disinfection and symptomatic treatment.

The method of operation will differ according to the size and form. If possible, excision is performed, and the wound closed with sutures. But if the surface is too large, the carcinoma may be circumscribed with the galvano-cautery knife at a distance of one centimetre; from this groove needles are passed under the tumor; the chain of the écraseur or the galvano-caustic loop is drawn after them, and the growth thus removed piecemeal from its base. In this manner hemorrhage is sought to be avoided. It must remain an open question whether it is not better to dissect off the tumor, tie the bleeding vessels, arrest the parenchymatous oozing by compression, and draw the healthy skin as far as possible over the wound, as in large defects of the mammary gland. At any rate, every case has its own peculiarities, according to which a special plan of operation may be needed.

Elephantiasis.

The vulva is a part of the body in which, even in the temperate zone, elephantiasis-like swellings are not rare. One labium alone, as well as the clitoris and the entire genitals may be hypertrophied. The clitoris is changed to a round or sausage-shaped cylindrical, smooth or uneven body. At times the elephantiasis-like proliferation extends to the anus, which is surrounded by a circle of cleft excrescences with intervening fissures. The form is very manifold; often clitoris, crura clitoridis, and labia minora can still be recognized, in other cases this is impossible. The cutaneous papillæ participate in various ways. It may happen that the whole tumor is papillomatous, resembling a mass of pointed condylomata, or else only part of it may be warty and rough. In other cases, hypertrophy occurs only in the deeper parts, leaving the surface of the tumor quite smooth. Secretion is usually present. In the depth, parts of the tumor rub together, the weight of a heavy swelling causes fissures which secrete fluid. The surface of the tumor sheds horny epithelia in a very active manner. These often mix with the secretion to form a smeary mass covering the tumor like vernix caseosa.

In other cases the tumor ulcerates, and larger raw surfaces exist. On section the tumor shows a jelly-like œdematous appearance. Microscopically, if the cutaneous papillæ participate, we obtain pictures showing the colossal development of papillæ and thickening of the cutis; while, when the surface is smooth, the subcutaneous connective tissue is hypertrophied in all its parts, and interspersed with numerous lymph-corpuscles.

Of etiological importance are said to be—onanism, irritations affecting the external genitals, at times also trauma. *A connection with syphilis can often be demonstrated;* not perhaps in such a way that the tumors are made up directly of broad condylomata, but that the chronic hyperæmia of the vulva, the syphilitic affection of the lymph-channels and the inguinal glands lead to lymph stasis and hypertrophy.

The *symptoms* are pain which originates from the fissures at the base of the swelling, as well as through traction exerted by the weight of a large tumor. The inconvenience of a larger tumor between the legs often is also quite considerable. Coition, if desired, may be practised *more bestiarum*, and may also be followed by pregnancy. Pregnancy and menstruation have little influence on the increase of the tumor, but there is a lack of sufficient observations. The general health does not suffer materially from the affection.

The course is variable; *tumors the size of a child's head may spring up in a year, and those the size of a hen's egg may have grown gradually in decenniums*. The ablation of a portion of the tumor, for instance, the labia minora, has a limiting influence on universal elephantiasis vulvæ.

On the other hand, good-sized tumors soon re-form from remnants of the growth. Elephantiasis is most frequent on the clitoris, then on the labia minora and majora.

The *diagnosis* is very simple, the hypertrophy of the vulva is recognized at first sight. Only in ulcerated elephantiasis may carcinoma enter into consideration. Excision of a small part would allow the diagnosis of carcinoma to be formed immediately.

The *treatment* consists in removal of the tumor. With good assistance we may proceed by dissection, tie the bleeding vessels, and, if possible, unite the wound primarily by suture. The practitioner, if alone, may apply the elastic ligature, cut above it, and ligate the pedicle in sections. Healing ensues without special danger. All ointments, caustics, etc., are to be discarded as opposed to the operation with the knife. At most the galvano-cautery may be employed.

However, if the tumor cannot be removed entire, for instance, if its base is too broad, spreading over the mons Veneris, and adherent, we remove as much as possible. Then portions are ligated *en masse*, as the vessels in the severed tumor cannot be grasped. If bleeding occurs in spite of the ligature, the hot iron is most to be recommended.

Rarer New-Formations on the Vulva.

Small tumors occur at the vulva which have been described as molluscum simplex. Twice I have cut off a tumor of that nature. One had an oblong, the other a round form; both were soft, not pigmented, without hairs, had sprung from the posterior third of the labium majus, and had caused no symptoms. Hildebrandt figures a case of four such tumors springing from the labia majora.

Myomata and *fibro-myomata* of the vulva have also been described. They are hid from sight or hang from the vulva as a pendulous tumor. Still rarer are *lipomata* which may attain a colossal size. Besides, cases of sarcoma, enchondroma, and neuroma are known.

On account of its rarity, lupus of the vulva has been but little studied, and is described under the name of *esthyomène*. Its slow course distinguishes it from carcinoma; its superficial extent, from elephantiasis; the quality of the ulceration, from syphilis or the phagedænic ulcer.

Lupus bears the strongest resemblance to elephantiasis. But lupus is more destructive, extends more energetically, and secretes more fluid. It is often barely credible how few symptoms are caused by the largest lupous destructions. All the soft parts from the anus to the mons Veneris may be changed into a uniform, moist, firmly œdematous, partly ulcerated, suppurating, fissured, fistulous mass without any complaint of pain and inconveniences.

According to the form, there have also been distinguished a lupus perforans, hypertrophicus, and serpiginosus. The treatment is usually hopeless. If the affection were to come under observation in time, the tumor or the ulcer should be removed most completely and carefully. In lupus, too, the suspicion of tertiary syphilis is often present.

Apart from varicose dilatations of the labia, very frequent in old women, there also occur congenital teleangiectases—nævus vulvæ. If they increase much in size, especially at puberty, their removal is advisable.

Cysts of the vulva have already been considered; they are mostly connected with the glands of Bartholin. But other cysts of unknown origin likewise occur. If they interfere with walking or coition, they should be opened. Here the differential diagnosis from hernia of the labium majus would be especially important.

The glands of Bartholin, too, may change into a carcinoma or a malignant adenoma.

D. Lesions of the Vulva (Perineal Ruptures).

Inasmuch as, unfortunately, it is not yet customary to repair all perineal ruptures after delivery, and as this operation in recent cases is often badly performed and fails, the gynecologist frequently sees cases of old uncured lacerations of the perineum. We distinguish complete perineal ruptures in which the sphincter ani is torn and the laceration extends a variable distance into the rectum, and incomplete lacerations in which often but a very thin membrane separates the vagina from the rectum. Nearly always a part of the primary laceration from above downward has closed, so that we see at the upper end of the injury a white mark—the cicatrix—extending upward into the vagina.

The etiology of the lacerations belongs into obstetrical text-books. In very rare instances the perineal rupture is the consequence of a non-puerperal injury. At times a bridge-like union also forms, leaving a rectal fistula above it.

The *symptoms*, often even in complete rupture, are quite insignificant.

We are astonished to see that, in the absence of the recto-vaginal wall to the height of three or four centimetres, continence of fæces still remains. This is possible by vicariating hypertrophy and function of the superior sphincter—the internal—as well as by the lower sphincter joining the cicatrix and contracting toward it, thus sufficiently closing the anus. In other cases only solid fæces are retained, the liquid masses passing spontaneously. Withal the patients usually believe that fæces and flatus pass per vaginam. Sometimes the rectum also develops a state of irritability, so that thin fæces are continuously passed. In incomplete perineal lacerations normal coition is impossible, and sterility is a frequent consequence. If the septum is very thin, fissures are developed in it which render defecation and coition painful.

As neither uterus nor vaginal portion is supported by the lowest part of the perineum, the perineal rupture has nothing to do *directly* with prolapse of the uterus. The posterior vaginal wall is the natural support of the anterior one. If the anterior vaginal wall is elongated, hypertrophied, prolapsed, and if, besides, its natural support, the posterior vaginal wall, is absent, the anterior wall may sink still lower. It drags on the uterus and pulls it downward, so that *indirectly* prolapse of the uterus may be connected with a considerable perineal laceration.

These processes are accompanied by the very annoying sensation of prolapse. The anterior vaginal wall crowding between the vulva causes subjective symptoms on the one hand, on the other it falls into a state of chronic inflammation. The hyperæmia in which the uterus takes part is also followed by various uterine diseases.

In complete perineal ruptures, small ectropia of the rectal mucous membrane into the vagina, or at least into the laceration, are not of rare occurrence. These ectropia, of vivid red color, are clearly demarcated from the paler vaginal mucous membrane, and make the impression of fresh granulating surfaces. In incomplete perineal laceration, both wound surfaces often have at once become widely separated, and the anterior rectal wall has prolapsed between them, so that a slight rectocele results.

The symptoms above described, and the injurious consequences, call for an operation.

We must first discuss the question when we should operate. Certainly soon after delivery. After about eight weeks, when the involution and the lochial flow have terminated, the operation can be performed even in nursing women. Here we have the advantage that the parts are still very hyperæmic, and consequently heal more readily. If the laceration be of our own making, we shall naturally be desirous to attain the cure as soon as possible. This is not supposed to mean that our chances are less in future. Even years after the injury the operation always succeeds. We may even operate while remnants of exudation are still present. In general the rule is accepted, to operate about one week

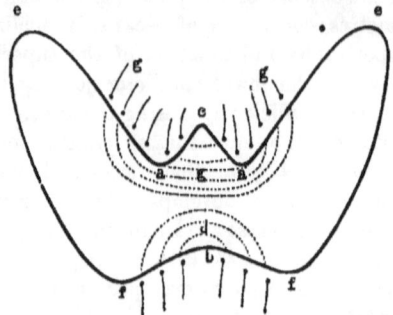

Fig. 47.—Diagram of the surface freshened by Simon and Hegar. *c*, central small vaginal freshened surface; *g*, vaginal sutures, approximating *a e* to *a e*; *b f*, rectum, *d*, sutures which are tied behind in the rectum; *e f* becomes the perineum.

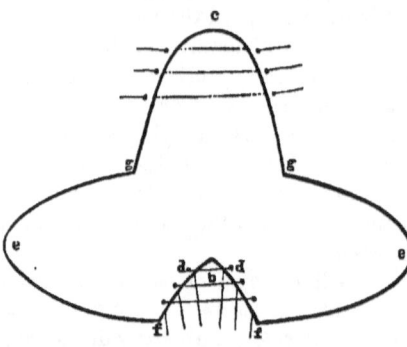

Fig. 48.—Diagram of the surface freshened by Hildebrandt. *c*, central vaginal flap; *c g e*, becomes the sagittal outline of the posterior vaginal wall; *f b*, rectum; *d*, sutures which are tied in the rectum; *e f* becomes the perineum.

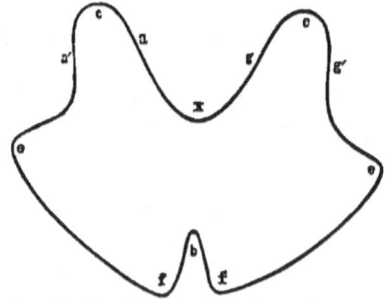

Fig. 49.—Diagram of the surface freshened by Freund. *x*, tip of the posterior rugous column; *c e*, the two lateral vaginal freshened surfaces; *a* is joined to *a'*, *g* to *g'*; *e f* becomes the perineum; *b f*, the rectum.

after cessation of menstruation. However, I have operated not rarely in the period of lactation without, as is feared, provoking by the operation a uterine hemorrhage—premature menstruation. Preparatorily the field of operation is to be disinfected by sitz-baths and vaginal irrigations for some days prior to the operation. (See Chapter IV.)

The modern methods of operation are those of Simon and Hegar (Fig. 47), Hildebrandt (Fig. 48), and Freund (Fig. 49). Hegar freshens a butterfly-shaped surface (Fig. 47). The freshening extends far forward to the labium majus, furnishing a broad perineum, but the union does not go far up the vagina. Hildebrandt, however, removes in the centre (Fig. 48, *g c g*) a rather broad flap. The details are made evident by the figure and its description.

Freund justly objected to this method that it did not consider the natural conditions; that the centre, the columna rugarum, hardly ever tears in a sagittal direction, but that the laceration extends upward on one or both sides, so that a tag remains in the middle. If that were cut away, or the union effected across it, the vagina would lose in expansibility. Freund, therefore, freshens as shown in

Fig. 49; x is the central, usually intact prominence of the rugous column. On both sides of this, $a' c a$ and $g' c g$, two tongue-shaped surfaces are freshened and each united by itself. Then the operation is continued in the usual manner, $g' e$ being joined to $a' e$, and $e f$ to $e f$.

I have operated in the following manner: The endeavor should be to reunite the parts in the same way in which they formerly were. Therefore I first trace the continuation of the defect into the vagina. Nearly always we see obliquely upward a white line—the cicatrix of the vaginal laceration. Not rarely the vaginal cicatrix extends high up into the vaginal vault. This cicatrix is the central line of the freshened surface. Then the freshening is commenced by the side of this cicatrix, as though we were to operate on a fistula. If the cicatrix is very long (Fig. 50, a), the freshening does not extend to the upper end of the scar. The upper terminus of the freshening must always be one and a half centimetres above the defect.

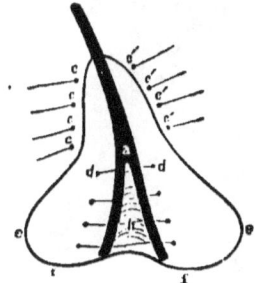

FIG. 50.—Diagram of the surface freshened by Fritsch. $c c c c$ united with $c' c' c' c'$ lies posteriorly, laterally, or medially in the vagina; c–e, posterior vaginal wall; $e f$, perineum; d, sutures tied in the rectum; b, posterior rectal wall; a, cicatrix.

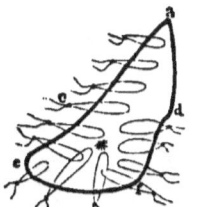

FIG. 51.—United perineal laceration. $a e$, vagina; $e f$, perineum; $d f$, rectum.

The defect is, besides, circumcised with scissors so that the rectal wall is also freshened. Portions of the rectum in a state of ectropion are removed, as they are usually inflamed and too readily permit the sutures to cut through. Intentionally the perineum becomes narrow, narrower than it has been. After being united, especially if the rectal sutures extend deeply, the sagittal outline of the joined posterior portion lies almost vertically downward from above. I have found too often that the artificial forward elongation of the perineum is useless. If the freshening be too broad above, the traction in the vagina is so strong that union takes place badly. A thin perineum, however, interferes with coition, ruptures during labor, and is absolutely without any influence in the prevention of prolapse. Therefore I make the perineum rather narrow. (See Fig. 51, $e f$).

The operation for laceration of the perineum is so easy that every phy-

sician should perform it. Of course, it is impossible to do this operation without having seen it several times, or, better, having assisted at it.

In order that every physician may be able to operate, the number of assistants should be quite small. To render this possible, I have devised the leg-braces represented in Fig. 52.

The illustration renders description unnecessary. These leg-braces can be screwed to any ordinary table. The rods carrying the legs are octagonal below and are inserted into octagonal sockets, so that they cannot be turned and the legs approximated. In this woodcut, engraved from a photograph, we also see the vagina exposed; my speculum holder

Fig. 52.—The patient ready for operation, fastened to the leg-braces.

drawing the posterior vaginal wall downward. A movable iron rod with corresponding curvature is inserted in a small bracket screwed to the table between the two leg-braces. To the upper part of this rod Simon's speculum for the posterior vaginal wall is attached. For the perineal operation this speculum is omitted. The anterior part of the vulva, however, is separated by Bozeman's speculum (Fig. 53). This instrument, being self-retaining, presses the soft parts of the vulva to the pubic arch in an excellent manner. Although the inventor applies it with the screw below, I have found it more advantageous to reverse the speculum so that the screw comes above the mons Veneris (see Fig. 52). Thus the perineum is readily accessible in every way. The median part of Bozeman's speculum for the upper vaginal wall is unnecessary, for the anterior vaginal wall cannot sink, because the two lateral wings greatly stretch the vagina in a transverse direction. In cystocele alone, the anterior vaginal wall sinks between the two lateral parts of the speculum. The patient being narcotized and fastened in the leg-braces, no skilled assistant is required. But one medical man is necessary to supervise the narcosis, and some one to irrigate the wound. Instruments, needles, needle-holders, etc., are in readiness on a chair, lying in a dish full of carbolic-acid solution.

Two fingers are now inserted into the anus, and the portion to be

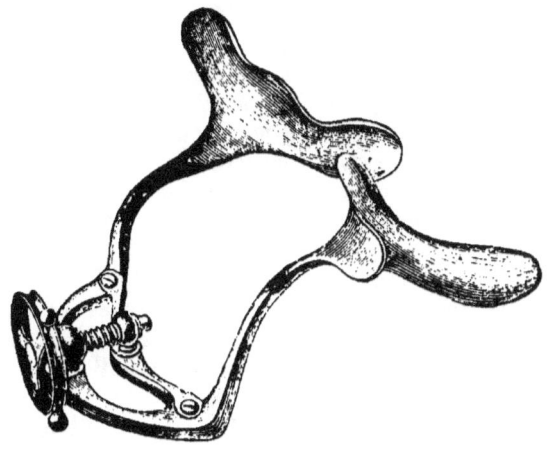

Fig. 53.—Bozeman's speculum.

freshened is bulged out. The entire figure is then marked out with the knife, and the mucous membrane is detached with the curved forceps and the knife. It is certainly more convenient if an assistant aids the operator by stretching and pressing, by passing his fingers into the anus, or pulling with tenacula, the American bullet forceps, or the forceps. But it can also be done alone without such assistance.

In order not to penetrate too far into the tissues, the edge of the knife is directed towards the flap. After one "flap" has been dissected off, the progress is more rapid. The beginner is to be advised not to freshen the entire surface from *one* line. In that way it is easy gradually to cut too deeply. It is better to commence at different points of the periphery, advancing toward the centre. Should the bleeding be too profuse, the loosened flap is pressed on, a sponge perhaps being superimposed, and the dissection is commenced from the opposite side. If a fenestra has been cut into the mucous membrane, the remaining remnant of the mucosa should be at once looked for. With my method we always get to the cicatrix in the centre, where it is often impossible to loosen the membrane. Then the superficial layer is cut off with Cowper's scissors The cicatrix comes into the depth of the wound, and therefore is of little importance. Below, where the incision extends into the external skin, we must not go too far beyond the still marked old anal margin, otherwise the traction becomes too great. After the whole surface has been freshened, I stretch the margin of the defect, and cut it out with scissors like a fistula. The particles of the rectum in a state of ectropion are cut away at the same time. The surface is now inspected while a stream of carbolic solution irrigates the wound. Small prominences and irregularities are cut off here and there with scissors. At the margin of the wound, the distance from the intact tissue to the freshened surface is rendered equal all around. Then follows the suturing. The material I use is silk, about which I never had any cause to complain. The needle-holder represented in Fig. 54 is largely used in gynecological operations. Grooves are filed into one side of the needle-holding jaws, the other side being roughened. Still better holders for the needles, but applicable to a more limited number of positions, are the needle-holders with grooves on both sides. The needles exactly correspond to these grooves. The fact that only certain needles fit these holders, and that but three or four different positions of the fixed needles are possible, restricts their use to the simpler operations, such as the perineal operation in question.

First the sutures (Fig. 50, c c') are inserted. In order not to cover the relations of the wound, they are left untied for the present. As the needle is passed, one finger is placed into the rectum with which the point of the needle is controlled. I take very strongly curved needles, and, on principle, I include as much tissue as possible, that is to say, I carry the point close to the rectum. In the centre of the freshened surface I cause the needle to emerge, but re-insert it immediately by the side of this opening, and on the right side again grasp as much tissue as possible. In this method, *including much tissue with strongly curved needles*, the results as regards certainty of healing and firmness of the cicatrix are better than if only the upper layers of the freshened surface are approximated.

FIG. 54.—Needle-holder.

Then the *rectum* is sutured. Here the sutures must be passed in such a way that they either enter in the rectum and re-emerge in the freshened surface one and a half centimetres from the margin of the rectal wound, or two needles must be carried on one thread. Then it is possible to lead both extremities of the thread from the wound-surface into the rectum. The rectal sutures are first carefully tied, the knots (Fig. 51) lying in the rectum from d to f. Then the remaining vaginal sutures (Fig. 51) from c to e are passed and united. In my method, with vertical or very slightly oblique posterior vaginal wall, this is very easy. The traction is insignificant. Then the remainder, $e f$, the perineum, is joined with three sutures.

I do not at all hesitate to *cross* the sutures exactly in the centre of the present united surface, at * in Fig. 51. Thereby the apposition of the wound surfaces becomes perfect, and union ensues more readily.

Before I had the necessary practice, when I still made the perineum as large as in Hegar's operation, Fig. 47, it often happened that $c e$, Fig. 51, remained ununited on account of excessive traction. Still the effect of the operation was a good one, as the rectal surface united. In the same way I have observed that the rectal suture failed to unite, but that of the vagina from c to e closed, and here also the desired result was obtained.

The after-treatment consists in vaginal irrigations (see Chapter IV.). The whole vagina and the anus are covered with benzoated or salicylated cotton to ward off impurities of the bed. If after two or three days the perineum is not swollen, union is certain. But even if swelling and suppuration ensue, the hope of success should not be abandoned. If the sutures but extend deeply, and grasp much tissue, union by second intention not rarely occurs.

The practice of locking the bowels with opium has been universally abandoned. But also the artificial production of passages soon after the operation is injurious according to my experience. If the bowels have been fully emptied before the operation, and if a restricted diet be prescribed, it is best to wait five days. Then the large intestine is filled with water by the irrigator, and a laxative is administered. After the first defecation, the most accessible sutures may be removed. We need not trouble about the other ones. If they cause any inconvenience, trifling suppuration, etc., and if they do not gradually come out spontaneously, they may be removed after some weeks.

When the rupture is combined with considerable descensus of the anterior vaginal wall and the uterus, more complicated methods of operation are necessary. On the other hand, if the laceration is incomplete, the operation is performed in the same way as in rectal laceration, but the rectal sutures are omitted.

CHAPTER VII.

DISEASES OF THE VAGINA.[1]

A. INFLAMMATION OF THE VAGINA (COLPITIS, VAGINITIS).

Etiology.

INFLAMMATION of the vagina often arises from *foreign bodies*. A tampon remaining in place too long incites inflammation. In the same way many pessaries very soon set up a quite acute inflammation producing bloody pus. Occasionally pessaries are worn for years, while in other individuals the most violent colpitis arises already after three or four days.

Most dangerous in this respect are old, previously used pessaries which, of course, no physician re-applies, but many a midwife does. I have seen most intractable inflammations which had arisen in this manner. Also wooden pessaries covered with leather and varnish, pessaries of rubber which is too strongly impregnated with sulphur, and inappropriate pessaries generally, provoke vaginal inflammations. With long-retained pessaries, the vaginitis may be so violent that sanious pus continuously escapes, and the pessaries become completely imbedded in granulations and cicatrices. Like pessaries, other foreign bodies, perhaps introduced for masturbatory purposes, cause inflammation. Even a sponge accidentally left in the vagina causes sanious discharge and contracting inflammation, so that in one case carcinoma was first diagnosticated. Furthermore, strong caustics with which the vaginal portion is treated cause vaginal inflammations, large cicatrices, and even stenoses.

Also from vesico-vaginal fistulæ an inflammation sometimes results. But it seems here that it has extended upward from the vulva; at least in cases in which the external parts are sufficiently clean, the vagina generally is not found inflamed. Under the influence of gravid hyperæmia, the investiture of the vagina hypertrophies, becomes very hyperæmic, single groups of papillæ swell considerably, become hard

---

[1] The malformations of the vagina will be discussed partly with the gynatresiæ, partly with the malformations of the uterus.

and painful (granulated vagina), many epithelia are cast off and lie in the vagina as creamy masses, often as scales and lumps.

Especially remarkable is the vaginal inflammation described as vaginitis adhæsiva, occurring in old individuals. This is not a universal, but a circumscribed form of inflammation. The epithelium disappears in places, granulating surfaces form in the vaginal vault which adhere to each other. Thereby the vaginal portion may *partially* adhere to the fornix, so that isolated cords can be felt; or *totally*, so that the vaginal portion cannot be felt at all, as in senile involution.

The etiology is very obscure. Why just in advanced age, where every injury and irritation are lacking, the fornix vaginæ should become inflamed, is difficult to understand. Vaginitis adhæsiva, especially, is not at all connected with the presence of pessaries, during the use of which this peculiar form is never seen. However, I have hardly ever seen a case of vaginitis adhæsiva without cervical catarrh, and it is possible that the secretion of the cervix present in the atrophying vagina, and removed neither by coition nor menstruation, has a macerating influence on the vagina. This maceration would be facilitated, especially, by the fact that the layer of pavement epithelium becomes gradually thinner with age.

Gonorrhœal infection, however, is to be considered the most frequent cause of a colpitis, concurring with all those hitherto named. In vaginitis, virulent infection is the etiological factor still more frequently than in inflammation of the vulva.

Moreover, an inflammatory destructive process may spread from the vulva to the vagina, so that in the grave forms of gangrene or diphtheria of the vulva, the vagina likewise perishes by sloughing.

Syphilis causes no specific vaginal affection. It will be impossible in a concrete case to exclude a gonorrhœa as etiology.

Infusoria and fungi have been found in vaginal inflammations. Oxyuris vermicularis, escaped from the anus into the vulva, may likewise lead to inflammation of the vulva and vagina.

Vaginitis occurs in scrofulous children, but, of course, the diagnosis is not quite certain on account of the impossibility to exclude cervical disease by careful examination.

Anatomy.

The secretion normally present in the vagina consists of cast-off pavement epithelium and cervical and uterine mucus. The vagina itself, the "mucosa" of which is really no mucous membrane but an epidermis, contains no organs secreting mucus. Although glands have been found in the vagina, this is by no means a constant condition; but these glands particularly do not impart the character of a mucous membrane to the vaginal lining. For this reason, too, a vaginal catarrh in the ordinary sense is impossible. Catarrhal secretion is a mass consisting of quantitatively increased normal secretion and pus-corpuscles. Therefore, as the vagina sheds horny epithelia, but normally possesses no secretion, a catarrh is as impossible as a catarrh of the external skin.

The superficial inflammation of the vaginal investiture, therefore, would rather possess the character of an erythema or erysipelas. After desquamative processes, of course, a copious suppuration of the exposed deeper parts is possible. In general, vaginal inflammations are not too frequent. "Fluor albus," by the ancients usually interpreted as vaginitis, generally consists in endocervicitis in which the vagina forms merely the place of passage or accumulation of the uterine secretion. The moistening of the vagina and complicating hyperæmia then lead to a plentiful separation of the superficial epithelia.

There is, in the first place, the *acute universal* colpitis. Here we have mostly to do with gonorrhœal infection. Noeggerath has asserted that in gonorrhœal infection the entire "mucous tract," from the vulva to the ostium abdominale tubæ, is always implicated. Such cases certainly do occur. A chronic gonorrhœa, however, may remain in the cervix, uterus, and Bartholin's gland, but the vagina usually does not become inflamed, and recovery soon results. Already the fact that the vagina is covered with pavement epithelium, and that the secretions escape from it spontaneously, is favorable. We know that, in the male, the gonorrhœal inflammation in the urethra becomes chronic, but the external cutis inflames acutely in the form of balanitis, but not chronically, and a similar course obtains in the female. The vagina inflames acutely; on the mucous surface of the uterus and the gland of Bartholin the inflammation remains chronic. Even the urethritis often complicating a gonorrhœal inflammation heals spontaneously in the female, disappearing without any direct treatment. In harmony with the close connection of the various uterine tissues, on upward extension of the colpitis, the uterus becomes affected in toto, that is to say, there occurs a quite acute endometritis, metritis, and perimetritis. Even more rarely than in the male, cystitis and nephritis may occur in connection with the urethritis. But a few such unquestionable cases I have seen.

In the speculum, in acute cases, the vagina is found vivid red, in some places there is even a deposit on the membrane resembling that found on the peritoneum in peritonitis. On advancing the speculum, isolated hemorrhages occur; the speculum scrapes off yellowish or brownish-red pus. In gravidæ especially, large masses of purely yellow pus are found, while the quantitatively equally considerable masses of pus from irritating pessaries are mostly tinged with blood and are more fluid.

On the other hand, however, a purely desquamative vaginitis occurs from pessaries. Without the occurrence of any symptoms, the entire pessary is found surrounded by a mass resembling inspissated vernix caseosa.

The worst cases of universal vaginitis are those running their course with very great swelling, œdema, and final gangrene of the vagina: the cases of diphtheria in infectious diseases. Here the tumefaction and swelling are relatively as great as those of the conjunctiva in blennorrhœa and chemosis. In these cases, gangrenous sloughs are cast off, so that even after recovery strictures or atresiæ remain behind.

A partial diphtheria is sometimes observed after unsuccessful fistula operations; the wound, after separation of the sutures, is first covered with deposit and hard, and does not clear until after disinfection. Usually the entire vagina is also reddened and painful.

In a great many cases, especially also in pregnancy, the inflammation confines itself to single papillæ or groups of them (vaginitis granulosa); they are infiltrated, hard, hypertrophied, and distinctly prominent.

Almost exclusively in pregnancy, or at least after recently preceding pregnancy, a peculiar affection of the vagina occurs—colpohyperplasia cystica. The vagina not being properly inflamed, this name is preferable to "colpitis emphysematosa." According to the latest investigations of Ruge, we here have to deal with air-cysts in the connective tissue, both in the papillæ and more deeply seated. Whence the air is derived, or what mixtures of gas these cysts contain, has not been determined. The condition cannot claim any practical interest.

As opposed to universal acute vaginitis, the above-described *vaginitis adhæsiva* is a circumscribed chronic vaginitis. Here there is only a circumscribed loss of epithelium; the inflammation has its seat merely above in the fornix vaginæ.

Cases of chronic vaginitis are rare, unless we include among them desquamative processes during pregnancy or in the use of pessaries. In most cases, aside from these causes, there will be merely a secretion derived from the uterus, which distributes itself in the vagina.

Under the name of *vaginitis exfoliativa*, rare cases have been described in which, during menstruation, fragments of skin are passed by the vagina. Such superficial exfoliations of the skin occur also during the application of too strong astringents.

A peculiar *paravaginitis*, an inflammation of the connective tissue surrounding the vagina leading to suppuration, has been described as paravaginitis dissecans. Through the suppuration here the entire muscular tube of the vagina may be separated, so that the vagina perishes by gangrene. In diphtheria of the vagina, too, a large part of it sometimes separates by gangrene. Then atresiæ and stenoses occur during the healing. The vagina may also be entirely occluded for some distance, so that it will be impossible to restore a canal even by operation. The consequences of these atresiæ at puberty are "retention cysts:" hæmatocolpos and hæmatometra.

A singular vaginal affection, colpitis gummosa, has been observed by Winckel. In a syphilitic person, the whole vagina was filled with friable masses of syphiloma. I saw a case of hypertrophy of the papillæ in which the touch was impossible, and the entire vagina was studded with villi.

Symptoms and Course.

The symptoms consist in pains and secretion. Particularly in gonorrhœal infection, the patients feel a burning pain and sensation of heaviness in the pelvis. Especially characteristic in this connection is the pain during urination—a symptom pointing to participation of the urethra. The touch or the introduction of a speculum may be impossible on account of the pain and œdematous swelling of the entire vaginal walls. Even the irrigation of the vagina is painful at first. Copulation cannot be borne on account of the pain. Especially if the vagina and vulva are eroded by ichorous discharges, the sensibility of the vagina may become so great that every touch will produce the most violent pain.

In the acute inflammations of the vagina, some fever is sometimes present. The whole abdomen, particularly movements of the uterus during combined examination, may be sensitive—a symptom pointing to implication of the perimetrium.

In diphtheria and spontaneous gangrene of the vagina, of course, high fever and ichorous discharge, mixed with gangrenous shreds, are present. The local symptoms, however, cannot be very severe, for in many gynatresiæ, decidedly of this origin, the history could not subsequently demonstrate any preceding local affection. It is possible, too, that the general and grave course of the causative disease, such as typhoid fever, may at that time have diminished the sensibility.

Both the colpitis and the participation of the bladder and the perimetrium may exert an unfavorable influence on the general condition, so that patients of that class may emaciate considerably.

The more chronic forms of colpitis have less influence on the general condition; in them, the complicating cervical catarrh is mostly of importance. Adhesive vaginitis causes as few symptoms as the emphysematous form; only in the former affection some sanguineous pus is usually discharged.

In the granulating inflammation of pregnancy, the vaginal touch is painful.

If the injurious cause is removed, such as an inappropriate pessary, the vaginitis soon disappears spontaneously. But if the inflammation remains untreated while the etiological factor persists, the inflammation may extend to the tube on account of the continuity of the mucous tract. If then it should be impossible to cure the uterine disease, the course of the vaginitis will also be a very slow one.

Diagnosis.

The report of the patient that discharge is present is not sufficient for the diagnosis. Only inspection of the vagina through the speculum permits of the formation of a diagnosis. But even if the vagina is everywhere covered with secretion, but is pale-red and not painful, the vaginitis cannot yet be diagnosticated. There must be pain, a general vivid-red or bluish coloration, and swelling, or the above-described tumefaction of single groups of papillæ. On the other hand, granulating and adhesive vaginitis can be diagnosticated by the mere touch. In the former, the hard granulations are felt; in the latter, the often quite friable cords between the vaginal portion and the vagina. Not rarely even the vaginitis adhæsiva *is felt*, while the cords can hardly be seen through the speculum.

Colpohyperplasia cystica is diagnosticated by scarifying and emptying the vesicles or demonstrating the presence of air under the surface of the skin.

In favor of gonorrhœal infection are—the acute, otherwise inexplicable beginning, a possibly complicating Bartholinitis, the presence of pointed condylomata, a painful sensation on firmly pressing the urethra against the symphysis, and tenesmus of the bladder, or burning during urination.

When treating children for scarlatinal diphtheria or other grave infectious disease, the genitals should be frequently inspected so as to guard against being suddenly surprised by an advanced destruction. I have seen cases in which unquestionably a diphtheritic ophthalmia and a diphtheritic vulvitis and vaginitis arose almost simultaneously from pharyngeal diphtheria.

Prognosis.

The more acute the case, and the sooner it comes under treatment, the better is the prognosis. In gonorrhœal infection, there is always the danger of an acute endometritis, metritis, and perimetritis. In chronic inflammations, the prognosis is bad because it is often impossible to remove the causative agent, for instance, cancer-juice. In diphtheritic gangrenous inflammations, small children almost always perish. But even if the momentary danger is past, the prognosis still is rendered doubtful by the sequelæ, adhesions, etc.

Treatment.

Vaginitis due to an irritating pessary or tampon usually disappears spontaneously after removal of the object. A few injections of lukewarm water hasten the cure. If a cervical catarrh simulates a vaginal secretion, the cervical catarrh must of couse be attacked.

For gonorrhœal infection, we may employ dry or wet treatment.

In the former method, the vagina is carefully cleansed as the speculum is inserted, while, on its withdrawal, the vaginal walls are dusted with the remedy. This is done with the pulverizer, with a bunch of cotton dipped into the powder, or by pouring the powder into the speculum, and pushing it in portions between the vaginal folds with the uterine rod as the speculum is withdrawn. The customary remedies are tannin and alum. The former is used pure, the latter is mixed with sugar or starch (1 : 10). Also the insertion of very large, long tampons the shape of a penis, dipped in strong astringent solutions of alum, tannin-glycerin (5 : 25) has often a good effect.

In a similar manner, the vagina may be painted through the speculum, using strong solutions of silver nitrate (1 : 20), tincture of iodine, which causes no serious pain, or alum. Especially in chronic inflammations, I have seen excellent results from tincture of iodine.

Another method is the bath of the vaginal walls: the astringent fluid is poured into a speculum, a part of the vaginal mucosa is allowed to remain in contact with the fluid, and the speculum is withdrawn *very slowly*, the rest being poured into a small dish held below.

Also the medicated suppositories in use already before the time of Christ, are still employed. The astringent is mixed with suppositories of cacao butter, or glycerin with gum-arabic. These suppositories have the disadvantage that in melting they drop out of roomy vaginæ, and soil the external genitals.

In gonorrhœal vaginitis a careful treatment is required. I have had the best results from a combined astringent-disinfecting method. Irrigation is performed every three hours, alternately with an astringent and a disinfectant. This has a better effect than if both, say carbolic acid and alum, are together mixed with the water.

For a disinfectant, I rather use a non-caustic remedy than carbolic acid.

Powders of three grammes of salicylic acid are dissolved in alcohol or Cologne water, added to one litre of water, and used as a vaginal irrigation. The water should be of about 25°–28° R. (88°–95° F.). The addition of alcohol has an excellent effect. After three hours an irrigation of alum is prepared (about a heaping teaspoonful dissolved in hot water, and added to one litre of water). This treatment is continued for three or four days. Should the discharge persist, it is best to paint the entire vagina with tincture of iodine or solution of silver nitrate.

Of course, the continued irrigation, by the immediate removal of the secretions, may deceive the patient as to their quantity or even about their presence. Therefore, previous to every renewed examination, the irrigation should be suspended for twenty-four hours.

In bad cases, vigorous tamponing of the vagina with dry benzoated cotton, continued for days, has also done me good service.

Colpitis adhæsiva requires no direct treatment.

The vaginitis of scrofulous children is best treated constitutionally and with salt-water baths.

In grangrenous vaginitis, frequent disinfecting irrigations and, after separation of all necrosed portions, the insertion of a tampon to prevent stenosis, is to be recommended. The tampon is coated with boracic acid ointment or carbolized oil.

In excessive secretion during pregnancy, cleansing irrigations with lukewarm water are permissible. If care be taken, the pregnancy is not interrupted. Granulating vaginitis seems to be made to disappear by the act of parturition and the involution of the puerperium.

B. Cysts of the Vagina.

Vaginal cysts vary greatly in size; they may be as large as a nut, or even so large as to interfere with labor and hinder the functions of the pelvic organs. As they usually cause no symptoms, they are not often recognized. But in thorough examinations, such as those of prostitutes, vaginal cysts are found in about one per cent of all cases.

These cysts are variously explained. They have been declared to be retention cysts of the still questionable vaginal glands. Veit thought it necessary to assume a dilatation of Gärtner's canals which persist in many animals. It is also possible that but one of Müller's ducts developed, and the rudiment of the other adjoins the fully formed one in the shape of a cyst. Others have made it appear probable that vaginal cysts originate from traumata or so-called thrombi (p. 77). The contents of the cysts are so variable that they permit no deductions. Clear serum and altered blood, thin or thick fluid has been found. In examined cases, the cysts were invested with cylindrical epithelium. But vibratile epithelium has also been observed in the cysts. Even multilocular cysts have been described.

The cysts may lie quite superficially, projecting even polypus-like into the vagina, or more deeply with a thick wall above them. Therefore, some authors distinguish superficial and deep vaginal cysts. Among the superficial belong also those occurring in colpohyperplasia cystica. The diagnosis is so easy that mistakes are hardly possible in combined examination from the rectum, bladder, abdominal coverings, and the vagina.

As these cysts refill after puncture, a good-sized piece should be excised from the wall. A knife is first inserted; into the opening thus made, a scissor-blade is introduced, the loose sac is lifted with the forceps, and as much as possible is cut off the wall. If the hemorrhage is profuse, we tampon. If the wall is very thick, the cyst-wall may also be united with the vaginal wall by the suture.

C. New-Formations of the Vagina.

Carcinoma of the vagina, secondary to that of the cervix, is frequent, but rare primarily. It occurs both in the shape of a tumor and in that of an ulcer. In the latter case, it may form a round, oval, or ring-shaped ulcerated surface. If situated posteriorly, the rectal wall is almost always found implicated, even if the growth be of small size.

The symptoms are the same as in uterine carcinoma—pains, discharge, and hemorrhage. Differing from uterine carcinoma, *vaginal carcinoma* usually attacks young individuals. Therefore, complications during labor are more easily possible than in uterine carcinoma.

The treatment is mostly symptomatic, as there is but little hope of operating in healthy tissue because of the rapid implication of neighboring organs or portions of connective tissue. Special methods have not yet been devised, owing to the rarity of the cases.

Tuberculosis of the vagina occurs only as a local manifestation of general tuberculosis, and even then but rarely.

Sarcoma of the vagina may occur secondarily in uterine sarcoma, and primarily as a circumscribed fibro-sarcoma. In the latter case it should be removed by operation. Or else it may represent a sarcomatous degeneration of the entire vagina, and be then inaccessible to the knife. Vaginal sarcomata have been observed in very young individuals. The diagnosis must be formed by microscopic examination.

Myomata of the vagina are not rare. When we consider the embryological relationship of the musculature of the vagina with that of the uterus, this is not surprising. Vaginal myomata, like those of the uterus, may be situated interstitially in the musculature, or project polypus-like. They may attain the size of a child's head, and then cause symptoms referable to pressure. But previous to that, difficult defecation and dysuria, various discharges, and pain during coition may point to some vaginal abnormality. The diagnosis is formed by examination, but it should be borne in mind that uterine polypi may enter into intimate connection with the vagina, so that on feeling the tumor adhere to the vaginal wall, we have no proof that it springs from the vagina. Only if, after operative separation, the empty os uteri is felt above, we can pronounce it a vaginal myoma. Uterine and vaginal myomata are also met with simultaneously. As vaginal myomata are more accessible than those of the uterus, their removal is also less difficult.

D. Vaginismus.

The peculiar symptom of vaginismus is that a painful cramp of the sphincter cunni prevents the entry of any body into the vagina. Thereby coition, the touch, or the introduction of a speculum is rendered impossible. Cases have been described in which merely blowing against the vulva or tickling with a feather, and even the bare thought of coition provoked the reflex spasm. Sometimes, besides the sphincter cunni, the adductors of the thigh and the levator ani participate in the cramp.

This group of symptoms is to be traced back to very different causes. In the first place, an injury may provoke the vaginismus. Should the introitus have been injured during the first attempts at intercourse, perhaps made awkwardly and against violent struggles of the woman, coition will be painful. The woman, as yet void of any feeling of voluptuousness, is alarmed, struggles, falls into violent nervous excitement, the legs tremble, and are pressed spasmodically together. As soon as the penis touches or lacerates the wound, the cramp of vaginismus occurs.

If the entire introitus is reddened by inflammation, the pain is equally great, and vaginismus ensues. Especially in gonorrhoic inflammation, with or without Bartholinitis, the complex of symptoms constituting vaginismus is set up. Should, as is often the case, all this concur; should a man during the attempt at coition with simultaneous injury infect the woman who is averse to intercourse, temporary vaginismus is easily explained. It is self-evident that very nervous, erethistic, hysteric persons suffer more readily from vaginismus than phlegmatic, quiet, sensible women. But this observation has led to the erroneous idea that vaginismus should be regarded as a local hysteria, a pure hystero-neurosis.

Temporary vaginismus is very frequent, especially in young married women. It often happens that the specialist is called upon to examine the genitals, because the conscientious woman is rather disquieted about a possible defect of her sexual apparatus. On examination pronounced vaginismus is observed, still the husband soon overcomes these obstacles.

Apart from these slight cases, graver ones occur. Sterility is the consequence on account of the impossibility of complete intercourse. It has been overcome by having the sexual act performed in narcosis. But after parturition the vaginismus returned. An over-thick hymen may also give rise to vaginismus. The abnormal thickness may be explained in some cases by the impotent man failing to effect perforation or rupture of the hymen with his not fully erect penis. However, the continual attempts lead to inflammatory infiltration and thickening of the hymen.

In other cases of long standing, small non-inflamed fissures were found, which provoked vaginismus on touch, and after the removal of which, cure was effected. The connection was thereby rendered clear. The condition recalls fissure of the anus. In such casss, owing to the impossibility to have normal intercourse, ill-feeling may exist between husband and wife; through that, sickness of the woman, and finally a reflex psychosis.

In the *diagnosis* it is essential to trace the cause of the well-marked complex of symptoms. If an inflammation is present, the case is readily understood. But if there is absence of any inflammatory symptoms, the entire vulva and the hymen must be examined in narcosis with the speculum, tenaculum, etc., to discover a fissure. With a thickened hymen the diagnosis is likewise clear.

The *treatment* in the temporary vaginismus after marriage is psychical. Attempts at coition are advised to be discontinued for the present, and in this way a redeeming word is often spoken for both parties. With increased mutual confidence, everything will be all right later on.

If an inflammation is present, it is treated as such. A fissure is to be cauterized—in narcosis—so that a large crust results. Silver nitrate or nitric acid is to be used. An imperforate hymen should be incised, or, better still, excised. As the fissure may also be situated behind the hymen, a complete excision is to be recommended. The two margins of the wound are united with a few interrupted sutures, both to arrest the hemorrhage and to cause rapid healing.

We then test by the touch if vaginismus remains; usually this is not the case, although it appears so because the patient dreads the old pain during the touch. By coaxing, the touch succeeds, and the patient is thus convinced of its painlessness. Should vaginismus still be present, the vagina is slightly stretched with dilators. Also rupture of the sphincter, analogous to the treatment of fissure of the anus, has been recommended. Of course, coition at first is still somewhat painful, but soon becomes normal.

Hildebrandt has described cases of spasm of the sphincter ani, mild forms of which are certainly not rare. In Hildebrandt's case, the penis was firmly held for some time during coition, hence the term "penis captivus." These forms have been described as painless vaginismus or vaginismus superior.

E. Recto-vaginal Fistulæ.

If a complete perineal rupture occurs during labor, at first only a part of the perineum closes, and above a recto-vaginal fistula remains. We often find that the vagina remains in a fissured state, the posterior wall at that portion being formed merely by the thin rectal wall. Then the fistula represents an opening in a thick membrane. Owing to the thinness of the surrounding parts, division and reunion are hardly possible with any prospect of success.

The symptoms vary according to the size. Smaller fistulæ may have a sort of valvular occlusion, so that they may exist without causing any inconvenience. In other cases, flatus and liquid fæces escape through the fistula, while the more solid matters pass by. If below the fistula not the entire perineum, but, as frequently happens, only a thin membrane has closed, symptoms of complete perineal rupture arise. Whenever a thin vaginal wall is covered with ulcerations or fissures, symptoms are complained of like those of fissure of the anus.

Vesico-vaginal fistulæ after operation do not rupture even during labor, for the anterior wall is compressed rather than stretched. On the other hand, rectal fistulæ easily reopen in subsequent labors. In one of these cases I first thought the posterior margin of the anus was the posterior vaginal commissure; the fistula, the os uteri; the thin rectal wall, the lower uterine segment. During the low engagement of the head, perhaps by the assistance of the midwife, the fistula had opened widely. Therefore I have cured two such cases simply by a perineoplasty, by taking no notice of the fistula and freshening a large triangle, and uniting it down to the labia. In the one case, a later labor proceeded quite smoothly, without any laceration. In the other case, I did not observe any subsequent labor.

Fistulæ may also arise from the pressure of a Zwanck pessary, from partial gangrene of the vagina in the puerperium, or from traumatism. Should these fistulæ be surrounded by thick layers of tissue, they are operated upon in the same way as vesico-vaginal fistulæ (see below).

The lesions occurring in carcinoma, of course, cannot be treated by operation.

Fistulæ of the vagina with the small intestine have also been observed. If the portion of intestine below the fistula has ceased to functionate, all the fæces pass through the vagina; otherwise, the greater part of the fæces pass through the intestine, a portion, however, also by the vagina. The differential diagnosis between rectal and ileac fistulæ is made by the quality of the characteristic chyle-like contents of the small intestine, and by injection of the rectum.

The operation is likely to be very difficult, and must be performed according to the rules prevailing in præternatural anus.

CHAPTER VIII.

DISEASES OF THE BLADDER AND URETHRA.

A. FISSURES AND ECTOPIÆ.

THE bladder is formed from the stem of the allantois; the upper end is the urachus, extending to the umbilicus; below the bladder opens into the vagina by the urethra. Aside from the excessively rare reduplication of the bladder, the existence of a sagittal or horizontal septum, and the congenital diverticula, the important malformations are traceable to urinary stasis. The latter again is the consequence of an atresia or the absence of the urethra. Possibly the malformations may even be traceable to an accumulation of fluid in the allantois, which may prevent the closure of the ventral fissure, or else the persistence of the ventral fissure may concur with amniotic ligaments or primary fissure-formations. In the latter case, ectopia of the bladder occurs, that is to say, a prolapse of the perfectly formed and closed bladder into or in front of the ventral fissure. More frequently the bladder is fissured in consequence of retention of urine. This is probably in those cases in which the centrifugal pressure in the abdominal cavity keeps the symphysis apart, so that a complete ventral fissure is found together with the defect of the anterior vesical wall. In that case the bladder was fissured even before it was supported by the abdominal wall, or the coincident ventral fissure facilitated the outward rupture. However, should there be above the closed symphysis, in a frequently small ventral fissure, a portion of bright-red vesical wall, it may be assumed that the rupture occurred even earlier, the ventral fissure subsequently closing everywhere except at the spot where the prolapsing bladder directly prevented it.

Likewise traceable to atresia or absence of the urethra are those cases in which the urine escapes from the umbilicus through the still patent urachus. Subsequently the urethra may yet become pervious, the urachus then remaining open, but no longer serving for the discharge of urine. It even happens that the urachus closes at the umbilicus and at the bladder, the accumulated secretions forming a cyst between these two spots.

In complete absence of the bladder and closed ventral fissure, the ureters empty directly into the urethra.

The external appearance of a *fissured bladder* is variable. The umbilicus is usually closer to the symphysis. In the worst cases, the entire abdomen is cleft from a spot close under the umbilicus down to and including the clitoris. The urethra is absent, the pubic bones are far apart, the anterior vesical wall is lacking, the posterior wall forms a bright-red, hemispherical tumor. The vesical mucous membrane covering this tumor is in a state of chronic inflammation. The orifices of the ureters project or are hidden in folds of the mucosa. The external abdominal integument terminates sharply at the tumor, so that we may insert a sound beneath this margin, as under that of a sinuous ulcer. Or else the integument irregularly blends with the vesical mucosa, some light-colored islands or proliferations of the epidermis are seen in the bright-red vesical mucous membrane.

In other cases, the fissure is situated farther downward or upward. Small ectopiæ also occur. Thus I saw a case in which, between the umbilicus and the closed symphysis, there was a round tumor, four to five centimetres in diameter, into which the ureters emptied.

Both in fistula of the urachus and in fissure of the bladder the *symptoms* consist in the inconvenience caused by the vulnerable tumor, and in the involuntary constant escape of urine. The prognosis is unfavorable, as operative treatment generally fails. As the entire malformation is traceable to the absence of the urethra, and as the sphincter is lacking in artificial urethra, complete " cure " will be impossible in many cases. In the first place, we shall seek to replace the prominent displaced bladder, then to separate large flaps of skin, and to unite them over the defect. Whether it is possible to form a urethra from the vagina, and to close it temporarily by a clamp or an instrument resembling a pessary, would have to be decided by a concrete case. But even without urethra much would be gained by covering the bladder with epidermis; in the first place, the vulnerable vesical mucosa is protected; secondly, the prolapse of the bladder cannot increase; thirdly, the dilatability of the bladder gradually becomes greater; and finally, if but a small opening remains, a urine receptacle or a temporary closure by an instrument is more easily applied.

The *prognosis* is best in fistulæ of the urachus. If a urethra is present, which is not rare in these cases, the operative closure of the upper opening causes no great difficulty, as has been demonstrated by experience.

In very rare cases of absence of the urethra, fissure of the bladder does not occur, but the bladder ruptures into the vagina. In the absence of the urethra, there were found more or less extensive vesicovaginal fistulæ from which the urine dribbled away spontaneously.

Outside of the ectopiæ, eversions or inversions of the bladder con-connected with fissure, inversions of the completely developed bladder are rare. There has also been some dispute as to whether, in inversion of the bladder, the mucosa alone is inverted, or whether all the coats of the bladder are affected in the same way. If we separate from the inversion the detachment of the mucous membrane in consequence of necrosis and the prolapse of the urethral mucosa, but few cases remain. Such a prolapse of the entire inverted bladder through the urethra, conjoined with prolapsus of the uterus, with otherwise intact pelvic organs, is preserved in the collection of the Pathological Institute, at Halle.

B. Inflammations of the Bladder.

The affections of the female bladder exhibit many differences from those of the same organ in the male. The absence of the prostate, the hypertrophy of which has so many grave consequences, protects the female bladder against a whole series of pathological processes. On the other hand, the frequency of neighboring inflammations—peritonitis, parametritis, and metritis—offers manifold dangers. The shortness of the urethra, too, permits an inflammation to advance inward from without more rapidly than in the male; while, again, the numerous dangers of chronic urethritis—stricture, etc.—do not exist at all in women. The accumulation of concretions in the bladder of the female is far less harmful than in that of the male, because removal is more easily effected, either by the urethra or by an artificial vaginal fistula.

Etiology.

According to what has been stated, the etiological factors of cystitis must differ from those of the male. In the first place, inflammation in consequence of the introduction of foreign bodies into the bladder is more common in the female. Intrauterine pessaries have been passed into the bladder by mistake. Furthermore, cystitis arises from penetration of the contents of neighboring abscesses. Cases have been described as cystic pregnancy in which the fœtus or parts of it have penetrated into the bladder from an adjoining suppurating sac inclosing an extrauterine pregnancy. Also hair and teeth derived from a dermoid cyst have been removed from the bladder.

All these foreign bodies irritate the vesical mucosa as much as it is done by fissure, inversion, and prolapse or by extensive fistulæ. Even pus, although not decomposed, may place the mucous membrane in a catarrhal condition. I have observed some cases in which pus from a perinephritic abscess had got into the pelvis of the kidney, or where a perimetritic accumulation of pus had perforated into the bladder. In these there was, as long as the flow of pus continued, tenesmus of the bladder and other symptoms of cystitis.

Pregnancy also predisposes to vesical catarrh, but it is not to be ascribed to the increased disposition to catch cold which the older writers have particularly blamed. It is rather the mechanical conditions which are of etiological importance. Should the head of the child press strongly upon the bladder and should the latter be hyperæmic like all the pelvic organs, a lesion of the mucous membrane by the pressure could readily be understood. Or else the hyperæmia of the vagina and an inflammation existing in it may travel along the urethral mucosa into the bladder. In the same way cystitis develops from acute gonorrhœa of the vagina. Inflammation of the urethra is a characteristic complication of gonorrhœal colpitis. This urethritis is sometimes succeeded by vesical catarrh. In general such cases are not frequent. The cystitis consequent on the introduction of bacteria has also travelled from without inward. Here we have mostly to deal with the transportation of noxious matters from without by means of the catheter. Such cases usually affect parturients. Stagnation of urine, according to the differences between man and woman above explained, is rarely the *sole* etiological factor of cystitis. Yet very grave inflammations of the bladder arise from the transportation of bacteria into a bladder from which the urine cannot escape spontaneously, for instance, in retention of urine in consequence of incarcerated retroflexed uterus.

Anatomy.

In acute vesical catarrh, where the mucous membrane alone is affected, the mucosa is bright-red, but soon becomes normal again. If the catarrh persists for a longer period and becomes chronic, the mucosa thickens, corrugates, and even acquires polypous prolongations. The vessels show vivid injection, are dilated, and easily bleed when touched. This form can be readily inspected in inversion of the bladder through extensive fistulæ or in vesical fissure with ectopia. The mucous membrane sheds its epithelia in large quantities; the vessels allow many leucocytes to pass through their walls. At the beginning in the acute stage, a number of red blood-corpuscles are also found in the urine. Bacteria are always found, though sometimes in small numbers. In the commencement the urine appears brownish-red, contains blood-corpuscles, but later it deposits a more or less thick, whitish opaque layer of mucus and pus mixed with bladder epithelium. Even after the complete cessation of all other symptoms, some pus continues to be present in the urine for a very long time. Such a chronic catarrh may exist very long without the implication of the musculature of the bladder. During post-mortem examinations, the mucosa is often found greatly injected in patches and even interspersed with some small hemorrhagic spots, so that the assumption of a partial affection of the mucous surface appears justified. Especially near the internal urethral orifice a partial inflammation frequently exists.

After the invasion of bacteria the affection is intense. Certainly the bacteria act not only by causing the urine to decompose, to become alkaline, and to put the mucous membrane in a state of irritation. They seem to penetrate also directly into the mucosa, and in severe cases to bring the tissues to a state of necrosis by their immigration. Although in the initial stage the total affection is the more frequent, it soon regresses and a circumscribed abscess remains behind. The abscesses are most frequent in the neighborhood of the trigone of Lieutaud; perhaps the mucous membrane is here less capable of resistance, owing to the absence of the submucous layer. The abscesses and their necrotic shreds lie in the bladder as foreign bodies, and become incrusted with urinary salts. Soon the musculature participates in the infiltration or inflammation; it thickens, becomes hypertrophic and hence paralyzed; the inflammation, therefore, becomes parenchymatous, is no longer a catarrh of the vesical mucosa, but a cystitis. Even if the abscesses are healed, by treatment or spontaneously, the hypertrophy of the vesical walls may persist for years. It frequently happens that the mucous membrane in these cases remains in a chronic catarrhal condition. Then after many weeks the thickened bladder may be felt like a recently delivered uterus, and at the same time the mucosa remains rough, thick, and liable to bleed.

However, should the condition not change into this at least partial recovery, the process will extend deeper and deeper, the peritoneum be exposed and become inflamed. Even then the infectious vesical contents may be kept from the peritoneal cavity by thickening of the peritoneum. In quite severe acute cases, however, the peritoneum is likewise penetrated; usually not rapidly by pressure from within the bladder, but by gradually progressive gangrene of the vesical walls. Then the urine escapes into the peritoneal cavity and a peritonitis brings about the fatal issue.

Still, in grave cases of mycotic cystitis, there occur not only partial softening and degeneration of the mucous membrane, but even total detachment of the entire mucosa. These cases have been called diphtheria of the bladder. The mucous membrane, rapidly necrosed through the invasion of bacteria, becomes detached from the underlying tissue, so that a second cyst lies in the bladder—the free mucosa. It has happened that this entire gangrenous mucous membrane is passed outward through the urethra and that new epithelium has formed on the denuded muscularis.

Often, however, the affection is so grave that death not rarely terminates its course.

Symptoms.

The symptoms of vesical catarrh consist in sensations of pressure and pain in the region of the bladder, frequently troublesome tenesmus, burning and pain during urination. In simple catarrh there is no fever, the urine appears turbid, at first brownish owing to the admixture of blood, later more opaque from the presence of mucus. In the test-glass or other vessel the masses of pus and mucus are deposited at the bottom. The quantity of sediment ranges up to one-half of the entire quantity of urine.

Soon the mucous membrane again becomes normal and all symptoms disappear. Still the affection lasts from ten to fourteen days. I observed the clinical picture of a simple acute catarrh of the vesical mucosa in a few cases in which an over-strong disinfectant had accidentally gotten into the bladder. I intended to give vaginal injections with a thin uterine catheter after perineoplasty. The catheter got into the urethra and bladder because I had passed it close to the symphysis in order to spare the posterior vaginal walls. Thus a two-per-cent carbolic solution flowed into the healthy bladder. Soon after, tenesmus of the bladder and hæmaturia occurred; the urine continued to show a great many leucocytes for four or five days. The subjective symptoms disappeared gradually. After ten days recovery was complete.

In chronic catarrh, the variability of the quantity of pus is noteworthy. Although differences in the quantity of pus are self-evident, according to the ingestion of fluid and the frequency of emptying the bladder, numerous recrudescences seem to occur on the other hand. This is particularly the case after highly spiced dishes, and more than all, after partaking of alcoholic drinks. With these recrudescences, the pains and inconveniences also increase. Not a few patients suffer thus for years with the chronic catarrh. But some pus is always found in the urine, even at the better periods.

In mycotic inflammation of the bladder, in cystitis, the symptoms as regards the general condition are much more violent; but, as is often the case in infectious inflammations, the local phenomena are more in the background. Thus, in diphtheria of the bladder, tenesmus is at times absent, while paresis and paralysis of the bladder occur. The latter depend upon serous infiltration of the detrusor urinæ.

On the other hand, in grave cases, the subjective symptoms may be very violent at the start. Of still greater importance is the quality of the urine, which is of brownish-red color, mixed with shreds, small coagula, and blood-clots, and of so cadaveric an odor as to impregnate the whole room. In it we find, besides red and white blood-corpuscles, an immense number of bacteria, single and in colonies. Should larger portions of membrane be detached, they get into the urethra at times, and then a colicky tenesmus may expel the shreds, or the physician diagnoses the cause of the tenesmus and removes the pieces by traction. The internal orifice of the urethra may also be blocked by the shreds, so that the full bladder is emptied with difficulty even by the catheter, either because the point of the catheter does not penetrate into the urine, but between muscularis and detached mucous membrane, or because the fenestra of the catheter is plugged by fragments of the mucosa.

Very grave typhoid symptoms arise whenever the parenchymatous inflammation attacks the entire substratum of the bladder. High fever, incipient uræmia, peritonitic pains, septic symptoms at once indicate the gravity of the case. Usually we shall find a neglected, misinterpreted retroflexion of the gravid uterus or a very severe, uncatheterized puerperal inflammation of the bladder.

Should a portion of the vesical tissue undergo gangrenous degeneration with coexisting ischuria, the urine infiltrates between the layers of the vesical walls and separates them from one another; the urinary infiltration leads to decomposition, the infectious urine penetrates into the peritoneal cavity with fatal results. We must denominate these cases *a malacia* of the bladder, a degeneration rather than a "rupture."

In many cases, the parenchymatous cystitis improves. Then a chronic cystitis remains behind, *i. e.*, a chronic inflammation of all the substrata of the vesical walls. An ordinary catarrh, too, though rarely, may cause a chronic parenchymatous cystitis through deep extension of the inflammation. The characteristic symptom is hypertrophy and paresis of the vesical walls. Thereby the general condition always suffers. Many cases run their course with continual pain and tenesmus, in others again the chief difficulty is the inability to retain larger quantities of urine or the involuntary escape. Invariably, however, the patient's general condition suffers so sensibly that relief is urgently indicated.

Diagnosis and Prognosis.

The diagnosis of vesical catarrh is formed from the subjective and objective symptoms. The tenesmus indicates the appearance of acute catarrh. In old cases, the history and the examination of the urine must suffice for a diagnosis. In retroflexion of the gravid uterus, the cystitis is the most dangerous complication. The symptoms of incarceration and the incipient peritonitis come pre-eminently into the foreground. Should a fatal result ensue, besides the diagnosticated peritonitis its cause will be found—partial gangrene and perforation of the bladder.

In chronic parenchymatous cystitis, the general condition is so poor that the suspicion of a malignant new-formation arises. Then examination must be made both from the bladder alone and combined with that of the abdominal coverings. An endoscope has also been devised, by means of which the mucous membrane may be viewed directly, but here the touch is of greater importance than the sight. In general, the diagnosis of chronic catarrh will be clear from the history.

The prognosis in the catarrhal inflammation is good; but in the graver cases of mycotic cystitis the prognosis depends on the time when the correct treatment is instituted.

Treatment.

In the non-febrile acute catarrhs of the mucosa, we confine ourselves to internal remedies. The formerly largely prescribed balsams mostly have but the one effect of disarranging the stomach. But the free use of milk is to be recommended. If the milk is not well borne, we add lime-water (25 gm. to 500 gm. of milk). If the milk causes diarrhœa, we give opium in small doses. A very good effect is produced by solution of potassium chlorate (5 : 100) or of sodium salicylate in the same proportion. For the tenesmus it is best to give morphine (0.05 : 10.0 of water) p. r. n., five to twenty drops every one or two hours. Belladonna may be advantageously added to the morphine solution. It is useless to inject the morphine solution into the bladder; in the first place, the morphine acts any way through the nerve-centre; and second, the bladder has very slight or no power of absorption. If the tenesmus is great, the patient should be put to bed, and the abdomen covered with hot cloths or Priessnitz's applications. After the symptoms of the acute catarrh are past, it is advisable to order a restricted diet for some period of time, especially abstention from alcoholics, and a course of Wildungen or Vichy water.

The urine must be frequently examined subsequently, to test the quantity of pus.

This simple treatment will not suffice in the grave, febrile, mycotic, or infectious cystitis. Here it is absolutely necessary to promptly institute local treatment. If this be done, we obtain excellent results even in this form. In the first place, we must provide for the evacuation of urine, especially in paresis of the bladder, or where there is great pain during urination. To this end I insert into the urethra a rubber tube fifteen centimetres in length and six to seven centimetres in diameter. The rubber tube is merely to keep the urethra patent, and is, therefore, not at all to be placed into the bladder. As soon as urine escapes during the rotatory insertion of the tube, it is left at that spot. Then, in febrile cases, the bladder is irrigated every two hours. Strong disinfectants are contraindicated on account of the vulnerability of the bladder. Carbolic acid should be used only in one-per-cent solution, salicylic acid in 0.3 per cent solution. The irrigation should be at blood heat. The point of the end-piece of an irrigator is inserted into the rubber tube lying in the urethra, and the injection is continued until the patient feels the need of passing water, when the mass is allowed to escape. This procedure is repeated two or three times to make sure that the entire vesical mucous membrane has come in contact with the disinfectant. If the double-current catheter be used for the irrigation, we are not so sure that the disinfecting fluid has reached every part. Moreover, the fenestræ of the catheter easily tear small fissures in the urethral mucosa. I have even observed such fissures after catheterization with an elastic catheter. They render the act of catheterization so painful that it is quite impossible to subject the patient to this torture every two hours. The fissures are also slow to heal. All this is avoided by choosing my simple method. Should the rubber tube-drop out, we select a thicker one. About every two or three days the tube must be removed and cleansed. Aside from that, the fever is carefully watched, so as to gain indications for the general condition. In the acute cases, too, besides a very bland diet, Wildungen water is exhibited. To every glass containing about three hundred grammes, we add 0.5 gm. of sodium salicylate and 0.005 gm. of morphine, and order daily at least three or four glasses. If there be no tenesmus, the morphine may be omitted.

As soon as the fever decreases, the vesical irrigations may also be given at longer intervals, perhaps four times in the twenty-four hours. If the urine is no longer offensive, and deposits but little pus, the bladder had better be left alone, and the plan of milk diet with Wildungen water commenced. Injections of astringents, according to my experience, do not materially hasten the recovery.

In the subacute, destructive cystitis in which membranes and shreds are cast off, we endeavor to remove these membranes. Hence it is first necessary to diagnosticate their presence. The urine is to be carefully examined; should it contain many small, especially incrusted particles of tissue; should the case not improve definitely; should the sanious quality of the urine always return, be mixed with blood, and the fever remain high, dilatation of the urethra, according to Simon, is justified. Only Nos. 1, 2, and 3 of the dilators are introduced, the bladder is well-filled, and during the escape watch is kept to see if a portion of the membranes gets into the speculum. Not rarely the membranes are pressed into the urethra; often during violent tenesmus they project from the external orifice, so that they may be touched directly, and removed.

Then we must likewise irrigate. Should pain or increased tenesmus occur with the above-mentioned disinfectants, a simple solution of common salt should be used, with which I have had good results in some grave cases.

In cases of long standing, it is often not so much the catarrh and the parenchymatous inflammation, but rather their sequelæ, the paresis and the diminished capacity of the bladder, which call for treatment. The paresis is best treated with electricity. A rod terminating above in a brass knob should be fastened in an elastic catheter. The bladder is first filled with fluid, then the pole is inserted into the bladder, the other pole is placed on the external skin, and the current closed. The effect of even a weak induced current is often quite surprising; continence sometimes returns after a single sitting. If this should not occur, the treatment is continued. In hypertrophy of the walls, slight dilatability of the bladder, continual dribbling of urine, and tenesmus, I have likewise employed the above-described treatment, that is to say, the bladder was stretched by injections of lukewarm water. As soon as the desire to pass water becomes urgent, the injection is interrupted, and the quantity of fluid noted which it was possible to inject into the bladder. Such expansions of the bladder are repeated once or twice daily, and usually it is soon found that the bladder gains the power of retaining larger quantities.

Of course, this treatment demands patience on the part of the physician as well as of the patient.

Cases of paresis occur also in acute catarrhs, *i. e.*, in very intense disease of the mucous membrane with infiltration of the detrusor. In these, the paresis disappears by the mere continued disinfection and improvement of the affection of the mucosa.

I have seen but little effect from the employment of astringents. Salicylic acid, solutions of common salt, as well as mechanical dilatation, seem to me to be more effective.

C. New-Formations of the Bladder.

The most frequent neoplasm of the bladder is the villous tumor, erroneously termed villous carcinoma. This new-formation is usually situated in the trigone of Lieutaud, and always in the lower part of the bladder. The villous tumor has a papillomatous structure; the single vascular papillæ are covered with cylindrical epithelium. This epithelium either covers only the single papillæ in the presence of real villi, or else it merely fills all the depressions, the apparently not fissured, soft mass conveying the impression of a solid tumor. The abundance of vessels proves that the tumor bleeds readily. The symptoms of such a neoplasm consist, first, in the interference with urination. The villi floating in the urine may sink in front of the urethral orifice, thus suddenly interrupting the stream. Second, frequent hemorrhages occur. Even the contraction of the bladder, but especially catheterization during the just described difficulty in urination, provoke hemorrhages. These may cause great anæmia, rendering the individual cachectic as in a large carcinoma. At first, the urine, even when not mixed with blood, is turbid, and the microscope may demonstrate numerous pus-corpuscles and cylindrical epithelium detached from the surface of the tumor.

Until recently, these tumors were looked upon as incurable by operation. But especially the circumstance that villous tumors are benign, in a clinical sense, makes the search for an operative procedure appear justified. Simon dilated the urethra, thus making a diagnosis, and opening a road to the tumor. Simon employed his urethral specula for that purpose. They are diminutive Mayer's specula provided with obturators. The thinnest instrument is first inserted, and gradually followed by the thicker ones. My dilators (Fig. 18, p. 41) can also be advantageously used for this purpose. If the external urethral orifice becomes too tense, it is incised. I consider it better to fix it with a sharp tenaculum, hooked backward from the urethra. After the largest number has been passed, the finger is insinuated. This always causes some hemorrhage. If the tumor can now be felt, it is carefully palpated, and its form fixed in the mind, to enable us to guide ourselves after partial destruction. Then the bladder is filled with 0.3-per-cent salicylic acid solution, through a thin catheter, and the tumor boldly scraped off with Simon's spoons. The middle finger of the hand, the index of which is inside the bladder, controls the scraping movements of the spoon from the vagina. Simon has also devised two other tong-shaped instruments (Figs. 55 and 56) with which larger particles may be grasped, torn off, and removed.

After the operation, to be performed under chloroform, the bladder is again washed out with salicylic-acid solution, and morphine given, if necessary, to diminish the vesical tenesmus. During the following

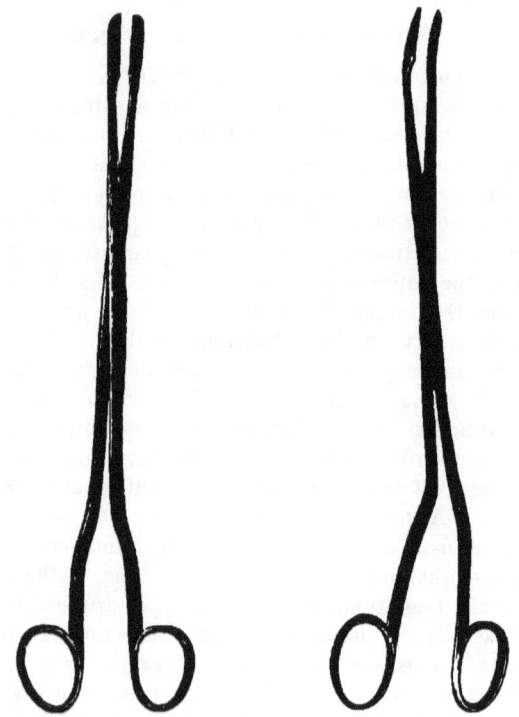

FIG. 55.   FIG. 56.
FIGS. 55 and 56.—Simon's forceps for operating on the female bladder.

days, catheterization is done with a very large catheter so as to gradually remove the remaining coagula and shreds.

Villous tumors being nearly always situated at the trigonum, the described method of operation is the most convenient. However, the inner surface of the bladder may also be made accessible by opening the organ. In order to invert the bladder completely downward, Simon first makes a transverse incision above in front of the anterior lip of the os uteri, followed by a central longitudinal cut extending downward. After that it is possible to invert the bladder completely, and to inspect its inner surface. Several extirpations of tumors of the bladder have been performed in this manner. Should an operation no longer be possible, the treatment is symptomatic.

In the villous tumor, a *carcinoma* may arise secondarily. A primary diffuse cancer of the bladder, followed by general carcinomatous infiltration of the entire walls of the bladder, has also been observed. Moreover, secondary vesical carcinomata very frequently occur in consequence of regionary metastasis from the uterus. In this way, carcinomatous degeneration ensues, and fistulous communications form through the disintegrating tumor. Superficial, epithelial carcinomata of the mucosa of the bladder are still doubtful. Sarcoma, fibroma, and myoma in the muscularis of the bladder have been very rarely observed. In one case, I found a teleangiectatic tumor in the bladder. The growth made the impression of a villous tumor, but consisted only of vessels devoid of cylindrical epithelium.

Tuberculosis of the mucous membrane of the bladder seems to be a rather frequent local manifestation of general tuberculosis or of tuberculosis of the genitals. Tuberculosis has been observed both in the form of small tuberculous ulcers and in that of a tubercular infiltration of the mucosa.

D. Lesions of the Bladder. Fistulæ.

Injuries of the vagina, the vesico-vaginal fistulæ, are almost exclusively due to parturition.

When the head of the child presses a part of the vagina or of the cervix uteri against the pelvic wall until this part ceases to be nourished, it subsequently separates spontaneously by gangrene. Should the gangrenous part be situated in the vesico-vaginal septum, a loss of substance occurs there, a fistulous communication between vagina and bladder, a vesico-vaginal fistula. The subsequent cicatricial contraction and the variable extent of the primarily gangrenous tissue will determine the size. The defect may range between one and five or six centimetres in diameter.

Besides, under certain circumstances the vesico-vaginal septum is completely bruised during operations. This is of rarer occurrence, for strong pressure, if of short duration, is much better borne than long-continued lighter pressure. It may justly be asserted that more fistulæ are due to operations having been omitted or too long deferred than to premature operations. In the present, more active direction of operative obstetrics which, at last, has demonstrated the nugatory character of theoretical temporizing, fistulæ will be of rarer occurrence.

Fistulæ also often arise from direct penetration of the bladder, for instance, by a splinter of bone projecting through a perforated head, or a faultily inserted sharp or blunt hook. I have operated on a peculiar case in which the forceps were applied in a roomy pelvis. No urine had been evacuated for twenty-four hours. Immediately after the rather difficult forceps operation, a large quantity of urine was said to have gushed from the vagina. Nothing further could be learned.

A fistula may also arise from an accidental trauma, as, for instance, from falling on a pointed object which penetrates into the vagina. Moreover, a number of fistulæ have come under observation which have been due to a Zwanck pessary. The latter rotates, and may perforate the bladder or the recto-vaginal septum.

Cases in which a uterine carcinoma has led to a communication between bladder and vagina will not be considered here, as they are incurable.

According to the position of the fistulæ, different kinds have been distinguished. An ordinary vesico-vaginal fistula perforates the vesico-vaginal septum between the uterus and the region of the upper urethral orifice. If the upper margin of the fistula consists of the vaginal portion, it is called a *superficial* vesico-utero-vaginal fistula. If the fistula extends higher up, that is, if a part of the vaginal portion is absent, we speak of a *deep* vesico-utero-vaginal fistula. If the communication is between the cervical cavity and the bladder, it is a *vesico-cervical fistula*. In the *urethro-vaginal fistula*, a passage leads from the urethra into the vagina. Even one of the ureters may open into the vagina: *uretero-vaginal fistula*. The latter fistulæ occur only where there is predisposition. For if the ureter be drawn close to the uterus by a parametritic inflammation, a violent lateral cervical laceration may completely sever the ureter. The lower end contracts downward and heals into the cicatrix, the upper end is drawn upward, but in consequence of the passage of the urinary secretion a point in the cicatrix remains open—a fistula is present.

These forms may complicate, two or three fistulæ occurring together. Then there have usually been extensive destructions, and the united shreds form bridges across the original large defect.

Besides the existing fistula, a number of cicatrices are often seen in the vagina, proving that other lesions were likewise present. In partial separations from the vagina, or in severe lacerations and gangrene of the vulva, cicatrices may contract the vulva so that the fistula cannot be rendered visible at first. Stenoses may also be present higher up, so that eventually, after curing a fistula, the persistence of the symptoms may demonstrate the existence of a second fistula above the stenosis. Or there may be complete atresia in the vaginal vault above the fistula. The uterus cannot be felt from the vagina. The loss of substance at times is so great that but little remains of the vagina, and the lower part of the uterus is also absent. The latter, under certain circumstances, cannot be found at all, being either imbedded in cicatricial remnants of exudation, or complete examination being impossible on account of the inaccessibility of the vagina.

Very small fistulæ are the uretero-vaginal fistulæ, situated laterally in the laquear of the vagina, and quite small, barely visible urethro-vaginal fistulæ occur. Very small fistulæ are often invisibly situated, deeply between, behind, or above folds or cicatrices of the vagina. Larger fistulæ are usually in the median line. If they are very large, the upper vesical wall finds no support on the rudiment of the vesico-vaginal septum, and prolapses into and through the opening, or even through the vulva; then the vesical mucosa is often greatly inflamed, vivid-red, or proliferated. The vesical mucous membrane may also adhere to the vagina, requiring separation previous to reposition. The anterior vesical wall may also be minus a portion. In almost total absence of the vesico-vaginal septum, we see in front the periosteum of the symphysis. Nearly always the openings are round with thin margins; but there may be callous, cicatricial margins, and oval, square, or irregular fistulæ. In all vesico-vaginal fistulæ which completely evacuate the urine, the upper part of the urethra gradually contracts. Not infrequently it may proceed to complete atresia situated exactly at the upper extremity. This atresia is only apparent; it may be overcome with the catheter by using some force.

Should, however, in more extensive lesions, portions of the urethra be lost, the lower remnant may terminate above in a cul-de-sac. It may even be distorted by a firm cicatrix, or be adherent to the bone. In these cases there is only a short inferior portion of the urethra left. Complete loss of the urethra may likewise occur. This proably is only the consequence of obstetric instrumental injury.

The *symptoms* consist in involuntary escape of urine. In extensive injuries, this occurs immediately; in gangrene due to pressure, after separation of the gangrenous part. At times, a large quantity of urine is retained in the vagina if the patient strictly maintains the dorsal decubitus, and if the vulva is swollen. In the lateral position, or if the thigh is raised, the urine gushes forth.

In urethro-vaginal fistulæ, the whole of the urine may pass through the fistula. In small fistulæ, the patient passes the urine into the vessel, while, at the same time, a portion flows down the legs.

In uretero-vaginal fistula, the urine from one kidney escapes spontaneously, that from the other is emptied from the intact bladder.

If the absolutely necessary cleanliness be neglected, vulvitis and vaginitis occur. Especially in the rima pudendi and around the anus excoriations form, and small, often condylomatous ulcers.

The involuntary escape of urine, the perpetual urinous odor of the atmosphere, the uncomfortable permanent moisture of the genitals and thighs, the impossibility of assuming any duties or having any enjoyments, torments the patients to such an extent that they deteriorate physically, and, if not relieved, lead a miserable life.

Diagnosis.

The diagnosis is usually clear even before examination. Escape of urine from weakness or paralysis is very rare, and occurs only after long-standing vesical catarrh, with thickening of the walls and diminution of the lumen of the bladder.

During examination, large fistulæ are readily felt and brought to view by a Sims speculum. Smaller, at first inappreciable fistulæ are discovered by pressing on the vesico-vaginal septum with a catheter. Or else fluid is injected into the bladder, and through the speculum its reappearance in the vagina is observed. For this purpose weak salicylic solution and the irrigator are best. It is necessary to color the fluid. Urethral fistulæ may also be recognized by sounding, and by observing the portion in question during urination.

Suspecting a small fistula at the vaginal vault to be a ureteral fistula, it is carefully probed with a very thin sound. The direction of the canal shows its nature. The bladder may also be well filled by means of the irrigator, thereby proving the non-communication of the fistula with the bladder.

Often the diagnosis of the nature of the fistula is very difficult, because at first it is impossible to employ specula on account of the stenosing cicatrices. Then a preliminary treatment is necessary with a view to make the vagina dilatable. During the diagnosis it is also important to test the patency of the urethra.

Prognosis.

Should the urethra be at least partly preserved, cure can be always effected in some manner. Despite cicatrices and other difficulties, despite lack of substance for the formation of a vesico-vaginal septum, cure is possible. If the fistula has once been cured, it does not reopen spontaneously. As the anterior vaginal wall, during labor, is rather compressed than stretched, closed fistulæ do not reopen even then. But if the defect is very great; if, as in one of my cases for instance, we are forced to stitch the vaginal portion one centimetre above the external urethral orifice, complete cure does not ensue. Still the patients are able to retain the urine at least one or two hours. With this result the patients are also contented.

The transverse obliteration renders the prognosis worse, inasmuch as coition, conception, and childbirth are thenceforth impossible. However, it may confidently be hoped that, with the great progress in the technique of the operation, the transverse obliteration will become one of the rarest methods. The prognosis of the fistula operation *per se* has nothing to do with the facts that peritonitis will occur from injury to the peritoneum, sepsis and pyæmia from infection of the wound, and hydronephrosis from ligation of a ureter.

Treatment.

A fistula recently discovered in the puerperium, if not too large, in many cases closes spontaneously under appropriate treatment. Any offensive lochia must be removed by frequent irrigations. Then the margins of the fistula, if they granulate badly, are to be rendered more prone to heal by being painted with carbolic-acid solution (1:25). *Besides, the bladder is drained, i. e.*, not as formerly by a catheter permanently inserted, but by a rubber tube ten to fifteen centimetres long and six to seven millimetres thick. *It is introduced only to the level of the upper end of the urethra.* The lower end of the tube is suspended in some urinary receptacle. The rubber tube is to be removed at least every two days, and cleansed from adhering crystals. By means of the same rubber tube, the bladder is irrigated in vesical catarrh. If the tube drops out continually, a thicker one must be substituted. In recent cases and small fistulæ, the hope to effect a cure in this manner is by no means slight. Of course, an opening the size of a dollar, with margins four centimetres apart, cannot close.

I would most earnestly warn, in both recent and old cases, against the *tamponade* which, though false both theoretically and practically, is still recommended. *In the first place, in stretching the vagina by a tampon, with increase of its lumen, of course the fistula must be made wider, i. e., its margins are separated.* Besides, the tampon irritates, even if disinfected. The secretions cannot escape and a vesical catarrh often arises rapidly. The nicely granulating wound becomes coated, even fever ensues, and there is no possibility of closure of the fistula.

If the fistula is old, *i. e.*, completely healed over, the bloody fistula operation must be undertaken. To be sure, even nowadays cauterization of small fistulæ is recommended, but incorrectly. In consequence of such recommendations, ignorant physicians feel justified in rendering the margins of fistulæ, by cauterizations, most unsuitable for future freshening. The only possible result is that secondary healing may occur. But where in modern surgery is secondary healing recommended as the *suitable* or *intended* method in plastic operations? *I should urgently advise to operate on principle only by the bloody method and to abandon cauterizations even in quite small fistulæ.*

Owing to the vascular condition of the vesico-vaginal septum, it is to be assumed, *a priori*, that a wound here will heal readily. Apart from experience which has recorded frequent failures, theoretically the healing of the freshened portions is very probable. The sutured wound is in a sheltered place, urine does not injure a wound, traction does not occur, for the other dilatable portions of the vagina may project into the abdomen and hold a good deal of urine. Hence the *properly* freshened and *well* united wound surfaces must heal easily.

But the difficulty lies just in the "proper" freshening and "well" uniting! The locality is inconvenient. All the later effective methods, therefore, have sought above all things to render the fistula so accessible that the operator may clearly take in all the relations and perform all manipulations, although with difficulty, yet with certainty.

In a strange misapprehension of the facts, often the greatest stress has been laid on the minor requirements—material and method of the suture, etc. *All this is unimportant as long as the freshening and approximation are done secundum artem.*

FIG. 57.—Front view of leg-braces for ano-dorsal position. Between them the speculum-holder.

Only since a method has been devised which permits the performance of fistula operations as safely and as completely as, for instance, that for harelip, certainty in results has been secured.

The preparations described on page 44 having been made, the patient is fastened in my leg-braces (Figs. 57 and 58) on a table opposite the window. With the leg-braces (p. 88, Fig. 52), of course, the anterior vaginal wall is placed so that the operator must rather look up from below. This is avoided by the ano-dorsal decubitus, in which the anterior vaginal wall is placed just opposite the operator, pointing almost vertically. In order to render the complete ano-dorsal position possible also in my leg-braces, I have devised two almost rectangularly curved rods extending backward. The thigh brace is applied under the thigh, that is, in legs bent backward, to the anterior side of the thigh which is thus secured. The leg is free and cannot hinder the fixation of the thigh. Fig. 57 represents a front view of this modification of the leg-braces specially adapted for fistula operations, and Fig. 58 the leg-braces and perineal holder (or speculum holder) in side view.

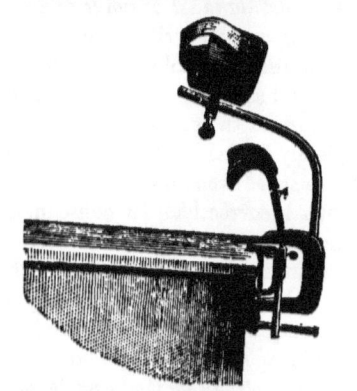

FIG. 58.—Side view of one leg-brace for ano-dorsal position and the speculum-holder.

Whoever can command a sufficient number of assistants, may, of course, as well have the legs held by them.

The vagina is then opened. If obstacles are encountered in the shape of cicatrices, many slight shallow incisions are made wherever the cicatrices are tense. As soon as the vulva and vagina are wide enough to admit a speculum, it is to be introduced. If some lacerations or even a perineal rupture are caused thereby, it is of little consequence; the wound is likewise closed after the fistula operation.

However, it may be necessary, with considerable cicatricial contractions, to institute a sort of preliminary treatment. Bozeman justly has placed particular value upon it. Suppositories and dilators are introduced into the vagina, cicatrices incised and excised, until the fistula can be thoroughly exposed so that the operation can be readily executed. Not till then is the operation performed.

In order to draw the perineum downward, I use a *short* speculum fastened to the table by means of the speculum holder (Figs. 57 and 58). It should be borne in mind that the vagina is to be stretched transversely. With long specula we directly oppose the object of shortening or transverse expansion. If the vulva and the lower part of the vagina are kept wide apart, this transverse stretching alone already diminishes the longitudinal extent.

FIG. 59.—The vesico-vaginal wall exposed in the speculum.

Long specula are absolutely unnecessary. The upper part of the vulva is best kept apart with Bozeman's speculum. For years I have labored to construct clamps intended press the soft parts within and without against the descending ramus of the pubic arch and have failed, and I have always returned to Bozeman's speculum. But I apply it in the reversed position. In the adjoining drawing (Fig. 59) it will be seen how I "engage" the fistula. I have purposely had the engraving made from a photograph, lest anything incorrect be represented. Any one inspecting this illustration must be convinced that by such exposure of the whole anterior vaginal wall as far as the anterior lip of the os, here likewise visible, every fistula is readily accessible without assistance. Should the fistula extend for some distance to the side, this portion must be rendered accessible with sharp tenacula (see p. 36, Figs. 13 and 14).

Having completed this exposure, during which many a little knack can be put to use, the freshening is commenced. For this purpose we need a number of long knives both with points and knobs. Those represented (Figs. 60, 61, and 62) have been devised by Simon. I hardly ever operate with angular, but with very sharp-pointed knives (Fig. 60).

For the freshening, too, different methods have been distinguished, certainly inappropriately. *The freshening must be adapted to the case.* First the margins are approximated with two tenacula or pincettes. Thus we learn in which direction the fistula can be most readily closed; this is not by any means always from above down or from right to left, but often obliquely. It must not be forgotten that the speculum stretches the vagina from right to left, and that the relations are unnatural as long as the speculum lies in the vagina. For this reason we sometimes take a narrow speculum, or screw the Bozeman speculum closer together, in short, observe the fistula as closely as possible in a natural position. This examination had best be done previous to the operation, with Simon's speculum, in the lateral position. Then we easily see how the margins of the fistula can be best approximated. Having thus determined which portions are to be joined, the freshening is done accordingly. Ectropia of the vesical mucosa are treated like the lumen of the fistula, *i. e.*, their periphery is freshened. Ectropia of the vagina into the bladder, however, should be folded outward, or be utilized in the freshening.

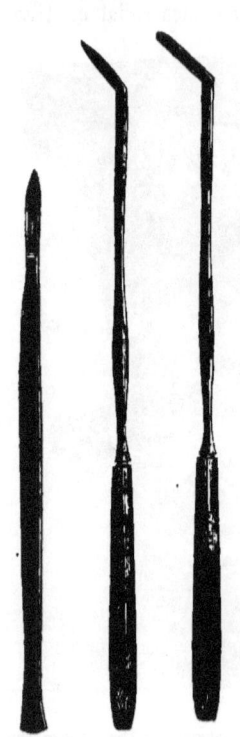

FIGS. 60. 61. 62.

FIGS. 60, 61, 62. — Simon's knives for the operation on vesico-vaginal fistulæ.

If the margins can be readily approximated, a belt to the width of one centimetre around the orifice is freshened. But if the fistula is so large that it would require too much traction to join the margins at all, the freshening is to be more funnel-shaped, preferably dispensing with a broad wound surface to avoid excessive traction. In a single fistula, one-half of the two corresponding wound margins may even be freshened broadly, the other flatly. If it be particularly difficult to approximate the margins, I have freshened laterally beyond the fistula. The freshening then extends to the posterior vaginal wall. Even an entire shortening ring may be stitched into the vagina. If then the sutures be tied first over the freshened surfaces away from the fistula, the fistulous margins can be easily approximated. Fig. 63 represents these conditions.

Little need be said about the method of freshening. Nobody will operate who has not seen similar operations. Certainly the vaginal surface must be completely removed, as described under perineoplasty, p. 89. The more profuse the hemorrhage the better the prognosis as regards the cure. During irrigation with water the unevennesses of the wound are best seen. It is levelled with knife and scissors and brought to a uniform depth. Often the freshening extends across a cicatrix. Whitish and bloodless it stretches close under the superficial skin, which is at once denuded by the pincette. These portions heal, only two cicatrices must not be brought together by suture. Injury to the vesical mucosa should be avoided if possible. For it is not impossible that the chances of healing are made worse by the incision and direct suturing of the mucous membrane of the bladder. The mucous membrane does not always close around the ligature, no matter what material be used. Small passages always arise, widened and irritated by the urinary incrustations on the thread, which, as it were, grow along upon it.

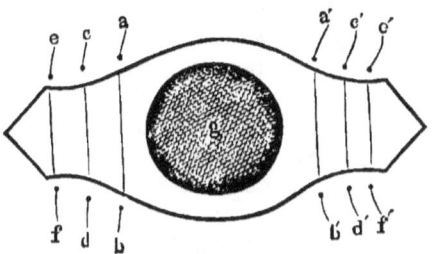

FIG. 63.—The freshened surface in difficult approximation of the margins of the fistula. In order to relax and approximate the margins of the fistula, a is first stitched to b and a' to b', then c to d, c' to d', e to f, e' to f'. Thereby the margins of the fistula are readily brought together.

The thread forms the nucleus of a small vesical calculus. In this way in the cicatrix, in the closed fistula, a small fistula arises in the track of the suture. Since I have repeatedly observed this, I prefer to introduce the needle close above the vesical mucosa. It is, besides, not improbable that vesical hemorrhage may be connected with the lesions of the mucous membrane of the bladder.

The material for the ligature is of no consequence if the approximation is thorough. After having tried all the suturing material recommended, I have returned to silk. Stiff silken threads are tied with difficulty, contain unreliable spots and tear easily, the only safe firm portion of a thread is often inconveniently short. Catgut has no advantage over silk, but tears more easily. Wire cannot well be tied, and in twisting it the operator has not the power to prevent the loop from becoming too tight and excessively compressing the tissues.

A needle-holder, to be appropriate for fistula operations, *first*, must hold the needle absolutely immovable lest it slip in passing; *second*, any direction of the needle must be possible; and *third*, the needle must be readily freed. These advantages are combined in the needle-holder represented in Fig. 64, which, besides, is self-closing on pressure.

Fig. 64.—Author's needle-holder.

During the passing of the suture the following must be kept in mind: Sufficient tissue must be included in the loop, otherwise the thread will cut through, and even in the most favorable case a small fistula arises in the closed fistula. Too many needles should not be passed, lest the included tissue be deprived of nourishment. Then we must reflect, before each needle is passed, how it is to lie, and thus avoid injuring too many capillaries by frequent reinsertions. The suture should never be drawn too tight, or else the entire included tissue will be deprived of nourishment.

The suturing is to be done with two equal needles at the ends of the thread, the direction being always from the wound toward its margin (Fig. 65). At *a* or *b*, the needle is inserted close to the vesical mucous membrane and its point carried around as deeply as possible, so as to make it emerge close to the margin of the wound at *a'*.

Only after presumably enough needles have been passed, at a distance of one-half to three-fourths centimetre, the tying is commenced. If then the vaginal investiture is not well joined, a number of fine sutures with No. 0 silk can be inserted. But the approximation in the depth is of more importance than that at the surface.

Should it be impossible to tie the knot with the fingers, the wire snarer may be used. Besides, sufficient dexterity is gradually acquired to tie a knot even in the narrowest and deepest places.

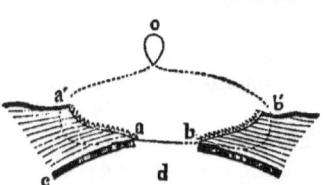

Fig. 65.—*a b*, points of entry; *a' b'*, points of exit of the needle; *c*, vesical mucosa; *d*, bladder; *e*, vagina.

The operation is not always so simple. Should the vesical mucous membrane be prolapsed, it must be held back during the operation with a stiff catheter. Should the wound be very irregular, it may be impossible to secure a straight line of union. Then single flaps must be formed and sewed together, producing figures like T or Y. In very difficult cases, a plastic operation may be attempted with the labia majora. They are movable and displaceable at their seat, so that for plastic purposes they may easily be shifted into the vagina.

In stricture of the urethra, the occlusion may be overcome by forcible passage of a stiff catheter. In the absence of a part of the urethra, the upper margin of the fistula is to be stitched to the upper end of the urethra. Even a few isolated fibres of the sphincter soon gain in power and functionate for the entire sphincter. If but one centimetre of urethra is present, there is a prospect of success. Above, the freshening may extend into the anterior lip of the os uteri, which will adhere readily. If the uterus is very mobile and displaceable, union to the urethral eminence is possible. Should an artery spout during the freshening, it may be left alone, at most while suturing regard may be had to make one of the sutures serve as ligature. Ligation and torsion will certainly diminish the prospect of primary union.

It has happened that a ureter has been incised during the operation. Simon mistook it for an artery on account of the escaping jet of bloody urine. Or else the suture accidentally grasped the ureter and closed its lumen. In the latter case, fever and anæmic headache, etc., rapidly set in, thereby permitting the diagnosis of occlusion of the ureter. In that event, nothing remains but to reopen the sutures at the point in question, or else to open a sufficient number of sutures to let the urine pass.

As the main point in a fistula operation is accessibility, the operation for vesico-cervical fistula is, of course, very difficult. Here alone the attempt at cauterization is justified. Previous to the operation, the cervix must be split open laterally, so as to expose the fistula. Should it be impossible to close the cervical fistula directly, the lips of the os must be stitched together. Then the menstrual blood flows into the vagina through the fistula; but the urine cannot escape into the vagina because of occlusion of the os uteri.

Formerly, when the technique of the operation was not so far advanced, many fistulæ were thought to be incurable. Then a modified cure was rendered possible by freshening a ring of the vagina below the fistula and stitching the walls of the fistula together. By this transverse obliteration, there was formed of the upper half of the vagina and the bladder a space containing urine, menstrual blood, and uterine secretion. Vesical catarrh did not occur, and aside from the impotentia coëundi, the patient felt well. In a somewhat modified form, in deep vesico-vaginal and cervical fistulæ, the posterior lip of the os uteri has also been stitched to the lower margin of the fistula.

Of course, there are cases in which we cannot do any better. But this helplessness is often confined to the operator. Simon has repeatedly reopened transverse obliterations made by other operators, and then operated directly on the fistula and cured it.

Should the entire urethra be missing, restoration is impossible, the unfortunate patient is incurable.

In uretero-vaginal fistulæ, cure has also been effected (Bandl). Should the lower end be completely absent, and the upper open into the vagina, the bladder is perforated into the vagina from the urethra, then a thin catheter is introduced through this opening and into the ureter. Now an oval surface is freshened and united over the catheter which remains for the present. After the wound has closed, the catheter is removed. The artificial passage is kept open by the urine. If the ureter is merely severed, a catheter is introduced into the upper and lower end, and union effected over the spot where the catheter is visible in the vagina.

Should this fail, transverse obliteration would have to be performed and a vesico-vaginal fistula perhaps made above it. The urine effused into the upper half of the vagina would then flow into the bladder through the fistula. The obliteration might then be oblique, so as to separate only the region of the fistula from the vagina.

It is one of Simon's great merits to have simplified the after-treatment of fistulæ. The patient must remain in bed at least one week. She is catheterized only if she cannot urinate and has tenesmus of the bladder. If possible, the sutures are not removed until urine can be passed spontaneously without difficulty; otherwise on the sixth or seventh day.

Should there be stricture of the urethra, a rubber tube is to be left in it. *I recommend drainage of the bladder as a procedure entirely free from danger.* It has the great advantage that we know, as long as the urine flows off, the fistula is closed. Moreover, by injections into the bladder we may at once control the healing. In a very large defect, however, the wound is certainly most perfectly at rest during drainage of the bladder. Should the fistula be above the trigonum, no urine will flow into the wound which can close undisturbed.

Should the wound fail to unite, this will usually be decided on the third or fourth day. In such cases I advise the secondary suture. In one case, in which I was very glad not to be obliged to operate again, on account of the many cicatrices present in the field of operation, I thus secured union. In connection therewith, of course, as above explained, the bladder must be drained. The process of healing is also disturbed by hemorrhages from the bladder; whether these are referable merely to traction and dilatation during refilling, to erosion of a vessel close to a suppurating suture track, or to injury during catheterization can often hardly be decided. But when the urine becomes brownish-red and tenesmus occurs, there is mostly a blood-coagulum in the bladder. Then, in a favorably progressing fistula, the urethra should be dilated with Simon's urethral specula or with my uterine dilators (Fig. 18, p. 41) and removal of the blood-clot attempted. Should it fail, the fistula will reopen under the tenesmus of the bladder. After removal of the blood, or diminution of the bladder, the hemorrhage ceases spontaneously. Should this not be the case, the vagina must be irrigated with cold, mildly astringent or disinfecting solutions.

Should a small fistula remain, for instance, in a suppurating suture, we must operate again in two weeks. Then the entire region is still very hyperæmic and we have a good prospect of success. In these cases, the second freshening should be laid vertically or at least obliquely to the first, to avoid the cicatrix of the closed fistula.

Communications of the bladder with other regions are found much more rarely than those with the vagina. These communications are generally the consequence of neighboring suppurations and inflammatory perforation into the bladder from the outside. Thus in parametritic accumulations of pus, perforations into the bladder occur. In extra-uterine pregnancy, too, perforation into the bladder has occurred, the fœtal bones being finally removed from the urethra. In the same way suppurating dermoid cysts of the ovary have discharged their solid and fluid contents into the bladder. Hair, teeth, fragments of bone, masses of pus, or colloid matter from ovarian adenomata, have also been passed from the bladder.

In a similar manner perforations from the pus cavity into the rectum occur at the same time. Then the urine may take a detour through the pus cavity to the rectum, and after closure of the abscess again find its old natural course outward. Or, in communication with the small intestines, thin chyle and intestinal gases may pass into the bladder.

Considering their great rarity, and the difficulties caused by their situation, we cannot as yet speak of a method of treatment of these communications. Dilatation of the urethra, dragging forward of the fistula, combined perhaps with a vesico-vaginal incision, would be the way to obtain cure in these cases.

Rupture of the bladder from excessive internal pressure in retention of urine does not appear to occur. But there are formed attenuations, diverticula, cystitis, gangrene, loss of substance, and secondary perforation. Death from peritonitis is the consequence.

E. Diseases of the Urethra.

The urethra inflames only in consequence of *gonorrhœal infection* or of injury. As gonorrhœal urethritis in the male soon loses its acute character even without treatment, so it is in the female. The tenesmus, the urgent desire to urinate, the pain on direct pressure on the urethra soon disappear. It is hardly ever necessary to institute treatment for the the urethritis of the female. The impossibility of retention of pus or formation of stricture seems to create in this instance conditions favorable to cure. Should this not be the case, a brush dipped in silver nitrate solution (1 : 15) is passed into the urethra.

More chronic are the urethritides referable to *injury*, especially to *catheterization*. Even with the greatest caution, small injuries are sometimes caused by the elastic catheter. These lesions have the same consequences as a fissure of the anus. During urination, particularly during repeated catheterization, often the most violent pains occur.

Often the locality of the fissure can be quite accurately diagnosticated. If the catheter be pressed to one side, and if pressure and traction on the fissure be avoided, the act of catheterization is painless.

In chronic catarrhs with paresis of the bladder, I have often seen such cases. But also in the puerperium, after a single catheterization, such fissures remain behind. At times, they cause perpetual desire to urinate. The absence of any turbidity of the urine, the impossibility to detect anything, cause such cases not rarely to be considered and treated as neuroses.

Should the fissure complicate a catarrh, I advise to drain the bladder, thus giving the fissure time to heal. But if the bladder is quite normal, rapid dilatation is performed with Simon's specula or my dilators (Fig. 18, p. 41). After that the troublesome symptoms quickly disappear.

According to my observations, such an irritated condition may also depend upon venous ectasiæ. The surroundings of the urethra appear dark-red, and on separating the urethral orifice, we often see thick ectatic veins close underneath the mucous membrane. We then commence general treatment with cathartics, make a small incision into the thickest vein with the scissors or the knife, and allow the vein to bleed. Cool compresses are also useful.

The peculiar condition of dilatation of the urethra is likewise looked upon as a consequence of *varices*. In these rare cases, the dilated, cystically expanded urethra filled with urine can be demonstrated by direct pressure.

Prolapse of the urethral mucous membrane seems to be equally rare. Its etiology is unknown. Retention of urine from prolapse of the urethral mucous membrane occurs more frequently in young girls than later in life. If not relieved, finally an ulcerated, furunculous, dark-red, painful tumor, three to five centimetres in diameter, lies in front of the urethra. The youthful age and the rapid development allow the diagnosis to be easily formed. It is especially necessary to find the urethra. This is not always easy. If an attempt at reposition promises success, it should be made. Otherwise, a thick catheter is introduced into the urethra, a ligature placed around the pedicle of the tumor as close as possible to the external urethral orifice, and the tumor severed close to the ligature, cutting toward the catheter. Thus I cured the only case which came under my treatment, a girl of eight years.

Among the affections of the urethra belong also the so-called *carunculæ of the urethra*. These are small, very vascular tumors, covered with pavement epithelium. They arise at different depths in the urethra, have therefore a varying length of pedicle, and are usually uniformly round or oval. Also lobulated and raspberry-like carunculæ occur. But rarely several carunculæ are found. However, in one case I have cut off in the course of three months, by degrees, as many as twenty-four.

Invariably after removal of one tumor another appeared after three or four days.

Urethral caruncles occur both in young and old women. They should not be mistaken for the small villi occasionally formed at the urethral orifice. It is always necessary for the *diagnosis* that the pedicle extend into the urethra. The bright-red color, markedly differing from the paler surroundings, readily permits the formation of the diagnosis.

The *symptoms* are variable. It happens that a caruncle is accidentally noticed, having caused no symptoms. In other cases, especially in older women, the caruncles are the cause of much suffering. Continual desire to pass water, pain during micturition, and prolapse of the caruncles from the urethra, greatly annoy the patients. It has even happened to me to have a patient believe herself to be suffering from a uterine polypus which always appeared during urination.

The *treatment* consists in ablation. This requires assistance. One hand must hold the scissors, the other the forceps; hence a third hand is necessary to keep the vulva apart. As touch with the forceps is not rarely extraordinarily painful, we must be quick with the scissors. Otherwise the patient presses her legs together, the forceps tear out, blood covers the field of operation, nothing can be done for the present, and the patient bears the second attempt at operation much less readily than the first. Subsequent cauterization is unnecessary. The hemorrhage soon ceases. Relapses do not occur. Cauterizations, ligatures, etc., do not give any better results than the ablation with the scissors.

Besides urethral caruncles, *myxomata* and *sarcomata* of the urethra have been described.

In long-standing *syphilis*, the surroundings of the external urethral orifice may ulcerate by syphilitic processes. If the vulva lies close below the symphysis, the penis in coition tears the urethra from its support and the urethral eminence from the urethra. Reunion is prevented by the syphilitic process and frequent coition. As the first indication of detachment of the urethra from the symphysis, we not rarely observe in old prostitutes quite oval urethral orifices or slight descensus of the posterior urethral wall into the lumen of the opening. With the complication of syphilis, frequent injury, and unfavorable (for coition) position of the introitus vaginæ, the entire external urethral orifice may be changed into one ulcer. The urethra, too, ulcerates upward, and strictures arise from partial cicatrization. If it be possible to prevent coition, and to treat the ulcers rationally, the *prognosis* is not bad. In more extensive destructions and detachments of the urethra, plastic operations must be performed. Schroeder has produced a new useful urethra in a large defect of that part by freshening and reunion.

CHAPTER IX.

UTERINE MALFORMATIONS, ARRESTS OF DEVELOPMENT, AND THE GYNATRESLÆ.

A. Malformations.

As in every other part, the malformations of the female sexual organs can be understood only in connection with the history of development.

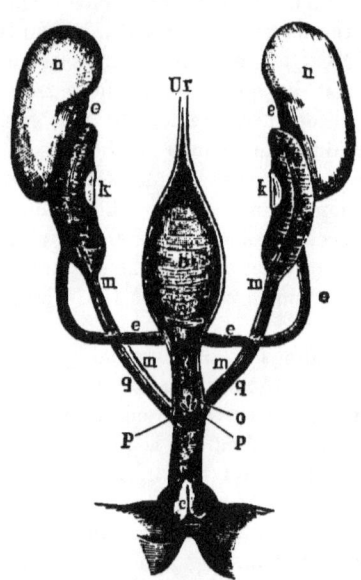

Fig. 66.—$n\,n$, kidneys; $e\,e$, ureters, emptying at f into the bladder b; $w\,w$, Wolffian bodies, primordial kidneys (parovarium); their efferent channels $q\,q$ terminate at $p\,p$ in the urethra d; $k\,k$, germinal gland (ovary); $m\,m$, Müller's ducts, adherent to the Wolffian bodies, terminate united at o; v, urogenital sinus; ur, urachus; c, clitoris, or genital eminence.

We therefore subjoin a brief description of the genesis of the female generative organs according to Henle. (Compare Fig. 66.) From the fourth to the fifth week of the fœtus, we find on both sides of the vertebral column the primordial or temporary kidneys, the bodies of Wolff or Oken (n). The name has been given them on account of the arrangement of the vessels, the presence of uric acid in their secretion, and the communication of their efferent channels with the urinary bladder. The efferent channel (q) is of no importance is the female, only vaginal cysts have been brought in connection with a partial persistence of this canal. In some animals, the efferent channel persists (Gärtner's canal). From the lower angle of the Wolffian body a ligament extends to the inguinal region, in the female developing into the *round ligament* of the uterus. At the median margin of the Wolffian body is the germinal gland (k), later the *ovary* or the *testicle*. From the anterior surface arises a cord, at first solid, *Müller's duct* (m) which becomes of the greatest importance in the female. Both efferent channels of the Wolffian body, as well as the two mutually united Müller's ducts, insert themselves into the lower end of the urinary bladder (*p, o*), at the point between the urethra above and the urogenital sinus below.

While the Wolffian body is retarded in its growth into the parovarium, and the function of secreting urine is assumed by the kidneys, Müller's ducts continue to develop. They join together, probably first at the point where later the *vagina* is united to the *cervix uteri*. The septum between the two ducts disappears so that a common canal arises. But the upper ends of Müller's ducts remain separated, diverging. The upper extremity becomes fimbriated, forming the *fimbriæ;* then follow the *tubes;*

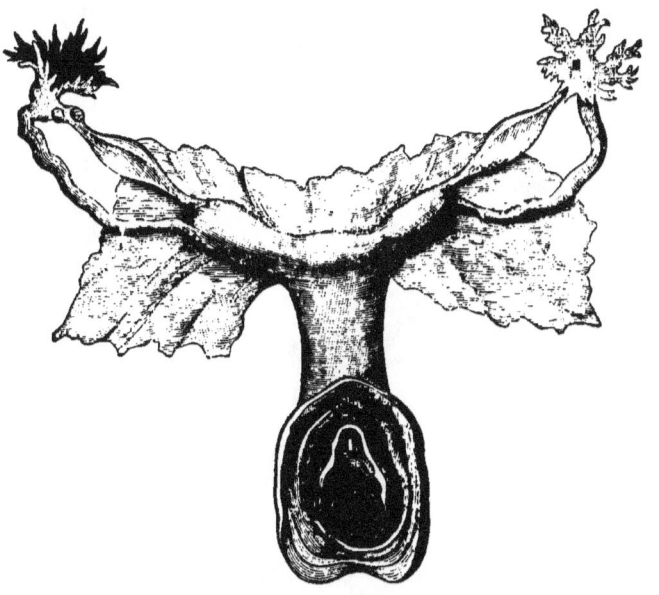

FIG. 67.—Uterus bicornis duplex, vagina duplex, after Böhmer.
If we were to assume the uterus to be externally perfectly normal, and internally divided by a septum, it would be a case of uterus bilocularis.

the next lower portions uniting form the *uterus*. The latter increases so greatly in size that not alone the trace of lateral cornua disappears, but that even a *fundus* projects in convex shape. Below the uterus, Müller's ducts form the *vagina*. The above-mentioned inguinal ligament approaches the upper edges of the uterus as the *round ligament*. The germinal glands, or *ovaries*, sink down from above until they lie beside the uterus. At the point where vagina and urogenital sinus adjoin, a fold arises—the *hymen*.

The *malformations of the uterus*, in connections with the conditions described, are easily understood. Assuming, first, that both of Müller's ducts are completely present, double formations result:

First. If Müller's ducts fail to unite (bicornuity).

Second. If Müller's ducts unite externally, but the septum fails to disappear (bilocularity).

Of the former, various forms occur, namely, *first,* an isolated development of the completely separated Müller's ducts—*uterus bicornis duplex separatus, s. uterus didelphys.* This double formation is usually observed in connection with other malformations in non-viable infants. Recently, however, three cases have also been found in adult women, in post-mortem examinations.

If the uterus is completely duplicated so that each club-shaped fundus diverges considerably, but the cervices below—though double—are closely adjoining, it is a *uterus bicornis duplex* (Fig. 67). In this the vagina may be single or double. A still slighter degree of duplicity exists if the uterus has two cornua, but a single cervix and vagina—*uterus bicornis unicollis* (Fig. 68). Among these belong the cases in which the bicorn-

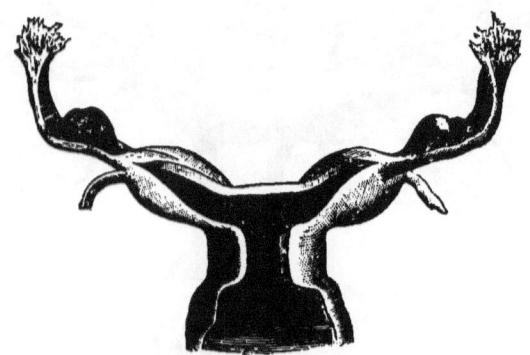

FIG. 68.—Uterus bicornis unicollis.

uity is indicated merely by the concavity of the upper outline—*uterus bicornis seu arcuatus*—or in which the fundus is extraordinarily broad—*uterus incudiformis.*

Of the second it may be remarked that the septum may be completely present—*uterus septus* or *uterus bilocularis duplex.* The septum in the vagina may either reach downward or be entirely absent, the separation extending only to the cervix (see Fig. 67).

If the uterus is not completely divided, the septum being merely rudimentary, it is called *uterus subseptus.* Still more frequently merely a fold of variable length depends from the fundus uteri. But only as an extreme rarity do we find two external orifices of the uterus—*uterus biforis*—with a single uterine cavity. These malformation are of greater importance to the obstetrician than to the gynecologist.

Third. Besides, malformations arise by one of Müller's ducts being altogether absent (unicornity) or only partially developed. In the former case, a *unicorn uterus* (Fig. 69) arises. It is usually rather defectively developed, with preponderating cervix, and deflected from the side where the defect exists. On the latter side, the ovary, tube, and round ligament are also absent. Otherwise a unicorn uterus functionates normally, even labor progressing with due activity and without abnormal symptoms.

If one horn be developed, but the other merely rudimentary, the latter may either closely adjoin the uterus or be dragged far away. The rudimentary cornu is formed both of a ligamentous solid accumulation of muscular fibres, and of small hollow body. Whether the latter be close to the uterus or at a certain distance, it is always recognized by the round

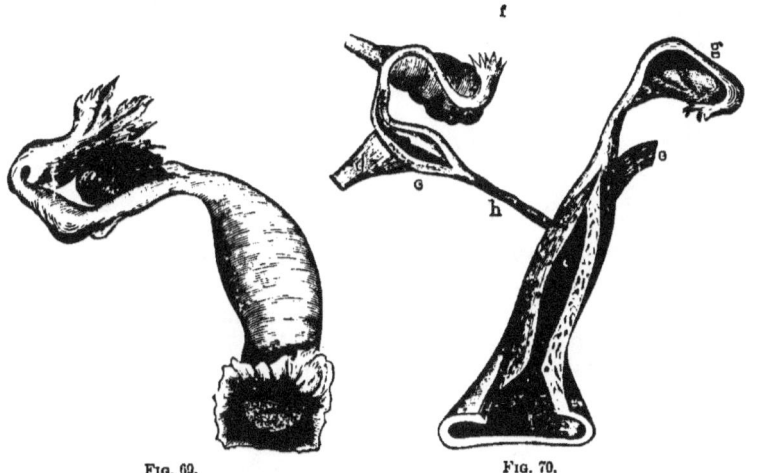

Fig. 69.
Fig. 70.

Fig. 69.—Uterus unicornis dexter.
Fig. 70.—Uterus unicornis sinister with rudimentary left cornu.
(Compare below the figures of the gynatresiæ).

a, vagina; *b*, unicorn uterus; *c*, rudimentary right cornu, representing a small hollow body joined by a solid muscular cord *h* to the uterus *b*; *d*, right, *e*, left round ligament, *f*, right tube and ovary; *g*, left tube and ovary.

ligament extending downward from it (Fig. 70, *e d*). The tube of the defective side may be a solid cord or be perfectly normal; the ovary likewise may be absent or present. In the latter case it is both normal and abnormal in that no follicles can be demonstrated in it.

Therefore, inasmuch as a rudimentary horn may be invested with mucosa and be provided with a normal tube and ovary, all functions are possible. The mucosa menstruates, the blood accumulates, a retention cyst arises, a hæmatometra in a rudimentary cornu. The tube may receive a fecundated ovum, the ovum may develop in the rudimentary cornu, and the latter rupture in consequence of the insufficient musculature, leading to hemorrhage and death as in extra-uterine pregnancy. We shall revert to these cases in connection with the gynatresiæ.

Fourth. There is series of malformations arising from the fact that *Müller's ducts, though joining and losing the intervening septum, and both being equally developed, still are in one portion of their length rudimentary to the same extent on either side, while everything above and below is perfectly normal. Thus the lower or upper half of the vagina or the uterus may be absent or in a rudimentary state.*

Most frequently that part of Müller's ducts may remain rudimentary the middle part of which forms the uterus. Hence on both sides tube and ovary, as well as the vagina, may be present, while the uterus is absent—*defectus uteri.* A complete absence of the uterus, however, has perhaps hardly ever been demonstrated with absolute certainty. Even if rectum and bladder were immediately adjoining, still a more minute examination discovered a slight accumulation of muscular fibres to be interpreted as a rudimentary uterus. Should the uterus be completely or almost completely absent, then ovaries and tubes are also lacking.

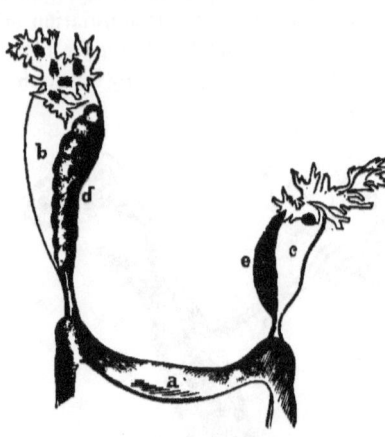

FIG. 71.—Rudimentary solid uterus (Kussmaul). *a*, muscular portion, representing the rudimentary uterus; *b*, right, *c*, left tube; *d*, right, *e*, left ovary.

But a rudimentary solid uterus is by no means rare; it may represent a small organ the shape of the uterus, or extend transversely from one tube to the other as a band of muscular fibres (Fig. 71). In that case

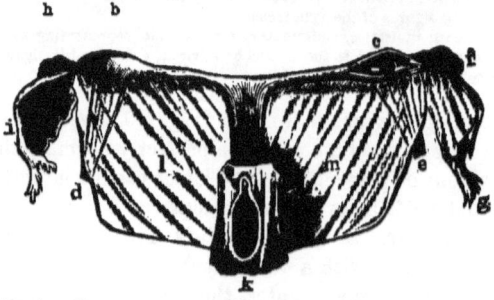

FIG. 72.—Rudimentary Uterus. Uterus bipartitus (Rokitansky).
a, the rudimentary middle portion; *b*, right horn of the uterus, *c*, left horn, cut open to show the small cavity; *d*, right, *e*, left round ligament; *f*, left ovary; *g*, left tube; *h i*, the same, right; *l m*, broad ligament.

the defect must be chiefly referred to the lower two-thirds of the uterus. Still more frequently the two cornua are uniformly developed; two small, widely separated, round or oval hollow bodies being present laterally, while the central portion and the cervix are lacking (*uterus bipartitus*. Fig. 72). In that event, the ovaries and tubes are not always absent. The opposite condition has also been found—a centrally located rudimentary uterus with absence of the two lateral cornua. Finally, there are those cases in which the uterus, with an increasing tendency to the normal state, is no longer rudimentary, but merely faultily or badly developed. As here the uterus is retarded in growth, and shows a fœtal or at least infantile form, this arrest of development is called *uterus fœtalis* and *infantilis*. An extraordinary narrowness—stenosis—of the orifices of the uterus is regarded as the slightest degree of arrest of development.

B. Defect of the Uterus.

The forms comprised in the fourth section of the preceding also acquire clinical importance. We have seen that the complete absence is difficult to demonstrate even anatomically. Clinically, all cases of uterine defect, and of functionally impotent, rudimentary uterus will have an equal importance. If we make the term so comprehensive, the *diagnosis* is not difficult.

Obviously, we must exhaust all possible methods of examination and then form the diagnosis by exclusion, that is to say, diagnosticate that nothing can be felt where the uterus should be. Such examinations had best be made in narcosis, so as to obtain *at once* nearly *absolute* certainty.

Even the anamnesis, the absence of menstruation, points to the defect of the uterus. As the rest of the body may be normally formed, we must not be surprised eventually to see the healthiest women affected with this condition.

In such cases, almost always only the lower portion of the vagina is present, ending in a cul-de-sac. It also happens that every trace of the vagina is absent. In that event, the vestibule terminates at the urethra, as the vagina otherwise does at the vaginal portion. Therefore in attempts at coition, the extremity of the penis will be directed toward the urethral orifice, will successively dilate it, and penetrate into the bladder. This dilatation of the urethra does not by any means always lead to incontinence of urine. Should every indication of a vagina be absent, the penis depresses the skin to a depth of five centimetres, causing surprise to find on touch a vagina, though short, while on exposing the parts and spreading the legs not the least indication of an opening can be seen.

Should the diagnosis be suspected, combined examination must be performed, first, from the vagina and the abdominal coverings; second, from the vagina and rectum; third, from the rectum and bladder. If it be difficult to penetrate into the bladder, dilatation must be made.

Especially the systematic palpation of the entire septum between bladder and rectum will be successful.

It will hardly be demonstrable with certainty whether the ovaries are present, even if bodies having their form be felt in the ovarian region. In that event a rudimentary uterus may be present on both sides (uterus bipartitus, Fig. 72), so that the oval, cylindrical bodies may be interpreted as ovaries and as uterine cornua.

C. Infantile Uterus.

At puberty the uterus, till then infantile, grows and reaches the normal size. However, should the uterus remain of the size appropriate to the child until puberty, the case is one of infantile uterus, a defect of development.

In the broad ligament the uterus lies as a small flat tumor, so that there is difficulty of finding the organ in the preparation. It has a complete cavity, but the palmæ plicatæ extend to the fundus, while otherwise, in the adult uterus, they reach only to the internal os. It is important that several investigators have found a decidedly infantile uterus—a hypoplasia of the uterus—in general hypoplasia of the vascular system—diminutive heart, etc.

Infantile uterus is diagnosticated in the living by the bimanual demonstration of deficient size of the uterus, as well as by measuring with the sound, if that be possible. Menstruation, and hence ovulation and fecundation, are absent.

Sometimes it also happens that the diagnosis of infantile uterus is formed in cases in which menstruation is present, but in which the uterus remains extraordinarily small. Then great dysmenorrhœa may force the patient to seek medical aid. Strictly speaking, this is not infantile uterus, as the organ possesses all the characteristics of a ripe uterus, except the size. However, the fact that the uterus is extraordinarily small, that menstruation occurs but sparsely and impregnation not at all, may well permit this condition to be designated as badly developed or infantile uterus, just as non-functioning uterine rudiments are included among defects of the uterus.

It is clear that uteri with badly developed musculature easily acquire flexions, or that the physiological flexions become more pronounced or acute-angled. Therefore, not rarely we find in infantile uterus anteflexion, or in anteflexion a very small uterus. For this reason we shall treat this matter more in detail in connection with anteflexion of the uterus.

D. Stenoses of the Uterus.

Stenoses of the External Os.

The external os has a variable form and width. That form is considered normal in which there is a small transverse opening, about one centimetre in length. Often, however, the transverse opening is much smaller, even consisting of a very minute hole not allowing the passage of the knob of a small sound. With this, *first*, the uterus may be perfectly normal and the cervical canal roomy; nay more, by the retention of the physiological secretion the cervical canal becomes very wide and contains vitreous masses which have been so inspissated that after dilatation they may be grasped and removed like a firm lump of gelatin (Fig. 73). *Second*, the vaginal portion is really too large. By counterpressure against the posterior vaginal wall it is flattened, mushroom-shaped, with a large surface in the exact centre of which is a small os (Fig. 76). And *third*, the vaginal portion is pointed, elongated, extraordinarily long, conical, the shape of a sugar-loaf. There is no lower surface. At the extremity of the vaginal portion is the small os (Fig. 78). In the second and third forms, the entire uterus is at times too weak in its musculature—infantile uterus.

Acquired stenoses of the external os arise especially from awkward cauterizations. Also puerperal gangrenous processes, or lacerations by

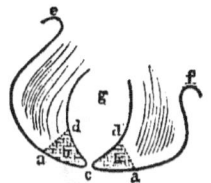

Fig. 73.—Originally normal vaginal portion, excepting the stenosis of the external os. *g*, dilated cervical canal; *e*, posterior, *f*, anterior vaginal vault; *a d c*, the portion *b* removed round about the os.

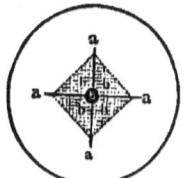

Fig. 74.—*a c, a c, a c, a c*, incisions carried from the os *c*, toward the outside; *b*, the four triangles removed.

obstetrical instruments lead to stenoses, though more frequently to defects. Furthermore, after plastic operations stenoses and even atresiæ are not at all so rare as is usually assumed.

The *diagnosis* of all stenoses is very easy. One look into the speculum, touching the conical portion, the attempt at sounding demonstrate the condition.

Owing to the variable form of the vaginal portion, different methods of operation are selected. In the first case (Figs. 73 and 74), the best result is obtained in the following mannner (Fig. 74): In Sims's lateral position the vaginal portion is seized from the inner side of the os with a sharp tenaculum (Fig. 13, p. 36). Then we insert a knife (Fig. 75) and cut in the direction opposite to the traction of the hook an incision one centimetre in length. The same is done on the other side, in front and behind. Thus four radial incisions extend from the centre (Fig. 74, *a c*). The four flaps (*b*) thus caused are seized one after another with the tenaculum and about half of each cut away. In this way the flaps (*b*) are removed (Figs. 73 and 74). The abscission may be done with the knife or long scissors. The os now has a funnel shape. After the operation a pledget of benzoated cotton is placed on the vaginal portion. It is removed after twelve hours and irrigations given. After this simple operation, often performed during my office hours, I have never seen any inflammation or after-hemorrhage. For antiseptic rules see Chap. IV.

The bilateral incisions, in order to secure an equally large opening, must be carried much deeper, bringing us into the dangerous neighborhood of the parametria. While in my method the cicatricial contraction renders the os rather wider, it becomes gradually smaller in the bilateral discission.

Should the vaginal portion have the mushroom form described sub 2 (Fig. 76), the above method will not suffice, for merely a piece of the narrow cervical canal would be cut off and the stenosis would be simply removed three to four centimetres higher. The method of operation here required was first described by Simon for the purpose of widening a narrow os. Let us imagine a short, thick, perhaps even mushroom-shaped vaginal portion (Fig. 76), then a wedge is excised round about the os (Fig. 77, $g\,h\,i$). If, after the removal of this wedge, g is stitched to i all around, the os and the lower end of the cervical canal must of course be drawn apart. To facilitate this excellent operation, the os is first cut open on both sides as far as the vaginal vault to e (Fig. 77). If this were omitted, the excision would present great difficulties. But as, after the lateral incisions, the two halves of the vaginal portion can be kept straight and drawn down singly, the excision of the wedges is not at all difficult. In this method, of course, we require no expansibility of the cervical mucosa, but stitch i—a part of the lower surface of the vaginal portion covered with vaginal mucous membrane—to g which is likewise covered with vaginal mucosa.

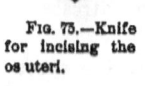

FIG. 75.—Knife for incising the os uteri.

But if the vaginal portion has the form described sub 3 (Fig. 78), the last method of operation is inappropriate because a large piece of the vaginal portion must be removed. The great length of the vaginal portion has different injurious consequences. First, disagreeable sensations arise either by the vaginal portion crowding apart and irritating the vulva, or by the vaginal portion, in resting on the floor of the pelvis, raising the entire uterus. An increase of the anteflexion may arise in this manner, as we shall see later on. Furthermore, conception is rendered more difficult; although I have seen several such cases in which conception occurred even without operation, still it may be assumed *a priori* that the great distance the spermatozooids have to travel from the fornix vaginæ around the entire vaginal portion is accomplished with diffi-

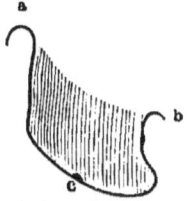

FIG. 76.—Stenosis of the external os, with thick, broad vaginal portion, *a*, posterior, *b*, anterior vaginal vault; *c*, os uteri.

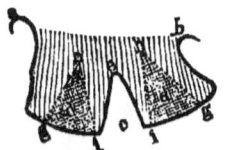

Fig. 77.—Conical excision of the vaginal portion. *a*, posterior, *b*, anterior vaginal vault; *g h i*, wedges excised; *d d*, the portion removed; *c e*, bilateral discission to facilitate the excision of the wedges.

culty. Inasmuch as the injurious consequences, described with the first form of stenosis, may besides complicate the stenosis with conical vaginal portion, an operation is indicated.

It is performed in the following manner:

The vaginal portion is first divided to the right and left as far as the

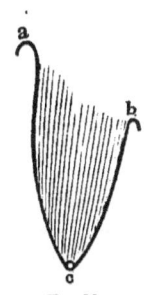
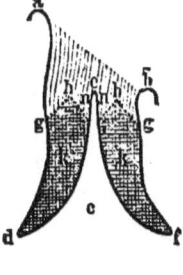
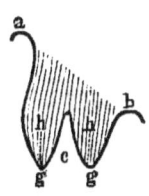

FIG. 78.   FIG. 79.   FIG. 80.

FIG. 78.—Conical vaginal portion with stenosis of the external os. *Col tapiroide*. *a*, posterior, *b*, anterior vaginal vault; *c*, os uteri.

FIG. 79.—*a*, posterior, *b*, anterior vaginal vault; *c*, the opened cervical canal; *d*, posterior, *f*, anterior lip of the os; *g h i*, the horizontal incisions which join at *h* at a right angle; *e*, upper extremity of the preparatory bilateral discission; *n*, point of insertion of the sutures; *k*, portions removed

FIG. 80.—Result of the operation. *a*, posterior, *b*, anterior vaginal vault; *c*, os uteri; *g h*, the united wounds, comp. Fig. 79, *g h* and *i h*.

vaginal vault (Fig. 79, *c e*) with the scissors, or, if the blade cannot penetrate into the vaginal portion, with a knife; then the two halves of the vaginal portion, *d* and *f*, can be separated anteriorly and posteriorly. This preliminary division facilitates all operations on the vaginal portion, by permitting all incisions to be made in a straight line instead of a circle. Then a wedge is excised from the lower lip by means of the incisions *i h* and *g h*. It should here be remarked that the transverse incision, crossing the inner wall of the cervix at *c*, begins about one-half centimetre below the end *e* of the two lateral preparatory divisions of the cervix. This is necessary, to permit of the ready insertion of the needles a little way above the margin of the wound, at *n*. If the portion has been divided in half transversely, *i h*, another transverse incision from without, *g*, meets the internal one so that both come together in a right angle at *h*. Then the wound is united by *i h* being joined to *g h*, giving rise to the form in Fig. 80. Thereupon the anterior lip is treated in the same way. The still remaining wounds on either side, due to the lateral discission, may either be united each separately, or, still better, be sewed together in such a manner that on each side the anterior lip is again joined to the posterior with one or two stitches. Then the vaginal portion has regained its previous form laterally, and no part of wound surface is exposed without mucous covering.

Stenosis of the Internal Os.

The character of stenosis of the internal os is thus far but little known. We may distinguish a dynamic, temporary, secondary, apparent, and anatomical form. *First*, the uterus contracts after repeated, especially awkward attempts at sounding. But if the entire uterus becomes hard, firm, and small, the internal os must likewise be permeable with greater difficulty. *Second*, secondary stenoses arise from small polypi, tumefactions of the mucosa, ovula Nabothi, or other tumors. *Third*, there are apparent stenoses, *i. e.*, the change of position does not allow the sound to enter. For instance, if the uterus is flexed at an acute angle, and the sound in a straight line, the latter cannot pass at the bend. *Fourth*, there are actual, anatomical stenoses. These may be acquired from cauterizations, or congenital—arrests of developments. They have been demonstrated in the dead subject. But in the living the demonstration finds great difficulties, apart from the possibility of mistaking the former three kinds for them. For in infantile uterus the internal os, of course, is also very narrow. Therefore it is essential to a correct diagnosis that we are sure of the normal condition of the uterus. Only if in an otherwise normal, correctly situated uterus, the sound encounters resistance always at the same place, both on insertion and withdrawal, it is possible to form the diagnosis of stenosis of the internal os.

If there really be this condition—and it is by no means frequent—the extent of the stenosis is important. The stenosed portion is rarely long, usually the knob of the sound passes it rapidly. But even in these cases we are surprised to see the stenosis disappear after a few soundings or a single dilatation with laminaria, when probably an ovulum Nabothi, an occlusion of the palmæ plicatæ with posteriorly accumulated secretion, or a small polypus will furnish the anatomical cause of the stenosis. The muscular substance, however, seems but rarely to be the cause of the stenosis. Certainly, there are cases in which the force of the musculature at the internal os cannot be overcome by laminaria. After the dilated laminaria tent is extracted by force, a deep groove is seen at the point occupied by the internal os, an impression, as it were, of the stenosis.

The symptoms calling for the formation of the diagnosis, and constituting the object of medical treatment, are dysmenorrhœa and sterility. If other causes of dysmenorrhœa or sterility can be excluded with probability, we may be justified to charge them to the stenosis and attempt its removal; but it is almost impossible to exclude with certainty *all* other causes of the above-mentioned symptoms! The diagnosis would be justified only if long-continued fruitless treatment forced the assumption of a stenosis as the cause of the suffering.

As always in therapeutics, we shall commence with mild measures and gradually continue with a more energetic treatment.

The symptomatic methods having been exhausted, we sound. If the sounding causes too much pain, anæsthesia is permissible; of course, all antiseptic precautions are necessary. We often have the pleasure of removing forever by a single sounding a suspected stenosis, or rather the group of symptoms usually taken for it. On the other hand, if repeated sounding has no effect, we institute dilatation with laminaria (comp. p. 38). To this end we must often scrape a solid tent very thin, because the *stenosis* exists. Anæsthesia is also often required for this manipulation. The removed laminaria tent will first permit of the formation of the diagnosis. If there be no groove, as above described, in the region of the internal os, there is no stenosis, and the treatment will have to seek other channels. But if it was possible to introduce a very thin laminaria tent, a thicker one is placed subsequently until the canal is really dilated to a diameter 0.5 cm.

We may also attempt to overcome the stenosis by dilating it forcibly with progressively thicker sounds. The method is described on page 41 (comp. Fig. 18). Furthermore, there are scissor-shaped instruments (Fig. 17, p. 41), the upper branches of which may be separated by compressing the handles. If the instrument be introduced closed and then opened, it presses the walls of the canal apart. The best of these instruments is that of Schultze represented in Fig. 17.

It has been sought to dilate the canal more rapidly but rather more harmfully with covered knives. Their application is not free from danger. As the uterine parenchyma gives way, we never know how deeply we are cutting. It is always an incalculable injury which is caused. Should the incision be too deep, we may wound the ureter which is situated in the parametrium. Should the incision not be deep enough, it will close readily and restitution *ad integrum* recur. But everything retarding or intended to retard the primary healing is connected with the danger of infection. Owing to the rapid extension to the peri- and parametrium, this danger is in this case very great.

I should like to know how often the internal os has really been touched with metrotomes, for to do that their point must reach to the fundus. If the internal os is wide enough to allow the instrument to pass, we cannot speak of a stenosis. But if the instrument does not pass the internal os, only the cervix is incised—a procedure which can be done much more carefully and with less danger in the above-described manner (p. 137). But if the intention be to inhibit the power of the muscular fibres, it will be much more rational, here as everywhere, to tear the sphincter than to cut it.

Besides, metrotomes are expensive, they often fail, and are not readily disinfected. The great body of physicians has never accepted the operation of the bloody dilatation of the os. And if we bear in mind that eminent gynecologists are of opinion that the stenoses in question are only spasms of the uterus, any operation coupled with danger is assuredly not justified. If we wish to act rationally, we must choose pre-eminently a method which cures positively, that is to say, after the employment of which the stenosis does not return. After numerous observations I am bound to assert that stenosis returns much more frequently after incisions than after rapid dilatation or dilatation with laminaria.

We shall again recur to dilatation of the os internum when speaking of the treatment of endometritis.

E. THE GYNATRESIÆ.

The escape of menstrual blood may be prevented by various causes, the blood accumulating in the genital canal, and either expanding all the sexual cavities or only a part of them.

The following kinds of gynatresiæ have been observed:

1. *Atresia hymenalis.* The hymen is not perforated, but completely occludes the introitus vaginæ. This pathological condition is congenital.

2. *Atresia vaginalis.* It is both congenital and acquired. If a part of the vagina be absent through *arrest of development*, *i. e.*, if Müller's ducts have not developed (p. 130), menstrual blood accumulates above. Or a transverse occlusion may be situated immediately posterior to the hymen. When, in these cases, the blood bulges out the transverse wall, it will touch the hymen. The latter may also adhere to the transverse wall, so that the vaginal atresia appears to be hymenic.

Acquired vaginal atresiæ occurs after gangrenous vaginitis, and the latter again, in consequence of a diphtheria or infectious disease (measles, cholera, scarlatina, etc.; comp. p. 95). Such inflammations are not sufficiently watched for in the general diseases of childhood, so that, when the latter consequence, the gynatresia, appears, we are often unable to decide with certainty whether, in a concrete case, we have to deal with congenital or acquired defect of the vagina.

3. *Atresia uterina.* Congenital atresiæ of the uterine orifices are very rare, the acquired are more frequent. By too oft-repeated cauter-

izations of the cervical canal (comp. p. 61), or by incautious application of the caustics, the os uteri as well as part of the vagina and the vaginal portion may become adherent. The same is liable to occur after gangrenous vaginitis in the puerperium or after serious injuries, such as laceration by obstetrical instruments. Atresiæ also occur much more frequently than is usually supposed, after amputations of the vaginal portion.

In old women, atresia not rarely occurs at the internal os. But as menstruation has long ceased with them, only some mucus and pus (pyometra) accumulate behind the atresia.

In a similar way, an adhesion occurs in the stretched cervix during prolapsus of the uterus.

4. *The gynatresia may be unilateral.* When, as above explained (p. 133), a rudimentary uterus, but invested with mucosa, exists (Fig. 70), this hollow uterine rudiment becomes expanded by menstrual blood.

Or else, the completely double uterus has a rudimentary vagina which adjoins the developed vagina for but a short distance, ending in a cul-de-sac. In this case, likewise, the blood accumulates, the blood-retention cyst implicating uterus and rudimentary vagina. The blood becomes inspissated to a fluid believed to be characteristic—tarry, chocolate-colored, often almost black. In rare cases, pus or mucus was found instead of blood. This might force us to assume the existence of constitutional amenorrhœa and uterine catarrh. But more correctly, this lateral pyometra is referred to secondary suppuration of the cyst-wall.

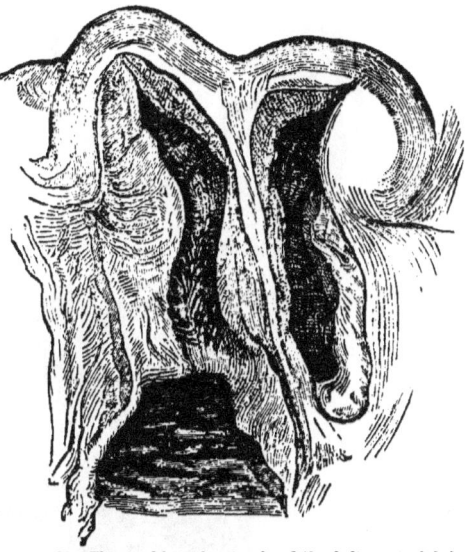

FIG. 81.—Uterus bicornis, atresia of the left os uteri, left hæmatometra.

In all cases, menstrual blood accumulates above the occlusion.

In *hymenic atresia*, the vagina becomes considerably expanded (*hæmatocolpos*). The uterus is situated above as an appendix to the blood-cyst. If a portion of the vagina be absent, the great internal pressure also gradually expands the uterus, the very attenuated walls of which form a part of the blood-cyst wall. Of course, the uterine distention is the greater if the vaginal atresia be situated higher up.

In atresia of the external os, or, more commonly, of the vaginal portion *together with* the laquear, the uterus alone forms the blood-cyst (*hæmatometra*). Also in double uterus, *uterus bicornis duplex*, one horn may be in a state of atresia, and the blood accumulate in it. Such a case is illustrated by Rokitansky, Figure 81 (lateral hæmatometra).

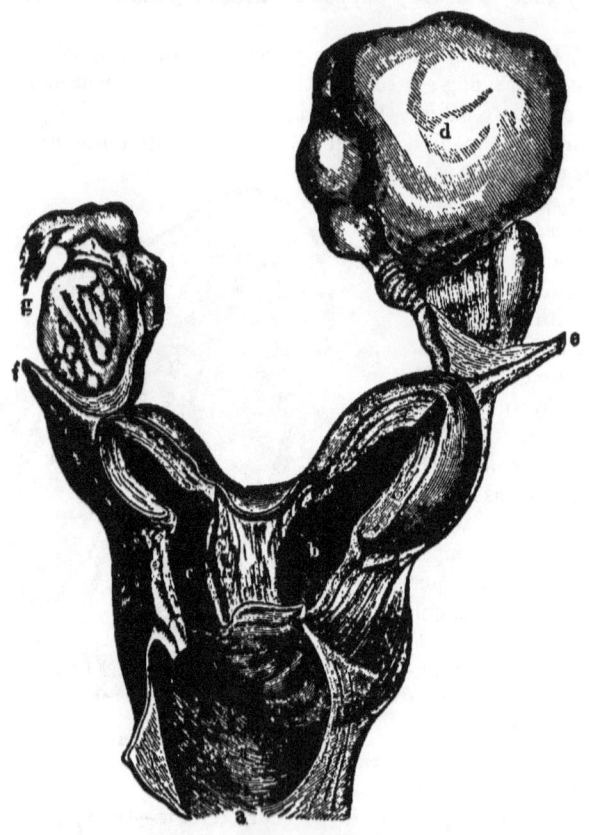

FIG. 82.—Hæmatometra and hæmatosalpinx (after Bandl).

a, vagina; *c*, the half of the uterus opening into the vagina; *b*, the half of the uterus closed toward the vagina, left hæmatometra; *d*, left hæmatosalpinx; *e*, left, *f*, right round ligament; *g*, right tube and ovary.

In not a few cases, the tubes likewise participate (*hæmatosalpinx*, Fig. 82). As the interstitial part was not found equally dilated, but usually occluded, the occurrence of hæmatosalpinx must not be explained by regurgitation. It would rather seem that, after repletion of the uterus, the centrifugal pressure on the uterine mucosa is so great that it does not bleed during menstruation, but that the tubal mucosa does so, vicariously. Of course, this blood may escape from the abdominal opening, but usually perimetritic adhesion of the tubes occurs, and a very large, quite thin-walled, sausage-shaped blood-cyst forms.

In accumulation in a rudimentary uterus and vagina (*hæmatometra et hæmatocolpos unilateralis*), the blood-cyst, if greatly distended, extends to the middle of the abdominal cavity. The depressed rudimentary vagina may so crowd against the hymen of the developed side that at first a simple hymenic atresia seems to be present.

We illustrate this case after a schematic figure by Freund.

Symptoms and Results.

The characteristic feature of all gynatresiæ is that the symptoms first manifest themselves with the beginning of menstruation. The accu-

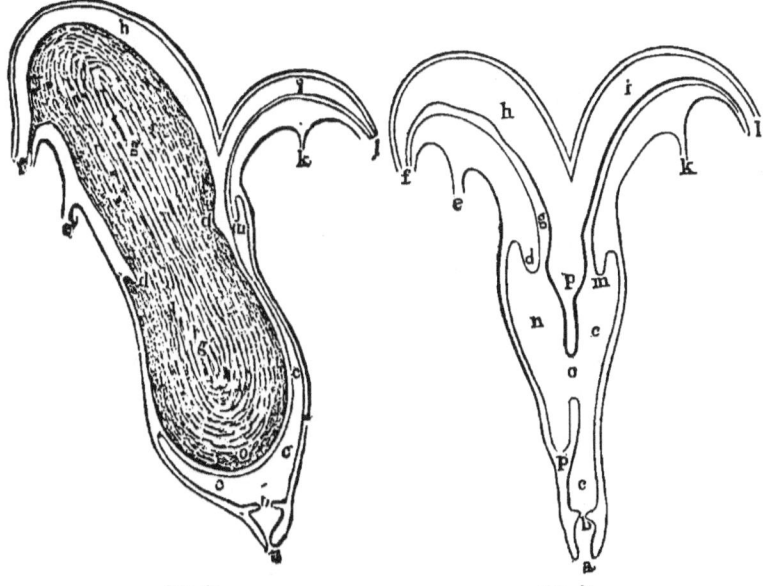

Fig. 83. Fig. 84.

Fig. 83.—Uterus bicornis, vagina subsepta. *a*, vulva; *b*, vagina; *cc*, the interspace between the blood-retention cyst and the hymen, disappearing on great accumulation of blood; *d*, right external os uteri; *e*, round ligament; *f*, tube; *g*, hæmatometra; *g'*, hæmatocolpos, separated from the open vagina, *cc*, by a vaginal septum springing from the vaginal portion; *h*, right uterus; *i*, left uterus, opening into the vagina, *c*, at *m*; *k*, left round ligament; *l*, left tube; *o*, point of incision whence the retained blood escaped.

Fig. 84.—The same case after operation. *a*, vulva; *b*, hymen; *cc*, vagina; *p p*, vaginal septum with the artificial opening *o*, through which the blood escaped; *n*, right vagina, previously hæmatocolpos *g'* of Fig. 83); *d*, os uteri; *g*, cavity of the uterus; *e*, round ligament; *f*, tube; *h*, right uterus; *i*, left uterus; *k*, left round ligament; *l*, left tube, *m*, left os uteri, opening into the vagina, *c*.

mulating blood, distending the genital tube, causes pain. At first these pains are slight, so as to be interpreted as menstrual molimina. Gradually, especially if the uterus is involved in the formation of the blood-cyst, the pains increase and may even become so intense as to depreciate the general health. After perimetritic deposits have formed, the pains become continuous. During the time of menstruation they are intensified to the most violent uterine colic.

The pyometra of old women, frequently accidentally discovered during post-mortem examination, runs entirely without symptoms. In them the uterus is expanded at most to a diameter of six to eight centimetres. A single dilatation to the size of a man's head is reported.

The *results* are very variable. In *hymenic atresia*, the hymen ruptures both spontaneously and in consequence of an accidental trauma, such as attempts at coition, during the touch, a leap, or a fall. In deeply seated vaginal atresia, this favorable result may also occur.

Should the upper part of the vagina be expanded (*hæmatometra and hæmatocolpos*), perforation into the bladder and rectum may occur. The upper part of the uterus does not rupture spontaneously, but an unfavorable result may arise from rupture of a *hæmatosalpinx*. Usually in consequence of an accidental trauma, the thin tubal wall breaks and the blood escapes into the peritoneal cavity. Should the effusion of blood be very considerable, fatal peritonitis generally occurs. In the most favorable case, a hæmatocele forms.

If the patient's health has gradually deteriorated through the long illness, it is possible that amenorrhœa may occur in consequence of the anæmia. In these cases, the growth of the tumor ceases and a kind of natural cure results. This may especially be the case in lateral hæmatometra in a rudimentary horn. Perhaps here—as in hæmatosalpinx—the normal cornu vicariously assumes the function of the rudimentary one.

Ordinarily, however, the symptoms gradually increase in severity so that medical aid is required.

Diagnosis.

The anamnesis shows the absence of menstruation in spite of occurring molimina, and the abdominal tumor which has arisen with accompanying pains. Preceding operations, cauterizations, difficult labors with prolonged suppuration, or infectious diseases during childhood are likewise important for the diagnosis.

By digital examination, the diagnosis of hymenic atresia is easy. Often the black blood shows through the attenuated hymen. Nor is there any doubt about the case in vaginal atresia. The combined examination demonstrates a fluctuation, or at least a distinct connection of the abdominal, tensely elastic tumor with the vagina.

Palpation must be done with the greatest care; any strong pressure may rupture the tube. Then fatal peritonitis is the usual consequence.

As we have generally to deal with nulliparæ, excepting the few cases of acquired atresiæ, the abdominal coverings are resistant, and render palpation difficult. In such cases, it is better to desist from closely palpating an indistinct tumor situated on the cyst. The uterus is most readily palpated. The tube extends laterally as an indistinct resistance. Even if merely suspecting a hæmatosalpinx, any strong pressure should be omitted.

If the external os be closed, or if the vagina end above in a cicatricial cul-de-sac, the blind pouch of the vagina must be carefully examined through the speculum.

The diagnosis is difficult in *unilateral hæmatometra*. The tumor, there forming, so greatly displaces all the pelvic organs that the knowledge of the normal relations does not help us. Amenorrhœa is also absent, for the non-occluded horn of the uterus menstruates normally. On examination, we feel a tense tumor adjoining the vagina. If the finger is still able to enter the vagina, the diagnosis will be easy after combinded examination, bearing the malformation in mind. But if the hæmatocolpos is so large that the finger cannot be introduced, the sound must demonstrate the presence of a lumen.

In more superiorly situated atresia, the os uteri adjoins the expanded cyst in semilunar form, conveying the impression that the side of the cervix is considerably distended. As the cyst is quite hard to the feel if the internal pressure be great, the diagnosis may hesitate between cervical myoma and lateral hæmatometra. The diagnosis is also difficult if the seat be very high or if even the rudimentary uterus be completely separated. Certainty is reached only after prolonged observation, by exclusion, and by oft-repeated careful exploration with all combined methods.

A very excellent, harmless diagnostic auxiliary is the exploratory puncture with a Pravaz syringe. In view of the importance of a correct diagnosis, the exploratory puncture is to be performed in every doubtful case previous to the operation. The presence of the thick black blood alone, which can be mistaken for nothing else, permits the recognition of hæmatometra. Bright blood is drawn from every fibroma by the syringe. It is self-evident that the skin and the syringe must be disinfected in the most careful manner. Employment of an unclean syringe may cause ichorous degeneration.

Prognosis.

The prognosis is good only in low vaginal and hymenic atresia. In all other gynatresiæ, the prognosis is doubtful. Without operation, the ever increasing pains greatly impair the nutrition, and the danger of an eventual rupture of a hæmatosalpinx is always present.

With correct diagnosis and treatment, *i. e.*, operative opening, the prognosis is likewise doubtful, owing to the danger of sanious metamorphosis and of rupture of the tube.

However, if the patient has happily recovered from the operation, and if a sufficiently large discharge opening has been made, the gynatresia is definitively cured.

Treatment.

The treatment consists in the operative production of a permanent opening for the escape of the menstrual blood.

In hymenic or vaginal atresia a puncture is made, and then a sufficiently large opening cut into the obstructing membrane with pincette and Cowper's scissors. Should the membrane be very thick, the union of the two surfaces by the button suture is to be recommended.

But if the lower part of the vagina be absent, great caution is necessary. Combined examination from the urethra and rectum is first performed to permit of an estimation of the thickness of the intervening connective tissue. If a thick cord be present, a transverse incision between urethra and rectum is first made, and the finger works upward toward the tumor, of course with continually repeated control from the rectum and bladder. Even if we fail to produce a real tube surrounded by muscular strata, a great deal has been gained by giving egress to the retained, and later the menstrual blood, by the natural passage.

However, should the urethra or the bladder lie immediately upon the rectum, so that even with the greatest caution perforation of the rectum or the urethra and bladder must be feared, it may be quite impossible to form a vagina. Although the simplest and apparently the easiest way would be to force in the trocar from the rectum, this must not be done on account of the penetration of intestinal gases and consecutive sloughing. The urethra must be dilated, and an artificial vesico-vaginal fistula produced, so that the retained and future menstrual blood flows into the bladder. Caution is also necessary here on account of the danger of injuring the peritoneum and the ureters.

Should the occlusion be high up in the vagina or in the external os uteri, the vaginal portion is to be sought from the anus, and we work toward it in the rudimentary vagina and plunge a long curved trocar into the blood-cyst. Its tube is first allowed quietly to remain. It is not removed until several days later, when the opening is to be operatively widened. If the locality is well recognizable, the use of the knife may be permissible for making incisions or excisions so as to gain a larger opening.

But if the accurate identification of the relations in the region of the occlusion be impossible, it is better to undertake the dilatation with blunt dilators. It is of the utmost importance in this operation to employ all antiseptic precautions (Chapter VI.).

If there be a lateral hæmatocolpos, a piece of the septum is to be excised (comp. Figs. 83 and 84).

Great difficulties must arise if the rudimentary horn, distended by retained menstrual blood, be at a great distance from the vagina. Here also we must operate if the diagnosis be certain. But insufficient experience makes it appear doubtful whether we should first seek to render the tumor adherent to the vagina, or whether it is not much more correct *to attempt to reach it by laparotomy and remove it wholly or in part*.

Experience teaches that the operations on hæmatometra are very dangerous. The dangers consist particularly in *sloughing of the cyst and in rupture of the tube*.

Sloughing often occurred because the operator hesitated to empty all of the retained fluid at once and to irrigate the sac with disinfectants. It was feared that the hyperdistended uterus would be unable to contract, and hence refill with fresh blood. It is questionable whether this fear is justified, and whether an irrigation with a double-current catheter (p. 48, Fig. 22) is not permissible. With this instrument as much fluid is injected as is returning, a remnant being left in the sac. This fluid can be renewed every two or three hours if the catheter be left. And in the mean time the cyst will be able to contract gradually, as the contents can escape from both tubes. Instead of a double-current catheter, the introduction of two rubber tubes may also be recommended, one for the admission, the other for the exit of the fluid.

The contents of the cyst are at times so tenacious and thick that they cannot be removed in this manner. Although real coagula or other formed masses do not occur, the tarry blood does not continuously flow from a catheter. Therefore, the following rules are valid: If there be only a demonstrated retention in the vagina, complete removal of the blood is permitted. Should, however, the uterus or even the tube participate, the fluid must be allowed to escape gradually and spontaneously. At most, the contraction of the uterus may be hastened by ergot or ergotin. But no pains must be spared to prevent decomposition during the gradual escape by appropriate prophylaxis. Independent of antisepsis before the operation, the genital region must be permanently irrigated while the fluid *flows profusely*. Should it escape slowly, a large wet carbolized compress is laid over the entire genitals. This compress is renewed every thirty minutes. Should the fluid pass merely by drops, the entire genitals should be surrounded by benzoated or salicylated cotton.

Sitz-baths are contra-indicated on account of the danger of rupture of the tube during movements of the patient.

We have seen that rupture of a hæmatosalpinx is frequently followed by fatal peritonitis. The rupture often occurred spontaneously. The tube must follow the contracting uterus, and the motion of the tube in this dislocation led to rupture. But especially every trauma, such as sudden change of position, vomiting or sneezing, pressure on and compression of the abdomen, lead to rupture. For this reason, pressure on the abdomen is to be strictly prohibited after puncture. The patient is to be most strongly impressed with the fact that the least motion may cause her death, and that hence she must lie perfectly quiet.

Theoretically counter-indicated is also the withdrawal by the aspirator or the syringe. In the first place, the opening of the aspirator needle is so small that it can hardly be found again in the depth, and is certainly dilatable with difficulty. Furthermore, if the uterine walls do not follow, the aspiratory force of the instrument acts like that of a cup, and thus additional blood is even sucked in.

It is to be observed that the diagnosis of the special form of the gynatresia is frequently possible only after the danger has passed by and the cyst has disappeared. Although the general diagnosis of "hæmatometra" may be clear, the positive differentiation of the peculiar anomaly of development does not succeed until subsequently, after the cessation of the displacement of all the pelvic organs caused by the cyst, during the then possible, harmless palpation, combined examination, and sounding. Thus it was in Freund's instructive case. Compare p. 145, Figs. 83 and 84.

CHAPTER X.

INFLAMMATION OF THE UTERUS. (ACUTE METRITIS, CHRONIC METRITIS.)

A. General Remarks.

A PROGRESSIVE inflammation of the connective tissue is likely to occur in the uterus as well as in other places, if there be a wound which becomes infected. Owing to the deep, sheltered position of the uterus, accidental wounds will be hardly possible. Only in the puerperium, in which there are always wounds in the cervix and at the os in consequence of parturition, where the puerperal fluids of the genital tract become easily infected and decompose, will a progressive inflammation, a metritis, arise. The hyperæmia of the puerperal uterus, and the specific dangerous quality of an infecting agent, cause the frequently pernicious and progressive character of the metritis in the lying-in period. Outside of the puerperium, it is usually the physician who causes a lesion during his manipulations on and in the uterus for therapeutic purposes. If this wound becomes infected, either directly by unclean instruments, or by decomposition of the wound secretions if not artificially removed, a connective-tissue inflammation, a metritis, must likewise arise. Outside of the puerperium, owing to the small quantity of blood in the uterus, these inflammations are usually not dangerous and become important only if they extend to the peritoneum or the loose parametric connective tissue.

That a *wound alone,* or the *decomposed vaginal and uterine contents alone,* do not cause metritis is proved, on the one hand, by the many primary unions after plastic operations on the uterus; on the other, by the cases in which offensive, decomposed menstrual blood gradually disappears without any untoward accidents.

These "traumatic" forms of metritis must always possess an acute character.

All chronic alterations of the uterine parenchyma are either secondary or they quite distinctly follow pathological conditions of the vessels. The close attachment of the peritoneum to the muscularis hardly permits of an isolated peritoneal inflammation without participation of the adjoining uterine layers. And in disease of the mucosa, especially in adenoid, deeply extending proliferations, the neighboring uterine parenchyma becomes likewise infiltrated and hypertrophic.

The idiopathic, chronic metritis, however, is the consequence of irregularities in the distribution of blood. After abdominal plethora, after disturbed puerperal involution, after excessively profuse or insufficient or abnormal menstruation, the uterine vessels remain permanently dilated. The thus effected, superabundant nutrition of the uterine parenchyma causes hypertrophy of the uterus, and this hypertrophy is denominated an inflammation—chronic metritis. Such a metritis, therefore, is always the consequence of *lasting* injuries, a *distinctly chronic* process.

To be included among chronic forms of metritis are the chronic affections of the mucous membrane, whether the endometritis is primary and the metritis secondary, or the reverse.

According to what has been stated, we distinguish: 1. *Acute metritis* (progressive inflammation of the connective tissue: 2. *Chronic metritis* (vascular dilatation, hyperæmia, hypertrophy of the uterus); 3. *Endometritis* (metritis mucosa, the pathological conditions of the uterine mucous membrane).

B. Acute Metritis.

Etiology.

Should an incision be made in the os uteri without sufficient disinfection, should a plastic operation be performed without antiseptic precautions, and should even sounding be done with an unclean instrument, a quite acute inflammation of the uterus may ensue. Especially often have ichorous sponge-tents and lack of antiseptic care in laminaria dilatation given rise to acute metritis. Furthermore, the application of intra-uterine or of ill-fitting vaginal pessaries may be followed by metritis. An excessively long stem penetrates the mucous membrane, the secretions decompose, and infection of the wound, re-inflicted with every motion of the uterus, takes place. After cauterizations of the uterus, with retained decomposing crusts or caustics, an acute inflammation likewise occurs. Gonorrhœal infection at times acts in a similar way. From these causes forms of metritis arise in which all the substrata of the uterus, peritoneum, parenchyma, and mucosa, equally participate, so that there exists acute endometritis, metritis, and perimetritis. According to the prominence of the one or the other, the diagnosis is determined.

Anatomy.

In acute metritis the uterus is often considerably enlarged. If we assume in non-puerperal cases the same anatomical conditions as in the puerperal, there occurs an infiltration of the tissues with serous fluid and a copious emigration of leucocytes. In harmony with the close connection of the substrata of the uterus, perimetrium and endometrium participate in the inflammation.

Symptoms and Course.

The symptoms of acute metritis consist primarily in fever. As in traumatic erysipelas, with which the disease is altogether identical, very high temperature and rigors occur. Moreover, there is distinct swelling of the uterus, which feels surprisingly soft, and which may reach the size of the uterus at the third month of pregnancy. The pains, the third symptom, often have clearly the character of peritonitic pains, forcing us to the conclusion that the peritoneum participates. However, cases occur in which there is at first no sensitiveness on pressure, but merely a dull pelvic pain. This is to be referred to the increasing hyperæmia, the swelling, and the pressure to which the uterine nerves are subjected. As in pregnancy nausea and vomiting occur with enlargement of the uterus, so it is in the present case. The bearing-down is also to be interpreted as a symptom of complicating perimetritis. If the metritis is the consequence of gonorrhoic infection, there is at the same time a quite considerable, often bloody discharge of pus.

As in erysipelas abscesses occur in the infiltrated tissues, so pus may form here in the uterine parenchyma. To be sure, without watching the case from the beginning, or without post-mortem examination, the differential diagnosis between perimetritic or parametritic and uterine abscess is impossible. The abscess may perforate in every direction, into the bladder, rectum, uterine or peritoneal cavity; in the latter case fatal peritonitis results. The diagnosis is certain if the knife open the abscess accidentally. Thus I have observed an acute metritis after bilateral discission. Later a dilating operation was again performed. In the course of it a teaspoonful of pus suddenly flowed over the surface of the incision.

Diagnosis and Prognosis.

The diagnosis is evident, especially, from the anamnesis; the pains on pressure, the discharge, the enlargement of the uterus, and the knowledge of any infection which may possibly have occurred render the case clear.

The prognosis is to be reserved, on account of the participation of the peritoneum, the remaining hypertrophy of the uterus, and the formation of an abscess.

Treatment.

The treatment is antiphlogistic. The vaginal portion of the cervix is engaged in a speculum and scarified. The larger the quantity of blood that escapes, *i. e.*, the greater the hyperæmia of the uterus, the more frequently should blood be drawn; but not oftener than once every twenty-four hours, as the entire manipulation is painful, owing to the implication of the peritoneum. The vagina is to be irrigated with lukewarm water every six hours. Caution is necessary in this, and, especially, the irrigator alone should be used. If the abdomen is very sensitive to pressure, an ice-bladder is applied, otherwise a Priessnitz cataplasm. The latter forces the patient to maintain a quiet position in bed. Should the vagina be inaccessible, leeches may be applied to the abdomen or the perineum. Their number is governed by the general condition; less than five leeches will have no effect, more than ten or twelve, however, are injurious.

If the pains are too violent, we must resort to morphine. It is very important to empty the bowels. Should constipation have existed for several days, we give calomel, 0.5 gm., and after profuse action exhibit some tinctura thebaica to quiet the bowels. Smaller, repeated doses of calomel cause salivation within two or three days, and therefore are strictly cautioned against.

After cessation of the fever and pains, rest in bed for several days, cataplasms, and eccoprotics should be used. Perhaps, during the after-treatment, injections of ergotin may be serviceable in causing the diminution of the uterus.

C. Chronic Metritis.

Etiology.

All conditions and circumstances leading to repeated hyperæmia of the non-gravid uterus or preventing the normal puerperal involution of the organ, are of etiological importance in chronic metritis. Normally we have, in menstrual congestion, a *stadium incrementi,* an acme with rupture of the vessels—the hemorrhage, and a *stadium decrementi*—the diminution and regression. When for any reason this normal course of menstruation is interrupted, disturbances of circulation result. Causes of circulatory disturbances are: Exposure at the time of menstruation, neighboring tumors, changes in the ovaries, and displacements or flexions of the uterus. In such cases, on the one hand, dysmenorrhœa arises—in infantile, small uteri—on the other, chronic hyperæmia, profuse menstruation, and relaxation of the uterine parenchyma. Moreover, in long-continued disease of the uterine mucous membrane, the inflammation will extend into the parenchyma. If there be a very intractable gonorrhoic endometritis, the musculature of the uterus always eventually participates. Again, in chronic inflammations in the surroundings of the uterus, especially in perimetritis, the uterus will gradually hypertrophy. This also is the case in those displacements which prevent the return of blood, that is, lead to stasis in the uterine parenchyma. Even a retardation of the circulation of blood in the domain of the inferior vena cava has an influence on the uterus, that organ taking part in the general abdominal plethora. In many uterine tumors, too, the seat of the growth becomes hyperæmic and hypertrophic.

The preponderating majority of the cases, however, occur in women who have borne children or aborted. We know that every peri- and parametritis disturbs the involution of the uterus. It has been demonstrated that the puerperal diminution of the uterus progresses slowly during every inflammatory pelvic affection. As in many cases the uterus is not only temporarily the path for the progressive puerperal inflammation, but permanently the seat of the inflammation proper, a slow diminution of the infiltrated, serously macerated, inflamed organ may be expected. However, not only after inflammations, but after mere disturbances in the lying-in period, for instance, hemorrhages, defective contraction, post-puerperal retroversions, etc., the uterus remains extraordinarily large. Abortions, too, especially if frequently recurring, are to be inculpated. If they pass unfavorably, they cause the same dangers as parturitions; if they run a favorable course, the women take too little care of themselves, thereby again causing dangers. Chronic constipation, early getting-up, exertions, etc., likewise disturb the involution. While normally, in tetanic contraction of the uterine musculature, the nutrition of the uterus becomes impaired and the musculature in consequence undergoes fatty degeneration, in disturbances of involution the uterus remains large. All its vessels are dilated, particularly those least influenced by the contracting musculature, the marginal vessels. As a result too much blood circulates in the uterus, it is too well nourished, the fatty metamorphosis does not ensue—a uterine hypertrophy exists. Masturbation and coition, i. e., the congestion following sexual excitement, have been likewise inculpated. However, during police examinations of public women, the group of symptoms of the uncomplicated chronic metritis is by no means frequent.

Anatomy.

The only facts positively demonstrated are, the enlargement of the uterus, the dilatation of the vessels, particularly the larger, and the frequently complicating peri- and endometritides. More theoretically, stages were formerly distinguished. The first was described as like the acute metritis, the second as like the chronic, the third as resembling the condition found in senile involution. An opinion is not very easy, particularly because no woman dies of chronic metritis and because the disease causing accidental death may greatly alter the relations. Thus an opinion respecting the anatomy has often been formed from pieces excised from a hypertrophied vaginal portion. But here the views of authors differ, some referring the hypertrophy to proliferation of the connective tissue, others to proliferation of the musculature, and others again to general hyperplasia.

Should a uterus affected with chronic metritis come accidentally under observation, we shall find, in the first place, multiform deposits on the perimetrium. Adhesions, pseudo-ligaments, cross the surroundings; the uterus appears much darker than in normal cases; the body is often soft, compressible, facetted, somewhat irregular in form; the shape differs in various dislocations. On section, too, the parenchyma is red, not white or pale pink like the normal uterus. At the margin of the uterus we find many severed wide vessels. In microscopic sections, a decided predominance of the connective tissue over the muscular fibres is evident. Furthermore, the vessel walls are greatly thickened and stiff, a section from the margin of the uterus exhibiting a distinctly cavernous structure.

Symptoms and Course.

All the morbid sensations and pains peculiar to the female sex occur in chronic metritis; there are no pathognomonic, characteristic symptoms belonging to this affection alone.

The symptoms can be most easily described if the hypertrophy is a sequel of abortion or the puerperium. To be sure, in many cases we have to deal with slight deviations of the uterus or affections of neighboring organs. Still, cases occur in which the uterus is in absolutely correct position, and we find the symptoms ordinarily ascribed to metritis. The women complain of pains during walking and standing, bearing-down, perpetual annoying sensation of pressure in the pelvis, even of symptoms present in incipient prolapsus. Incidentally, there are phenomena resembling early pregnancy. For, anatomically, the processes in the uterine parenchyma are similar. The patient complains of vomiting, loss of appetite, constipation, tenesmus of the bladder caused by the pressure of the heavy uterus resting on that viscus, pains in the mammæ, the stomach, the whole abdomen, the back, etc. With this occurs hysteria in the most comprehensive sense, perpetual peevishness in consequence of the unceasing pains or at least the abnormal sensations. Owing to the pronouncedly chronic course of the affection, accidentally superadded injuries are self-evident. Invariably the women, whether of the upper or lower classes, neglect to take care of themselves. Should the injuries be of a nature to provoke hyperæmia, to lead to congestions, the condition is rendered worse. On the other hand, rest, abstention from coition, and appropriate treatment improve the state. Renewed injuries again cause exacerbations. *Thus there is the possibility of recovery, but actually the cure is very rare on account of the continued occurrence of injuries.* Even the menstrual congestion alone may each time effect an exacerbation of the condition. In the course of chronic metritis anomalies of menstruation occur. Dysmenorrhœa, enormous but painless hemorrhages and quite atypical discharges of blood are observed. These symptoms are referable rather to the affections of the endometrium. The uterine mucosa from the fundus to the vaginal portion becomes diseased and secretes pus or normal secretion in abnormal quantity. The glandular organs of the cervix proliferate and cover the external surface of the vaginal portion in adenoid new-formations.

The uterus is always felt to be thickened; pressure upon and movement of the organ on combined examination are painful. If sterility be present, it must be referred to eventual complications—adhesions, flexions of the tube, displacement of the ovaries. In non-complicated cases of chronic metritis, conception frequently occurs. However, after long-continued observations, I must side with the authors who bring frequent abortion in connection with chronic metritis.

The *course*, as appears from the description of the symptoms, is chronic; many exacerbations produce a retrogression in the cure. By the influence of the chronic hyperæmia, the time of the menopause is postponed. Often the symptoms even become worse in the climacteric period. And even after the definite onset of senile involution, a slightly purulent secretion of the endometrium persists often for decades. If hysteria was the consequence of metritis, it may remain in its full extent after the menopause.

Diagnosis and Prognosis.

In nearly all affections of the uterus a chronic metritis must be diagnosticated. Thus the peculiar hypertrophy exists in carcinoma and fibroma as well as in the displacements of the uterus. Pre-eminently, however, in extensive pelvic peritonitis, the uterus is always in a condition which admits of the diagnosis of chronic metritis. Formerly, before physicians knew how to differentiate the several diseases of the uterus, chronic metritis was exceedingly often diagnosticated. But if we first exclude all other pathological alterations, few cases will remain of simple, uncomplicated, chronic metritis. Its diagnosis requires an enlargement of the uterus, sensitiveness both spontaneous and on pressure, and abnormalities in the secretions, whether consisting merely of menstrual anomalies or whether discharges of any nature demonstrate the participation of the endometrium.

In the differential diagnosis, pregnancy in its first months enters pre-eminently into the consideration. If it be not suspected or if it be denied, a mistake is possible, owing to the almost identical subjective symptoms. As the treatment necessary for chronic metritis may be followed by abortion, in every doubtful case a careful examination, and if need be a prolonged observation will be required.

The *prognosis* is unfavorable. Especially in cases dating from the puerperium, the alterations, vascular dilatations, are so considerable that a *restitutio ad integrum* is impossible. But if the last cause remains, the symptomatic treatment will have little effect. If the chronic metritis is the complication of another, non-removable cause, cure is likewise impossible. However, even in the cases in which cure lies within the bounds of possibility, carelessness of the patient, incompleteness of the cures, disregard of the physician's directions, abortions and labors, new metritides and other injuries, always again provoke exacerbations which, though admitting of temporary amelioration, actually and fundamentally render the patient an invalid. The reflex effect of the continual illness on the mind and body undermines even the strongest constitution, making the patient debilitated, sickly, and but slightly resistant toward other intercurrent diseases. With long-continued rational treatment and the absence of additional injuries, a condition is possible which is equal to cure.

Treatment.

According to our delineation, metritis is very frequently complicated with other affections, or is the consequence of other diseases. For this reason, the point of attack for the treatment is seldom the uterus alone. The manifold inflammatory processes around and within the uterus make it appear necessary to commence first an antiphlogistic treatment.

The patient must be careful of herself in the most comprehensive sense. Particularly such injurious influences as incite a congestion of the internal genitals are to be avoided. Dancing, horse-back exercise, carriage riding and railway travelling, climbing stairs, physical exertions, etc., are to be interdicted. Pre-eminently, however, coition must be abandoned, because the condition is made worse partly by the hyperæmia during the orgasm, partly by eventual pregnancy and abortion. Besides these, defecation is to be regulated. The choice of means depends upon the variable circumstances. As the constipation is usually of very long standing and obstinate, care should be had to employ such measures as will be borne for some length of time. Often it is necessary to change the remedies repeatedly, lest the patient become gradually accustomed to excessive doses. Drastics are by no means contra-indicated; in moderate doses they often have a better effect on appetite and digestion than the salines. Owing to the anæmia, iron is preferably combined with the cathartics, as in the form of pill. aloë. et ferr., two to three per diem. In very anæmic patients, ferrated waters are to be employed, but strict watch must be kept lest the appetite suffer or constipation occur. The soluble saccharate of iron and the pyrophosphorated iron water are preparations which are very well borne. Good results are also secured by a well-directed cure with laxative spring waters —Elster Salzquelle, Franzensbad, Eger, Marienbad, Kissingen, Carlsbad, Tarasp ("the Carlsbad of Switzerland"), etc. Not a few patients even feel so well after regulation of the digestive apparatus that they believe themselves cured. If salines are not well borne, the whey and grape cures are likewise to be recommended. The family physician here has a wide field for painstaking activity, and especially with unreasonable patients his task is to insist with great severity upon regular diet and the avoidance of injurious influences.

The local treatment consists in first removing everything pathological. If the vaginal portion be very large and hyperæmic, scarifications are made from which one to two tablespoonfuls of blood escape. After the scarification, the hemorrhage is arrested with a glycerin tampon. The glycerin abstracts moisture from the tissues, thus having a desiccating effect and diminishing the swelled vaginal portion. In more recent cases, a large vaginal portion, after a single scarification and glycerin tamponade for three or four days, may have diminished one-half. As this treatment is very simple and harmless, it is to be urgently recommended. The abstraction of blood is repeated at most six or eight times in the intermenstrual period, while the glycerin tamponade may be continued for weeks. The glycerin may be mixed with iodine or the vaginal portion may be previously painted with iodine. Also inunction of the vaginal portion with an iodoform ointment has been recommended. However, I should lay the greatest stress on the scarifications and the glycerin tamponade. Incidentally, of course, the affection of the internal surface of the cervix must be looked after. Should the vaginal portion remain large in spite of the treatment, and if this enlargement is not caused by easily evacuated ovula Nabothi, a wedge-shaped excision of both hypertrophied lips is to be commended, after the method described on page 139 (see Figs. 78 to 80). As, after amputation of the hypertrophied vaginal portion of a hypertrophied uterus, the latter frequently diminishes in general extent, amputation has been recommended as nothing less than the most rational treatment in chronic metritis.

In cases of uncomplicated metritis I have repeatedly undertaken dilatation with laminaria for therapeutic purposes, have irrigated the uterus, swabbed it, and painted it with iodine. Immediately after the procedure, one or two syringefuls of ergotin solution (1 : 8) were given hypodermically. In this way we often effect a quite material reduction of the uterus, improvement of the endometritis, decrease of the hemorrhages, and thereby an improvement in the general condition.

Should, however, as is frequently the case, the chronic metritis be complicated with pelvic peritonitis, the treatment is directed more against the perimetritis. Here Priessnitz cataplasms, not over-warm sitz-baths, etc., are in place. Irrigations of the vagina, too, during the entire duration of the sitz-baths, often have a good effect.

Watering-places nowadays are frequently visited without the advice of the physician; saline, mud, and steel baths being most in favor. A direct influence of any definite bath on chronic metritis must be decidedly denied. A spa is to be selected according to the temperament, the social position, and the entire habit of life of the patient. The enforced idleness, the great physical and mental rest often have more effect than the quality of the spring. Not rarely it will occur that of two patients suffering from the same affections, one will return from the spa almost cured, the other rendered decidedly worse. In general it may be said that forest life, sea-baths, carbonic-acid cool baths, and cold-water cures are effective in nervous individuals without appetite; that saline baths and mud-baths should be recommended in more recent peritonitic affections; and that in anæmia, ferruginous waters, best in conjunction with laxatives, are to be employed. Not rarely, the prolonged, warm, full baths are exciting; the patient, therefore, should be told that after a few baths she must feel *better*. Should the reverse occur—nervousness, sleeplessness, anorexia, and great excitement—only local applications and sitz-baths must be employed. *The first rule to be adhered to is, that a patient should never drink the waters or take the baths of any spring without being under the observatory treatment of a spa physician.*

Should the metritis be complicated with or be the result of displacements, the latter are the objects of treatment.

Consequently, in a whole series of affections of the female genital organs, we shall revert to the treatment of chronic metritis.

CHAPTER XI.

DISEASES OF THE ENDOMETRIUM.

A. ACUTE ENDOMETRITIS.

ACUTE endometritis is a concomitant phenomenon of acute metritis (described on page 152). Especially in gonorrhœal affection, but also from infectious or irritating intra-uterine pessaries, as well as after intra-uterine medication generally, the affection of the endometrium manifests itself by copious secretion. The mucous membrane in such cases becomes distinctly sensitive. A similar process occurs physiologically *before and after* menstruation, and displays itself by slight, watery, mucous, sanguineous secretion of the internal surface of the uterus. The importance of acute inflammation of the endometrium as well as its treatment are identical with that of acute metritis.

B. CHRONIC ENDOMETRITIS.

Much more frequent are the chronic pathological conditions of the uterine mucosa, in which it is possible that the entire mucous membrane from the external os to the fundus, or merely the cervix, or the uterine cavity alone, is morbidly altered. The most frequent is cervical catarrh, then comes the affection of the entire mucous surface, and the rarest are pathological conditions confined to the mucosa of the body, as in endometritis fungosa.

Etiology and Anatomy.

Considering first the *catarrh of the entire internal surface of the uterus*, cases are of frequent occurrence which must be called hypersecretion rather than catarrh. Particularly in anomalies of menstruation, in chlorotic, anæmic individuals, the uterus produces a large quantity of glairy mucus. If the external os be narrow, and crowded close to the vaginal wall as in acute flexions, the mucus cannot escape, thickens to a firm lump of jelly, and gradually dilates the cervix. Even the whole uterus can be expanded, thick sounds passing the internal os without hindrance. If we remove the mucus from the cervix, the ever succeeding mucus indicates the participation of the endometrium in the hypersecretion. Even in small uteri, we daily find one or two teaspoonfuls of mucus in the fundus of the vagina.

Similar conditions also develop after abortions and labors, yet rarely in robust, plethoric women. On microscopic examination of the mass of mucus, epithelia from the internal surface of the uterus are by no means frequently discovered. Therefore, it seems to be simply a hypersecretion of the uterine glands.

Should the condition of hypersecretion exist for a longer time, it seems to lead to alterations in the circulation. The entire mucosa thickens, menstruation no longer runs a normal course, and becomes especially copious in women who have borne children. In nulliparæ, dysmenorrhœa may occur.

After prolonged existence, pus is often mixed with the mucus, particularly when the cervical catarrh predominates and there are erosions around the os. The uterine secretions irritate the vaginal walls upon which pus then likewise forms. But cases also occur in which pus flows from the uterus. This is more frequent in advanced age. The positive demonstration may be furnished by the test tampon (Schultze). A tampon dipped in glycerole of tannin is laid in front of the vaginal portion, and after twelve hours a tubular speculum is introduced through which the tampon is carefully withdrawn. The uterine secretions then can be plainly seen on its upper surface.

In old syphilitic subjects and in the presence of chronic gonorrhœa, the endometrium often secretes a thin-fluid pus without any admixture of mucus. Here then we must assume an alteration of the mucosa in which the glandular activity is placed in the background. And in fact, in old women, glands are frequently no longer found on microscopic examination.

A peculiar form of endometritis, limited to the mucous membrane of the body, exists if the whole mucosa is hypertrophic, *endometritis fungosa* (Olshausen). In this form, cases occur in which there is a uniform velvety swelling, as well as such in which there is a villous condition of the mucous membrane. The *etiology* is obscure. On microscopic examination portions are seen exhibiting throughout nothing but uniform hypertrophy of all parts of the mucous membrane. At times, however, the vessels are ectatic or the glands dilated. The latter may possibly be the sole cause of the hypertrophy, for in some preparations they are not dilated, but elongated, interlaced, closely packed on account of the displacement of the connective tissue of the mucosa, and nearly devoid of any lumen. Should this be the case, the condition approaches that of adenoma. Endometritis fungosa is often very obstinate, the treatment taking years.

The so-called *membranous dysmenorrhœa* likewise belongs among the diseases of the mucosa of the body. In this affection, at every menstruation the entire superficial layer of the mucous membrane becomes detached and is expelled with intense, labor-like pains in the shape of shreds of membrane. Bearing in mind that in membranous dysmenorrhœa the expulsion of a membrane at *every* menstruation is the characteristic feature, the diagnosis is easily made. Such a membrane must always be carefully examined with the microscope; in remnants of abortion, we often find chorion villi, frequently large decidual cells, glandular lumina without any or with epithelia in fatty degeneration; while the dysmenorrhœal membrane shows the normal uterine mucosa punctured with red and white blood-corpuscles, together with but slightly enlarged connective-tissue corpuscles and portions of glands.

Symptoms and Course.

In chronic endometritis the chief symptom is the increase and the qualitative alteration of the uterine secretions; besides, the anomaly of menstruation, the menorrhagia. The latter is the characteristic symptom of endometritis fungosa. With long-continued profuse discharge from the uterus the general condition suffers, though often the hypersecretion is referable to the anæmia, rather than the anæmia to the hypersecretion. Owing to the complication nearly always present with metritis, we have in endometritis also all the symptoms described on pages 156 and 157. The course, as its name expresses, is pronouncedly chronic.

Diagnosis and Prognosis.

The *diagnosis* must not by any means be formed simply from the statements of the patient. On principle, the practitioner must credit a statement only when he has convinced himself of its truth. Patients often give the most exaggerated descriptions, and call "the whites" a quantity to be considered strictly physiological. Hysterical women, who frequently suffer from leucorrhœa, even on purpose lie to the physician. On the other hand, a patient using the irrigator daily may completely deceive herself about the quantity of the discharge, or the contracted os does not allow the mucus to escape, the cervix being dilated like an ampulla. For all these reasons, a digital and a specular examination is necessary.

The differential diagnosis between the various discharges is made partly by the inspection of the stains in the linen (pus causes yellow, mucus pale-gray stains), partly also by direct inspection. That the entire uterine cavity is affected is learned, *first,* from the dilatation of the uterus; *second,* from the non-effectiveness of the treatment of the cervical catarrh ; *third,* from aspiration of the uterine contents with Braun's syringe.

The *prognosis* is favorable under appropriate treatment. But especially the milder forms of hypersecretion are harder to cure than cases of copious purulent or sanguineous discharges. Of course, the prognosis depends also upon the complications, particularly metritis.

Treatment.

The treatment must be general and local. Both must act together in order to remove or mitigate this chronic affection.

Regulation of the bowels, building up of the constitution by iron, good nourishment, country life, or change of scene are necessary. If there be no suspicion of lung disease, which is often the case, sea or mountain air is to be recommended first of all. There are not a few cases in which the local affection disappears with the improvement of the general condition. Especially in virgins, a hasty local treatment should be warned against. The great difficulties in exploration as well as in every manipulation in the depth of the vagina greatly impede *complete* local treatment; the psychical excitement caused either by the pain and aversion to having the genitals touched, or, on the contrary, the sexual excitement and the inclination to repeat the excitement independently, is incalculable in its consequences. I believe I do not exaggerate if I assert that in many a psychosis the local treatment of some uterine affections, in itself slight, was the first cause.

If the condition does not improve by constitutional treatment, we first try mild local measures. We order vaginal injections with salt or alkaline water. Should the discharge return even then, local treatment of the mucosa is indicated. In women with roomy vaginæ, we might decide on local treatment sooner than in virgins.

Before commencing the local treatment, the patients must be given to understand that an affection of many years' standing cannot be removed by a single cauterization, etc. We must emphasize the importance of the *prolonged treatment*, of perseverance, and conscientiousness, and warn against wilful interruption of the treatment.

In hyperæmia, scarifications are useful. When, during the scarification of the vaginal portion, the blood escapes in large quantity, repeated abstraction of blood, glycerin tamponade, and painting with iodine are indicated. The inner surface of the uterus must be carefully cleansed previous to every local treatment. If the internal os be narrow, it must be dilated; usually this is not necessary. With the uterine catheter (Figs. 23 to 25, p. 48) disinfecting irrigations of the cavity are given. Then with the uterine rod armed with dry cotton (Fig. 32, *b*, p. 62) the uterus is repeatedly wiped off until no more glairy mucus adheres to the cotton. If hemorrhage ensues, a uterine rod is armed with cotton, inserted into the uterus and allowed to remain. Then a second rod, wrapped with very little cotton, is dipped in tincture of iodine and, after removal of the tamponing rod, quickly introduced into the uterus. The latter rod is allowed to remain one or two minutes. The readily diffused tincture of iodine is thus brought much more certainly in contact with the entire uterine cavity than if injected with Braun's syringe (Fig. 35, p. 68). After removal of the rod, the superfluous tincture of iodine is carefully wiped off, and two or three daily lukewarm vaginal irrigations ordered.

In cases of simple hypersecretion, much better results are secured with the above-described harmless treatment than with the recently recommended heroic methods of cauterizing and curetting the uterus. For the uterus is no fistulous track covered with unhealthy granulations and slow to heal. A partial destruction of this diseased mucous membrane has the therapeutic value of a direct abstraction of blood merely. If the mucosa is to be incited to a healthy secretion, nothing can be gained by taking away portions of it. It is much more important to remove the secretions and then to bring the *entire* mucous membrane in contact with an alterative remedy.

If the internal os be narrow, we are not sure of reaching every part of the uterus with the instrument. In that case, often the pressure with my catheter suffices to dilate the os. Particularly appropriate are my uterine dilators (Fig. 18, p. 41). No. 1 can be passed even into the nulliparous uterus without special effort or pain. Immediately after its withdrawal large masses of retained mucus often well forth. The object having been the insertion of the rod armed with cotton so as to swab out the uterus, a greater dilatation is for the present uncalled for. Others prefer laminaria dilatation from the beginning. Indeed the dilatation of the tent in the uterus and cervix is needless if the internal os alone is to be stretched. Still it must be admitted that the antiseptic dilatation with laminaria is the most harmless in its consequences, and withal the easiest method of dilatation. A lasting effect may be ascribed to laminaria dilatation; the expanded, serously infiltrated uterus often diminishes surprisingly afterwards, especially if the procedure be concluded with a few ergotin injections. The uterus having been dilated with laminaria, it is irrigated, swabbed, and painted. If ineffectual, this method may be several times repeated.

Should we secure absolutely no effect, or should there be from the beginning a purulent rather than a glairy discharge, a more energetic dilatation of the uterus is indicated, for the purpose of palpating the internal surface of the uterus. This is often effected with surprising facility; if the external os be wide, the finger bores into the uterus without encountering much resistance. In this the other hand must assist externally, that is to say, press and impale the uterus over the finger. Should the control from without be difficult, we operate in Sims' speculum, and insert into the anterior lip of the os, or into both lips, a Muzeux tenaculum forceps (Fig. 15, p. 36). Then the finger penetrates into the uterus while an assistant holds the speculum. We may also operate without speculum, penetrating with one hand and holding the Muzeux forceps with the other. Schroeder has recommended, in difficult cases, to incise the os on both sides. This has the advantage that, when the lips are turned over, we can see the internal surface, and that the penetration into the uterus is greatly facilitated. In nulliparæ, however, I believe it to be less dangerous to insert laminaria tents first and dilate afterwards, either with Schultze's (Fig. 17, p. 41) or with Ellinger's instrument. These instruments are best inserted into the uterus in the lateral position (Fig. 11, p. 34) through Sims's speculum (Fig. 12, p. 35), the vaginal portion having been fixed with tenacula.

But here, too, I earnestly caution against force; there certainly are uteri of nulliparæ which cannot be dilated sufficiently for the finger to reach the fundus. However, for the formation of the diagnosis it will do to introduce a sharp spoon (Figs. 26 and 27) or a curette (Fig. 38, p. 70). If these instruments can be readily moved to and fro, if they everywhere encounter the smooth internal surface of the uterus, without, perhaps, scraping off too much in their downward course, there can be present neither pathological contents nor a largely tumefied mucous membrane. The recognition of that fact suffices. But if the contents are pathological, the entry is usually not difficult. Thus, where the entry is very difficult, we may suspect *a priori* that there is nothing in the uterus, and the reverse. Exceptions to this rule are small myomata, in the presence of which the cervix is at times quite hard and firm.

Having formed the diagnosis that the mucous membrane is diseased without being hypertrophied, and that, perhaps, no tumor is present in the uterus, the cavity is treated with caustics. Instead of tincture of iodine, nitric acid may be used. The entire uterus is cauterized with a uterine rod. A mixture of carbolic acid, tincture of iodine, and water has also been proposed. Quite excellent results are often secured with the harmless wood vinegar. While an assistant holds the Sims speculum, a teaspoonful of pyroligneous acid is poured into the vagina in front of the vaginal portion, after the uterus has been wiped clean. Then my uterine rod, armed with some cotton, is inserted ten or fifteen times into the uterus, so that the acid is repeatedly pumped into the cavity. After the procedure, another uterine rod armed with a large tuft of cotton removes the superfluous vinegar from the vagina, lest it flow out on turning the patient. Strong solutions of alum, tincture of sesquichloride of iron, and many other astringents and caustics are also in use. Liquids are to be preferred to solid remedies.

Not rarely an indication for therapeutic measures is derived from the hemorrhages alone. Both menorrhagia and metrorrhagia weaken the organism so that aid must be rendered.

Of course, a large number of different diseases may cause hemorrhages, and the cause, *e. g.*, a myoma, often is quite evident. But cases occur in which the nature of the pathological process must be sought in a mere relaxation and repletion with blood of the entire uterus and especially the mucosa. Then a treatment is instituted having for its object the arrest of the hemorrhage. Here, again, all the measures which have an invigorating influence on the organism in general, and attack the chronic metritis directly and indirectly, find application.

If we intend to combat the hemorrhagic tendency of the mucous membrane, it is best to begin with the injection of tinct. ferri into the uterus by means of Braun's syringe. The procedure is described on p. 67. Previous to it, the vagina, the vaginal portion, and the cervix must be disinfected, lest any crusts possibly escaping from the uterus might decompose in the vagina. To prevent this, the vagina is irrigated, besides, for two or three days after the procedure. Should the mucosa alone be the object of treatment, a preliminary dilatation is unnecessary. However, should the antecedents of the case be known, and there be a suspicion that the uterus incloses abnormal contents, or should combined examination disclose a peculiar softness and enlargement of the uterus, the cavity must be explored.

In this case, too, we first endeavor to enter without dilatation. This failing, rapid is to be preferred to slow dilatation. *First*, the long-existing hemorrhage always prepares the conditions for rapid dilatation equally well as is done artificially by laminaria; *second*, the tent may easily bruise a polypus which becomes gangrenous, and endangers life by sepsis and peritonitis. I have made the post-mortem examination of a case in which, after the employment of sponge tents and destruction of a large mucous polypus, peritonitis could be demonstrated to have developed by way of the tube. Immediately after rapid dilatation the uterus is palpated. Should the finger encounter foreign contents in the uterus, their nature must be ascertained. In the above-described (p. 162) hypertrophies of the mucous membrane, removal from the uterus is indicated. This is done with the curette or the sharp spoon, with careful control from without. Sims's lateral position, speculum, and seizing the anterior lip of the os with tenaculum, greatly facilitate the procedure. The uterus is scraped out thoroughly in every direction. After completion of the operation the uterus is irrigated with three-per-cent carbolic solution. No purpose can be served by cauterizing subsequently. It is better to favor the diminution by ergotin. After removal of the mucous membrane, the hemorrhage usually ceases with moderate labor-like pains for several days.

If no distinct limitation of the mucosa can be felt in the depth, or if ragged tissue be deeply penetrated at a circumscribed spot, a malignant tumor is present. The latter must be diagnosticated by microscopic examination of the scrapings.

After all procedures of that nature, the patient must keep as quiet as possible. Simple painting of the-uterus with tincture of iodide or pyroligneous acid when the organ is relaxed may be performed at the office. For dilatation the patient must always be abed, and if laminaria tents are inserted, rest in bed for three or four days is necessary, or for a longer time where there is a hemorrhagic tendency. If only my dilators Nos. 1

to 3 have been passed, the patient may get up after twenty-four hours, provided she feels well. After churetting of the uterus, the temperature must be watched for three or four days, and the surroundings of the uterus be examined as to their sensitiveness or the formation of tumors. The patient may get up only when no more blood escapes and the uterus is small and hard. Oft-repeated cleansing, disinfecting vaginal irrigations are evidently in place after every intra-uterine manipulation.

For membranous dysmenorrhœa, dilatation and strong cauterizations of the uterine cavity are to be recommended.

C. The Pathological Conditions of the Cervix Uteri.

Etiology.

If we divide the uterus into two parts at the internal os, we have seen that disease common to both halves (metritis) is frequent. It is self-evident, too, that if the body of the uterus is diseased, the vaginal portion eventually becomes involved, just as in long-standing cervical catarrh with hypertrophy of the vaginal portion, not only the whole uterine mucosa, but also the entire parenchyma of the uterus will take part. Still there are some causes which may effect disease confined to the lower half. In the first place, the vaginal portion is exposed to many injurious influences; it becomes lacerated in parturition, the cicatrices in some parts cause congestion and hypertrophy, in others anæmia and atrophy. Should larger portions remain ununited, the granulating surface becomes covered with atypical proliferations of epithelium. Cylindrical epithelium seems to possess much greater energy of growth than pavement epithelium. The latter is displaced, while the former acquires territory in which it is not found primarily. The different anatomical quality of the glands is also important; those in the cervix are acinous, easily proliferating; the uterine cavity contains only tubular glands. In the next place, the altered relations of pressure enter into the consideration. If, in a nullipara, the vulva is closed and the vagina narrow, the entire pelvic floor forms a part of the abdominal walls. But in a multipara with relaxed, roomy vagina, with descensus of the anterior vaginal wall, the intravaginal pressure is less. By the descent of the vagina the vessels are dragged down, congestion ensues, the stasis prevents involution in the puerperium, the vaginal portion remains large, thick, plethoric, and irregular by the proliferation of cylindrical epithelium.

During pregnancy the intimately anastomosing glands of the cervix have also become hypertrophic. In the puerperal uterus we find an almost adenomatous structure of the cervical mucosa. This hypertrophy also takes place pathologically, or remains after the puerperium. Perhaps there are individual differences regarding the quantity of the glands. Or else, an inflammatory irritation, which renders the tissue of the vaginal portion hypertrophic, causes in some cases a peculiar energy of growth of the glands. Gaping of the cervical canal, besides, frees the cervical mucosa from the pressure of the opposite side, and the bright-red mucous membrane projects directly into the vagina. If excessive nutrition of the entire uterus be superadded, or gonorrhœal infection, or erroneous, too irritating treatment, there ensues enormous hypersecretion of the cervical mucosa mixed with pus.

In nulliparæ, however, it is possible, on the contrary, that the mucus cannot escape from the greatly contracted os externum and so expands the cervix, as we have repeatedly mentioned.

Anatomy.

Only very rarely is one of the three characteristic affections of the vaginal portion—hypertrophy, superficial cervical catarrh, and glandular hyperplasia—present singly. Nearly always we find, although one predominates, that the other two affections complicate, or else *one* affection is the cause or sequel of the *other*.

Regarding, in the first place, hypertrophy; purely hyperplastic conditions of the connective tissue, and of the musculature without participation of the glands, rarely occur singly. Although they are not congenital—for they are never found in the fœtus—they are based on a defective transformation of the fœtal uterus into the ripe organ. Such a hypertrophy is more frequently a concomitant phenomenon of chronic metritis, of hypertrophy of the entire organ. Or else the hypertrophy is the consequence of congestion, and the latter the sequel of dislocation, of descensus of the anterior vaginal wall—a condition which we shall consider separately.

Isolated cervical catarrh is more frequently observed. That herein again the internal surface of the body of the uterus often takes part in the production of mucus we have explained on p. 154. In isolated cervical catarrh, the mucosa is thickened, spongy, and bleeds very readily. The mucus is mixed with pus, or consists of pure yellow pus, in gonorrhœal infection. In the latter part of the lying-in, pure pus often escapes in large quantities from the vaginal portion. Withal the external surface of the vaginal portion may be quite intact; only where the cervical mucosa begins, a bright-red mass is seen when the external os is pulled apart.

Inasmuch as this catarrh has its essential cause in an affection of the muciparous organs, it is natural that an increase or enlargement of the organs, the glands, is nearly always present.

The pathological processes in the glands lead to what has been of old recognized as a concomitant phenomenon of uterine or cervical catarrh—erosions. A larger or smaller reddened portion, nearly always connected with the cervical mucosa and extending to one or both lips of the os, with irregular surface, was seen through the speculum. This part clearly had an investiture differing from the normal. As the surface was at times apparently granulated, it was called granular or papillomatous erosion, and when these "papillæ" were greatly tumefied, it was denominated cockscomb excrescence. If some follicles, ovula Nabothi, retention cysts with mucous contents, were observed in the growth, this condition was specified as follicular erosion. A careful investigation of the nature of this affection was not made; partly because the material was not sufficient, partly because the interest was centered more in the treatment. The several erosions were separated more in reference to the appropriate caustics than to the anatomical substrata.

It was generally assumed that the epithelium disappeared first, and that then the underlying tissue began to hypertrophy, that, therefore, the surface of the erosion was formed of denuded papillæ. An erosion in this sense occurs occasionally in the vagina and on the vaginal portion, which equals the vagina with reference to its surface (Fig. 85).

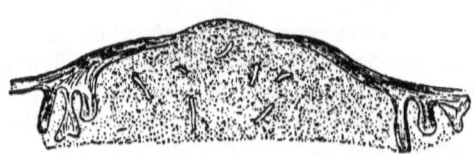

FIG. 85.—Erosion, abrasion; right and left papillæ. In the centre, loss of epithelium. Protrusion of the subepithelial strata, infiltration of the tissues, after Fischl.

In the great majority of cases, however, the condition is altogether different.

It was Veit's and Ruge's undeniable great merit to have approached this question with careful patho-anatomical studies. The result was very important—the above-named investigators demonstrated that in all cases the "erosions" were covered with cylindrical cells. These cylindrical cells were said to have originated from the deepest layer of the pavement epithelium overlying the papillæ of the vaginal portion. The deepest layer of these cells, in form cylindrical, but of the nature of young pavement epithelia, were said to be incited to independent proliferation by some irritation; they formed both depressions and projections. The depressions again might form retention cysts—follicular erosions—or the projections, papillæ—papillary erosions—which, of course, were something quite different from the normal papillæ or a hyperplasia of the latter.

Without detracting from the merit of these authors, this explanation might be received with some doubt. How an "irritation" was to change young pavement epithelium suddenly into cylindrical epithelium was hard to prove despite the developmental relationship of the epithelia. But the key to this discrepancy is soon furnished. Acinous glands occur normally at the vaginal portion, both on the external and especially on the internal surface. When an abrasion of the normal pavement epithelium had been effected, the glandular epithelium proliferated and covered the parts denuded of epithelium (Fig. 86). It is even possible for the cylindrical epithelium to detach the pavement epithelium, to crowd be-

DISEASES OF THE ENDOMETRIUM.

tween the underlying tissue and the epithelium. We may also represent to ourselves that in total enlargement of a gland the spreading orifice displaces the pavement epithelium, crowds it one side, so that a spot covered with cylindrical epithelium in midst of the vaginal portion consists of protruded and proliferated glandular lumina. Should there be an especially great proliferation of the cylindrical epithelium or of the glands, protuberances will arise—papillomatous erosions—which are really glandular lumina which have proliferated outward (Fig. 87). The process essentially recalls the puerperal state. In the latter, the uppermost layer of the uterine mucosa, the decidua, disappears—abrasion—and from the remaining glandular fundus the cylindrical epithelium overspreads the entire inner surface of the uterus. It is not impossible

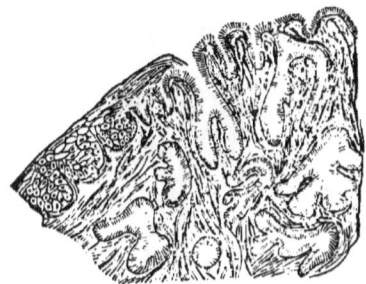

FIG. 86.—Follicular erosion (after Veit and Ruge); on the left, normal papillæ with pavement epithelium; adjoining, adenoid proliferations; in the depth, many glands in section.

FIG. 87.—Papillomatous erosion. Depressions of newly forming glandular tubes, with intervening masses of tissue simulating papillæ (after Veit and Ruge).

that in the puerpera, regeneration of the uterine mucosa the glands of the cervix also take part, that the new uterine epithelium partially springs from them and proliferates upward. In favor of this is the great hyperplasia of these glands in pregnancy.

Not rarely there is occlusion of a part or of a whole gland. Inspissation of the secretion, as well as compression of the efferent duct by swollen periglandular tissue, by very great distention of the vessels, by neighboring follicular glands, by great proliferation of the epithelium between the efferent ducts of the glands, hinder the passage of the glandular secretions. If a gland cannot discharge its secretions, a retention cyst is formed which is called ovulum Nabothi. Its genesis will be hindered by great resistance of the surrounding tissues. Where there is no resistance, as at the surface, therefore, the formation of such cysts will be effected most readily. Should such cysts be opened, either spontaneously by internal pressure or by any trauma, the cylindrical epithelium may be laid bare. Should an erosion form at that spot (a portion covered with cylindrical epithelium), and, with the abundance of acini in the acinous glands, should other cysts crowd upward into the erosion, the appearance of follicular erosion is presented.

But if the tissue is too firm to be pressed apart by the glands, the cyst, as it were, elevates itself from the parenchyma. It depends from the vaginal portion, usually from the os, in the shape of a small polypus (Fig. 88, *b b*). This polypus may be single-chambered, hanging pearl-like out of the os, and gradually acquire a longer stem, or several sections of a gland may proliferate, and freed from the pressure of surrounding tissue, rapidly increase in size in the vagina. Should the polypus proliferate outward from the cervical cavity, its surface of course will bear the same cylindrical epithelium as the retention cysts. It seems to be possible, however, that the cylindrical changes into pavement epithelium when the polypus has remained very long in the vagina.

FIG. 88.—Vaginal portion with erosion *c*; ovula Nabothi of varying size *a a a*; small mucous polypi formed of ovula Nabothi *b, b*.

FIG. 89.—Two long pedicled mucous polypi depending from the cervix, after Billroth.

Such polypi differ largely in form. In Fig. 88 we see two small polypi which should be designated ovula Nabothi. In Fig. 89, two long pediculated polypi are represented depending out of the vaginal portion. But they occur also in flat form. The larger ones are always flattened by the vaginal walls, being composed merely of structures of the mucous membrane, and hence very soft.

However, very considerale hyperplasiæ of the glandular portions underlying the pavement epithelium of the external surface of the vaginal portion also occur. Whether these glands gradually grow from the internal surface of the cervix, or whether they were originally present close underneath the pavement epithelium, will be hard to decide. At all events, so great a hyperplasia of the glands ensues that an entire vaginal portion may consist of nothing but small cysts. The vaginal portion is transformed into a cavernous tumor, the connective tissue is entirely displaced, a section cuts through innumerable small retention cysts filled with mucus. Among them is here and there a larger one measuring one to one and a half centimetres in diameter. In this cystic degeneration the vaginal portion is much enlarged; it always secretes considerable quantities of mucus. Often the cysts have opened here and there, so that in one place the crowded cysts shimmer through, blue and glassy; in another, the surface presents a cribriform appearance like a decidua.

But the cysts are not always immediately under the skin; often during scarification we accidentally evacuate deeply seated ovula Nabothi, at times we also encounter a single large cyst down near the internal os during amputation.

If the glandular hyperplasia affects but a single gland or a small group of glands, the surface of the vaginal portion becomes prominent; and after the enlarging gland has overcome the pressure of the tissues of the vaginal portion, it rapidly proliferates into the lumen of the vagina. Then the growth may attain enormous proportions; the minute acini become tubes the thickness of the finger, filled with tenacious mucus; the hypertrophic part changes the form of the lip of the os whence it springs. In the course of years, tumors the size of the fist are formed, their surface having a tonsillar, irregular appearance. These large polypi were also called—follicular hypertrophy of the vaginal portion. Although anatomically identical with the above-described mucous polypi, their appearance is different, as the parenchyma of the vaginal portion and its investing pavement epithelium take part in their formation. It has also happened that the cylindrical epithelium of a retention cyst proliferated papillomatously, then we find a simple papilloma springing from the lumen of a gland, covered with the pavement epithelium of the vaginal portion (Ackermann).

All the described erosions and polypi, then, are referable to pathological processes in the glands of the vaginal portion. Whether there be simple erosion with cylindrical epithelium, or the cylindrical epithelia proliferate so as to produce papillomatous forms; whether a small retention cyst lie in the erosion, or depend from the os on a pedicle; whether a mucous polypus depend from the vaginal portion, or the latter be entirely in cystic degeneration—it is nothing but a different anatomical arrangement of the same formations, the hyperplastic cervical glands.

Symptoms and Course.

Hypertrophy of the vaginal portion, whether involving the connective tissue merely or due to cystoid degeneration, causes a number of symptoms encountered likewise in chronic metritis: sense of pressure in the pelvis, tenesmus of the bladder, pain on defecation, a disagreeable sensation when sitting down suddenly, pain during coition.

The erosions often maintain a quite considerable discharge of a muco-purulent nature. Besides, the menstruation is nearly always much increased. The erosion is so plethoric that blood escapes during coition and the touch. If the erosion be merely freed from the counterpressure of the vaginal wall by the insertion of a tubular speculum, points are seen from which drops of blood exude. The loss of blood and the production of abundant secretion weaken the general system.

With mucous polypi, at times a purely serous, quite thin-fluid secretion escapes in such quantities that in the morning, after the fluid has accumulated during the night, from thirty to fifty grammes gush forth. Some blood is frequently lost during coition, especially during pregnancy, in which the polypus is hyperæmic. In follicular hypertrophy, too, whether it affects an entire lip of the os, or whether a polypus depends from the vaginal portion, bloody and mucous discharge occurs.

In cystoid degeneration, the discharge is usually very considerable, like the discharge symptomatic of cervical catarrh.

The course is chronic; during pregnancy, before and after menstruation, the secretion becomes larger in quantity. Spontaneous recovery ensues in *this* sense, that with decrease of the general hyperæmia the secretion likewise becomes less.

The polypi usually grow remarkably slowly. They occur in the married and the single, while erosions follow with especial facility on parturition or abortion.

Diagnosis.

The diagnosis is made by digital and specular examination. In the speculum, the ovula Nabothi are often seen to gleam through. The expert discovers the small protuberances even with the finger. Mucous polypi are not at all easily felt. With surprise we sometimes observe in the speculum a large, bluish-red polypus, which is so soft as to escape everywhere from the finger, and not to be felt. The polypi projecting from the external surface of the vaginal portion have a peculiar wrinkled appearance, are of irregular form, but may also represent perfectly symmetrical, oval formations. Their consistence insures the differential diagnosis from sarcoma and carcinoma. Their entire appearance separates them from papilloma. On section, the large retention cysts with glassy mucus appear striking.

With larger erosions, the differential diagnosis between carcinoma and erosions enters pre-eminently into the consideration. The erosion frequently succeeds the puerperium, occurs in young subjects, does not at first depreciate the general health, is soft, covers a larger extent of the hypertrophied vaginal portion, forms no circumscribed hard tumor, and, especially, disappears on appropriate treatment.

In the affections described, the prognosis is good.

Treatment.

We start with the treatment of the cervical catarrh. As these affections are of an essentially chronic nature, we must not expect to effect the cure rapidly.

To begin with, the vaginal portion is very thoroughly cleansed with my uterine rod, but *cautiously*, lest a hemorrhage disturb the observation. Then a dry, benzoated cotton tampon is laid in front of the vaginal portion, and removed after twenty-four hours, thus gaining information as to the quantity of mucus produced. Then the vaginal portion is scarified, in order to evacuate the visible and invisible ovula Nabothi on the one hand, and to produce a depleting effect by the hemorrhage on the other. With a small sharp spoon (Fig. 36, p. 70) the cervical canal is scraped out, beginning at the internal os. Thereby the mucous membrane is to be depleted through the hemorrhage, accidentally present polypi are to be destroyed, and larger tumefactions removed. After the bleeding surfaces have been washed during the hemorrhage with three-per-cent carbolic solution by means of the rod, and about two to three tablespoonfuls of blood evacuated, a glycerin tampon is laid in front of the vaginal portion. The tampon is removed after twenty-four hours, the entire vaginal portion again cleansed with wet carbolized cotton, and if the scarifications bleed, some more blood is allowed to escape. In the intervals the patient may take disinfecting injections; painting of the vaginal portion and the cervical canal with tincture of iodine also has a diminishing influence. Should the hypertrophy have diminished after these measures—often not until a week after the first procedure—the erosion and the cervical canal are most carefully cleansed of mucus. Should the cervical canal bleed, a tamponing tuft of cotton is inserted, and the cauterization of the cervix performed immediately after the removal of the tampon. With my uterine rod (Fig. 32, *a*, p. 62), I take up a single drop of nitric acid, and first touch the erosion; every part after being cauterized becomes covered at once with a white crust. The erosion is cauterized carefully in every direction, sparing normal portions, but penetrating especially into fissures and into angles of the os. Then we proceed into the cervix, commencing as close as possible to the internal os. By pressure against the walls, the nitric acid is expressed from the cotton. We enter from six to ten times, so as to reach every portion. Brownish masses flow from the os. They are wiped off, the vaginal portion is inspected for some time to see if any red uncauterized spot be left, and finally a tampon of benzoated cotton moistened with glycerin is laid in front of the vaginal portion. The whole procedure is performed at the office. The cauterization is not painful. Should the patient complain of burning, this will pass off in a few minutes.

To be sure, the whole procedure—scarification and cauterization—may be crowded into one sitting, but then the result is by no means as certain as when done in the manner described.

Vaginal injections are to be used for three or four days, then the exploration is repeated. In recent cases, especially late in the puerperium, healing is now seen to progress. The swelling of the vaginal portion may fall quite rapidly. But should the bright-red erosion be again seen of equal size, it is cauterized with nitric acid four or five times in succession, if possible *once daily, while the cervical canal is left at rest*. For the erosion is to be *destroyed*, the cervical mucosa *merely alteratively influenced*; were we to cauterize here in rapid succession, a stricture or atresia would have to be feared (comp. p. 64).

I can affirm that I have positively and definitely cured the most extensive erosions, covering both lips of the os, in this harmless manner.

In one case, however, the result is not permanent, namely, in the so-called follicular erosion or in cystoid degeneration of the vaginal portion, that is to say, where the ovula Nabothi are so numerous that the evacuation of the superficial ones has no great influence. In this case, in which after every renewed scarification fresh masses of mucus escape, oft-repeated scarification will be of use. Two days after thorough scarification, the vaginal portion may be reduced from the size of the fist to the normal. But definitive cure is usually not secured by scarification.

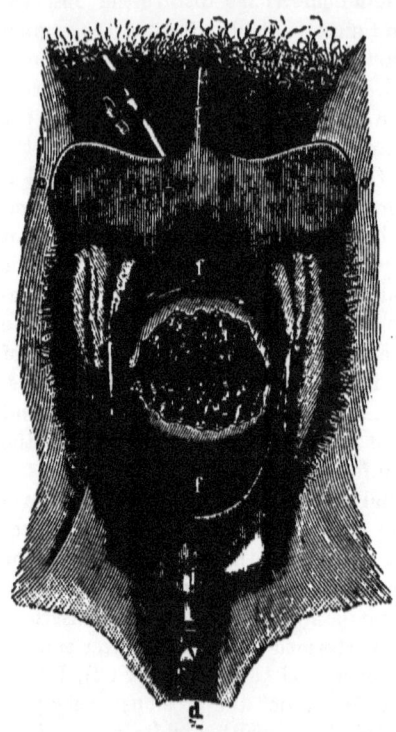

FIG. 90.—*a, b, c*, gigantic speculum; the wings *c c* are intended to crowd the pudendal hair and the labia as far as possible out of the field of view. At *b* the speculum is inserted into the upper handle of Simon's speculum; *a*, tube to which the irrigator tube is fastened; *e e*, labia majora; *f f*, vagina; *d*, lower speculum, fastened below into the perineal or speculum holder *d* (comp. Figs. 57 and 58, p. 120). Between the specula appears the vaginal portion, both lips of which are covered with a fissured, papillomatous erosion.

Such a vaginal portion must be amputated. It is just in these cases that we obtain quite a wonderful result by the amputation. Although the entire diseased portion is not removed, it seems that the loss of blood and the cicatricial contraction or compression bring about obliteration of the glands. Cervical catarrh disappears in such cases at once. The operation is performed in the manner described on page 139, Figs. 78 to 80.

Fig. 90 shows the specular appearance of a vaginal portion thus covered with erosions. The upper "gigantic speculum" is very short, like the lower one (see also Fig. 58, p. 120). For if the vaginal portion is to be drawn down, it is incorrect to use long (Simon's) specula. In order to do away with the lateral holders, I have had the lateral parts *e e* of the lower speculum made very high, as will be seen by reference to the figure. The lower speculum can be fastened to the table, so that but a single, and not even a skilled, assistant is required for holding the gigantic speculum. Chloroform narcosis is unnecessary for this operation. The gigantic speculum is made double; at the upper end *a*, the disinfecting fluid which permanently irrigates the field of operation escapes.

If more cysts are observed within the surface of the incision, the wound had better be stitched together.

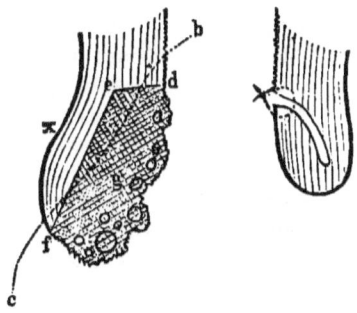

FIG. 91.　　　　　FIG. 92.

FIG. 91.—Schroeder's method of operation for the removal of erosions. *d e*, transverse incision, met at *e* by an incision from below, *f*; *g*, the erosion, removed with the soil whence it starts; *b c*, suture; *x*, the point whence the incision toward *e* would be made in the wedge-shaped excision.

FIG. 92.—The opposite lip of the os after completion of the operation.

For extensive erosions Schroeder has proposed another method of operation which he has often performed with success. First the lateral incisions are made, in order to separate the lips of the os. Then a transverse incision is carried at a right angle to the uterine axis across the internal surface of the vaginal portion (Fig. 91) *d e*. It intersects another incision *f e*, so that *g*, the erosion, is removed. The suture *b c* then unites *d* to *f*. Therefore the external surface of the vaginal portion is folded upward and inward, the cervical canal "lined" with vaginal mucosa. But this method, too, is really only a modified wedge-shaped excision. If more of the vaginal portion be removed, so that *f* lies more externally, say at *x*, we have a wedge-shaped excision. I am inclined to place but little value on the lining of the cervix, for the whole cervical mucous membrane is not removed; at most, if we go very high up, two-thirds of the mucosa. The value of this operation consists in the influence exerted on the chronic metritis by the amputation of the vaginal portion, and in the obliteration of the other glands to be secured also by the partial removal of the muciparous glands. If there be a suspicion that an erosion is of a cancerous nature, it might be more advisable to remove as much as possible by the wedge-shaped excision, rather than to detach the mucosa superficially.

Should the vaginal portion be misshapen in any other way, for instance, should the os be too narrow, the treatment is that laid down on p. 139, Figs. 76 and 77.

Should the os be quite irregular, for instance, should a thick ridge project into it, it must be excised. Should there be a laceration in the vaginal portion extending into the fornix, the laceration may be united. Emmet referred a whole series of injurious consequences to such lateral lacerations and united them. The cicatrized margins of the wound were freshened and stitched together with wire sutures, thus restoring the former shape of the vaginal portion. Where there are two lacerations, one of the lips is not rarely so atrophied that after bilateral freshening hardly any room remains for the os. However, these lacerations are so frequently found without any symptoms that I am altogether sceptical as to the necessity for the operation and the certainty of a result.

Mucous polypi are abscised. If they are too small, they are scraped off with the sharp spoon, or destroyed with the hot iron. Owing to the ample blood-supply of these polypi it is necessary to observe the patient carefully, to tampon thoroughly, or to cauterize the point of origin with the hot iron.

Fig. 93.—Scissors for the abscission of mucous polypi.

Peculiarly appropriate scissors with which the operation can be performed also through a tubular speculum are represented in Fig. 93.

In polypi springing from the external surface of the vaginal portion or in follicular degeneration, it is best to sew the wound margins carefully together. Should it be altogether impossible to secure assistance, a ligature might be applied and the polypus cut away under it. The remnant of the polypus disappears spontaneously.

Galvano-cautery and écraseur are not necessary.

CHAPTER XII.
DISLOCATIONS OF THE UTERUS.

On page 11 we have discussed the position of the uterus and its normal mobility. A pathological condition exists if the uterus so alters its location that none of the described positions are maintained; or, if fixation occurs in some position and the normal motions can hence no longer be executed.

We distinguish *flexions*, in which the body bends from the cervix forward or backward; and *versions*, in which the fundus of the non-flexed uterus sinks forward or backward. In versions, therefore, the vaginal portion moves upward when the body sinks down; while in flexions the vaginal portion preserves its position, although the fundus is bent downward anteriorly or posteriorly.

By the term descensus or prolapsus we understand the displacement in which the entire uterus or a part of it sinks downward.

A. Anteflexion.

In anteflexion of the uterus the body is bent forward from the cervix at an acute angle. The angle may be so acute that the upper half runs parallel to the lower. Among the cases of pathological anteflexion are also included those in which neither the filling bladder raises, nor the menstruation stretches, the uterus; in which, therefore, the angle is constant, and never obliterated.

Etiology.

Anteflexion is often a defect of development, that is to say, not an arrest of development in the ordinary sense, a defect of fœtal life, but rather of the development of puberty. An extraordinarily small uterus (infantile uterus), great anteflexion, narrow os, disproportionately long and thin vaginal portion, occur as complications. Here it must be assumed that not only the uterus, but also the peritoneum has been somewhat retarded in its development. The transformation of the infantile to the mature uterus has proceeded with little energy. In consequence, the uterus stands higher, is suspended farther upward.

On the other hand, should the abnormally long vaginal portion (comp. Fig. 94) extend farther down into the vagina, it must take the same direction as the latter. But with firm abdominal coverings the intestines keep the body of the uterus pressed tightly against the bladder. Thus a very considerable anteflexion arises more by the vaginal portion bending toward the body than by the body flexing on the vaginal portion. Fig. 94 represents these conditions.

Should the uterus maintain this position, the physiological dislocations are not easily possible. The fundus inserts itself between the bladder and cervix, the bladder does not lift the body, but pushes past it in an upward direction. The angle of flexion, however, must be badly nourished in the presence of permanent compression, thus causing secondary atrophy of the flexed portion. The corresponding upper wall likewise becomes atrophic by traction, so that the uterine wall above and below at the angle of flexion is rendered abnormally thin. In the menstrual congestion, the blood cannot penetrate into the compressed or stretched vessels, erection fails to occur, the blood stagnates in the body, which becomes hyperæmic. The final result is that the body is movable at the angle as on a loose hinge. Thus retroflexion easily arises from anteflexion. By these processes the circulation, and consequently the normal course of menstruation, is disturbed. In this way pathological processes occur which lead to chronic metritis on the one hand; on the other, to hyperæmias in the surroundings of the organ which incite perimetritidis. The originally uncomplicated anteflexion becomes complicated in consequence of the long duration and the sequels of the menstrual congestion.

FIG. 94.—Anteflexion with *col tapiroid*.

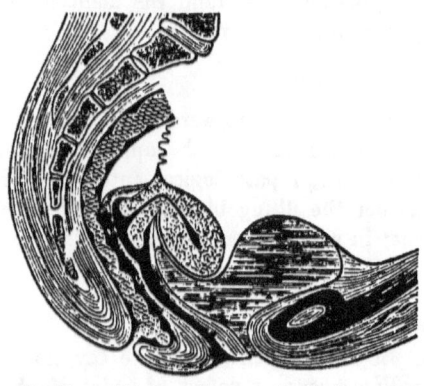

FIG. 95.—Anteflexion with posterior fixation. Mushroom-shaped vaginal portion.

While here, then, the complications were the consequences of the flexion, it is the reverse in a second series of cases. First there is a peritoneal or subperitoneal inflammmation posterior to the lower part of the uterus. That condition may arise from the extension of an inflammation of the cervical mucosa to the cervical parenchyma, the adjoining connective tissue, and the neighboring peritoneum. Moreover, in gonorrhœal infection, the mucous membrane of the entire uterus and the tube becomes diseased. Should some of the irritating secretion escape from the abdominal orifice of the tube, it flows into the fossa of Douglas, the lowest part of the peritoneal cavity, and there produces inflammation and adhesion of the posterior wall of the uterus to the opposite peritoneum. Besides, other causes may lead to perimetritis and contraction of the fold of Douglas, and to fixation of the uterus at the angle of flexion. But if this part of the uterus be drawn still farther upward, the fundus must sink lower. In Fig. 95 we see the posterior fixation represented by the zigzag line. The full bladder is unable to raise the fundus, but extends upward past the uterus.

In these anteflexions, the vaginal portion is pressed against the posterior vaginal wall, producing a mushroom shape, just as when a soft clay cylinder is pressed against a hard plate. After prolonged existence of the flexion, the angle will become atrophic, so that in this second form the same final result may occur as in the first form. Should—perhaps in consequence of reckless treatment—the perimetritis in-

FIG. 96.—Retroposition, retroversion with anteflexion.

crease behind the uterus, not only the angle of flexion, but the entire posterior wall of the uterus will become fixed. And should the perimetritis be intense, the subperitoneal layers will likewise participate in the inflammation. Then the peritoneum may be fixed to its substratum, and the uterus again be adherent to the fixed peritoneum. In this way the uterus becomes immovably fixed to the posterior wall of the pelvic cavity. Should a strong anteflexion have existed previously, it will continue. There will, therefore, be formed a retroposition, or a retroversion, with anteflexion. Such a case is represented in Fig. 96; in the latter, as always in these cases, elevation of the uterus succeeded, *i. e.*, the entire uterus was drawn somewhat upward.

Should pregnancy ensue—a rather rare occurrence—we often find the uterus after the close of the puerperium just as much anteflexed as before. On the other hand, the changes of pregnancy and parturition may be able to cure the anflexion, atrophy of the angle of flexion does not occur, and the uterus thereafter lies in its proper position and functionates normally.

Through small myomata within the body of the uterus an anteflexion may also be brought about or be increased. Of course, it is immaterial whether the myoma be situated in the anterior or the posterior wall. The important element is the increased weight and the descent of the fundus. As physiologically it rests on the bladder, it here must sink still lower.

Symptoms and Course.

Besides the manifold morbid phenomena depending on the complications, it is mainly three symptoms which are ascribed to anteflexion.

First, the dysuria. A glance at Fig. 95 shows how the low-lying fundus prevents the normal distention of the bladder. If the latter cannot raise the uterus, and if the womb prevents the uniform filling of the bladder, vesical tenesmus will be felt sooner, especially at the time of menstruation, during which, first, the hyperæmia renders the nerves more susceptible of irritation, and second, the uterus is heavier and its pressure greater.

The other two symptoms, dysmenorrhœa and sterility, have led, especially of late, to many scientific discussions, and we must consider them more nearly, if for no other reason, because they are of the greatest importance for the principles of the treatment.

In most recent times, the mechanical explanation of dysmenorrhœa stood at the head. It was stated that the internal os partly was really constricted in the way of complication, partly was narrowed by the uterus being flexed at these points. When blood is effused, it cannot escape from the uterus, thereby uterine contractions are excited, and this was said to be the dysmenorrhœa. As a proof of the correctness of this assertion, the following was accepted: If sounding be done previous to menstruation, the latter is often painless; if the flow of blood be established, the pains likewise cease. After labors, the dysmenorrhœa usually does not recur. But that occlusion of the internal os with retention of the secretions can cause violent colic is shown by the cases in which uterine colic occurred in consequence of foreign bodies, or where, for instance, a small polypus, swelling during menstruation, prevented the escape of secretions.

Against this view we have, *first*, the fact that the pains have not the pronounced uterine character, are by no means confined to the uterus, but are felt in the whole abdomen and the thighs. *Second*, very frequently the pains are present even before there is a drop of blood in the uterus and do not disappear with the hemorrhage. In the worst cases, moreover, so little blood is extruded that it could not possibly incite the uterus to contractions. *Third,* the liquid blood is not to be identified with a foreign body. Nor does the uterus react by dysmenorrhoic pains on the formation of a polypus, on the sound, on accumulation of mucus, or on intra-uterine pessaries.

The pains are to be explained rather in the following manner: We know that the mucous membrane is greatly swollen and the uterus thickened during menstruation. The resistance of the uterus against this thickening and swelling, as well as against the hyperæmic condition generally, causes contraction and sensitiveness. As soon as the hemorrhage is established, *i. e.,* the swelling of the mucosa falls, the pains of course diminish. A flexed uterus decidedly will oppose much greater difficulties than a normal one to the thickening of the mucosa and the *uniform* expansion. Should the uterus be lax, like one previously impregnated, the expansion of the organ by the swelling mucosa will of course be easy, and hence the dysmenorrhœa must cease after childbirth. If this explanation be valid for uncomplicated cases, *i. e.,* such as are free from any inflammation in or about the uterus, the complicated cases can be understood still more easily. During existing inflammation or inflammatory irritation of the uterine nerves, in perimetritis, etc., the menstrual congestion will and must cause an exacerbation. A process completed painlessly and physiologically in the normal uterus, will cause symptoms and run a pathological course in the diseased organ.

But where the internal or external os is narrow, the mucus cannot escape at all or with difficulty and thus the swelling of the mucosa is hindered. The same effect is produced by a swelling polypus.

Anteflexion has also been largely enumerated as of etiological importance for sterility. Although the cause of sterility can never be ascertained with certainty during life, it is true that anteflexion and dysmenorrhœa are in many cases complicated with sterility. But by no means invariably. I can positively assert that in great anteflexions with narrow external os, where painless sounding was impossible, with long conical vaginal portion and considerable dysmenorrhœa, conception took place; just as it may fail to occur in the normal uterus, although it is not possible to find any reason for it.

Narrowness of the os is unjustly accused. For the spermatozooids are so small that they may enter through the narrowest os if they get at all into its neighborhood. But where the vaginal portion is as long as in Fig. 94, the semen is poured out above the os, and where the vaginal portion is strongly pressed against the posterior vaginal wall, the pressure likewise will prevent the approach of the semen into the region of the vaginal portion. That the position of the vaginal portion is important we see by retroversion with descensus. In that position pregnancy often occurs with surprising facility.

Apart from the fact that the angle of flexion becomes gradually more atrophic, and that thus the symptoms—the dysmenorrhœa—increase, it seems also that the irritation *repeatedly* affecting the nerves renders them both more susceptible and perhaps alters them. At least there are cases in which finally the painfulness increases more and more and lasts always longer, without any change in the condition of the parts. Eventually the ovaries participate, ovulation and menstruation become irregular, the reflex neuroses of hysteria appear. The parenchyma of the uterus suffers by the irregularity of its functions; with the uterus the mucosa becomes affected and secretes excessively. The chronic metritis thus arising keeps up the nervous irritation, so that finally no time is free from pain, and the sufferings become intense. The continual pains, the feeling of being always sick, the impossibility of bearing exertions or enjoying pleasures, the disturbances of digestion caused by constipation, the mental depression due to the unsatisfied desire for conception, the dread of the pains of approaching menstruation, finally so reduce the system that the patient becomes a confirmed invalid, a useless member of society.

Should an incautious treatment lead to inflammations in the surroundings of the uterus, death may result.

Diagnosis.

The diagnosis of anteflexion is formed by careful combined examination. It is essential, in view of the treatment, to determine whether any disease in the neighborhood of the uterus complicates the anteflexion. Therefore the uterus is raised, drawn forward, pushed backward, and its outlines are most carefully palpated externally and internally. It is of special importance to learn whether the uterus is fixed posteriorly. Even an increased resistance in the surroundings or a sensitiveness at a certain spot or during a certain motion, may give rise to the suspicion of inflammation and furnish indications for treatment. Should there be exudations at the uterus which complicate the diagnosis, these exudations will be the first objects of treatment. It would be altogether erroneous, if there be neighboring inflammations, to use the sound to ascertain the direction of the uterine cavity. On the other hand, if we desire to commence intra-uterine treatment, sounding is necessary, if only for the purpose of ascertaining whether the ora are wide enough to permit the introduction of instruments. The gentlest way of sounding in anteflexion is that in Sims's lateral position. The anterior lip of the os is to be seized with a tenaculum and carefully drawn forward and downward. Then the sound is passed as near in the ascertained direction of the uterine body as possible.

Treatment.

The old dispute, whether the flexion or an eventual inflammation is the main point, is to be decided to this effect, that it is the object of treatment to remove whatever is pathological. Should we find anything pathological in the surroundings of the uterus, the peri- or parametritis ought to be removed by the methods to be described below. Should there be nothing pathological except the anteflexion, the latter will have to be attacked directly, and even if we had any result from it, the treatment would be purely symptomatic. Of course, with the causal, the symptomatic treatment will not be neglected from the beginning. Withal, we must never forget that a long-standing affection is rarely cured by any single measure, that on the contrary weeks and months will be required before there will be any result, that not rarely the tentative treatment will guide us in the right direction.

The plan of treatment is the following. In the first place, we seek to improve the general nutrition and to regulate the digestion. Not rarely the dysmenorrhœa is much less important after the constipation has been removed. The bowels should always be kept empty before and during menstruation by mild laxatives.

We then look to the removal of the complications, especially perimetritis and parametritis. Should the symptoms remain unchanged, we attack the uterus itself. If there be *hypersecretion*, we secure an unobstructed passage for the flow. The secretions may be mechanically removed with the uterine rod, or, if necessary, the external os is dilated. Should the cervical mucosa be greatly swollen, it is scraped off with a small sharp spoon, cauterized, and, if the vaginal portion be hyperæmic, the swelling is sought to be diminished by scarifications and painting of the vaginal portion with tincture of iodine, glycerin tampons, etc. Then we try to sound a few days previous to menstruation. The cases ameliorated by sounding are by no means rare. I should most urgently recommend the dilating sounding with my dilator No. 1. In a whole series of cases, I have, by a single or repeated introduction of this instrument before respective menstruations, completely cured dysmenorrhœa with uncomplicated anteflexion. The introduction is done in Sims's lateral position, the anterior lip of the os being fixed with a sharp tenaculum.

In these cases, perhaps, the swelling of the mucous membrane is possible only if the uterine secretion can pass away, when the mucosa, in its expansion, can force the mucus out of the uterus. And the first requisite for the cure of the diseased mucous membrane is the removal of the secretions. Furthermore, the premenstrual artificial stretching of the uterus facilitates the menstrual normal erection of the organ. After the stretching of the organ, however, the causes of the disturbed circulation are lacking, as well as the partial prevention of the swelling of the mucosa. Hence the entire process passes with less disturbance and without injurious consequences.

Should the succeeding menstruation be as painful as before, and should the case be equally uncomplicated, it is permissible to make the effect of the sounding permanent. As the sound cannot be allowed to remain, we insert an instrument corresponding to the uterine part of the sound—an *intra-uterine* or *stem pessary*. The latter consists of a stick of ivory, rubber, wood, or metal, four millimetres thick and six centimetres long, and is intended to maintain the uterus free from flexion. Figs. 97 to 99 represent such instruments. Those with flat bases, Figs. 97 and 99, are most useful. Figs. 98, Amman's stem pessary, can be employed only when the instrument easily glides down. Several pledgets of cotton laid in front of or behind the plate fix the pessary.

As there are dangers connected with the wearing of such instruments, this treatment is the last resort in anteflexion. The condition of the patient without treatment must be such as to justify a hazardous cure. In the same way as we decide on capital operations, so we are justified in instituting a risky treatment in grave disturbances of health. But it would be criminal to commence the hazardous method without the most urgent necessity, before all mild measures have been tried.

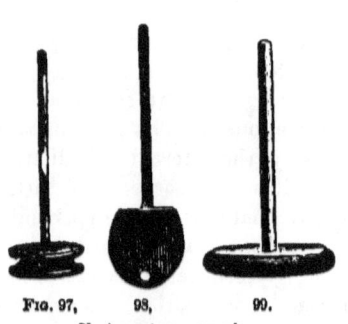

Fig. 97, 98, 99.
Various stem pessaries.

Having decided on the introduction of a stem pessary, the mobility of the uterus should first be tested. Existing inflammations, painful resistances, and fever contra-indicate the mechanical treatment. Should the uterus be mobile and the surroundings painless, sounding should be performed. Thereby the length of the uterus is ascertained. The pessary must always be one-half centimetre shorter than the uterine cavity, lest it cause a lesion by penetrating into the mucous membrane. For when the uterus becomes dislocated, for instance, during a difficult defecation, pressure may be exerted against the lower end of the pessary.

There are patients who react alarmingly even on the sounding; they scream with pain, they faint, and urgently request that all attempts in this direction be abandoned. This strong reaction, however, is by no means a counter-indication. But it calls for the greatest caution. The instrument may even be inserted in narcosis if it be too painful without. Especially where the vagina is narrow, these manipulations are so painful and exciting that we gladly anæsthetize.

The pessary is most readily inserted in the lateral position. After a most thorough cleaning of the vagina, the disinfected pessary is immersed in carbolized oil or vaseline. We then take a number of sounds, give them curvatures, and sound with successively straighter instruments; usually this is readily effected. After the almost straight sound has entered the uterus, we seize the stem pessary, with the tenaculum pull the anterior lip of the os strongly downward and forward, and allow the pessary to glide into the cavity. This method is by no means difficult of execution, much easier and less dangerous than the introduction in the dorsal position of the patient.

After the pessary is in place, it is watched for a while. Not rarely it gradually slips from the uterus. In such cases, the examination must subsequently be frequently repeated, lest the pessary, reaching merely to the internal os, lie in the vagina. Should the pessary be too long, and should the plate fail to closely adjoin the os, the stem must be shortened.

Should the pains continue in spite of morphine, or should they even increase in severity, the instrument is removed. The most scrupulous control of the discharge, of the external and internal sensibility to pressure of the abdomen, and of the temperature by the thermometer, are positively recommended.

Much more frequent are the uteri which bear the instrument well from the beginning, or soon become quiet. Even in this favorable case, the patient must remain abed three or four days, then at first be up or walk about but little, etc. Never should the patient be without medical control, even if there be no difficulties from the beginning. More especially should every patient be positively directed to send at once for medical aid should there be pain. The entire abdomen and the position of the instrument must be persistently examined from time to time. By no means should we rely on the statements of the patient, and especially should the removal of the instrument not be made dependent on the subjective sensations of pain. For there are women who bear high fever for days, in whom an exudation the size of a fist is formed, before they send for the physician! Nearly all untoward results are due to carelessness in the after-treatment.

During menstruation the instrument may remain. If no leucorrhœa arises, and if the condition of the patient continues good, the instrument is allowed to remain, the longer the better, for eight to twelve months. But should hemorrhage or profuse leucorrhœa set in, the pessary must be removed. We wait for some time, and introduce a second instrument only if the treatment was ineffectual thus far. Of course, all these manipulations must be strictly aseptic. Since this is the rule, accidents are hardly ever reported.

From the stem pessary, or from the removal of the flexion and the straightening of the uterus, we expect a higher degree of the result which I have ascribed to a single dilating sounding, p. 183.

The application of a stem pessary is always an experimental treatment. Despite the fulfilment of all the preliminary conditions and the *a priori* justified assumption that the instrument will be useful, the desired result is not always realized.

Should the symptoms, after an ineffectual trial of the treatment thus far described, be so grave as to justify a still more dangerous treatment, *castration* is indicated. After that, menstruation ceases completely and the case is radically cured. Whoever has for years treated cases which gradually went from bad to worse, in whom dysmenorrhœa is succeeded by hysteria, local hystero-neuroses, reflex neuroses of various kinds, and finally by a psychosis, will not doubt for a moment that castration is justifiable in such desperate cases. However, it is often impossible of performance. Inasmuch as in just these cases perimetritis complicates the anteflexion, it is downright impossible to remove the ovaries from the firm adhesions, in the unfavorable conditions of nulliparæ. We shall subsequently return to the subject of castration.

Besides the intra-uterine, vaginal pessaries are also largely employed. Pessaries have been constructed with the view to lift the anteflexed *body* directly. Although the pessaries lift the uterus *in toto*, an influence on the body alone is unfortunately impossible. Thus, in the pessary illustrated in Fig. 100, the upper spring part is usually firmly pressed down upon the lower, but there is no such thing as an elevation of the body of the uterus. On the other hand, it is correct that, when the uterus is elevated, the upper adhesions are no longer dragged upon, and that, therefore, one cause of the pains is removed. Thus many a case improves by a simple elevating fixation of the uterus from the vagina, that is to say, by the introduction of a small round rubber ring.

Fig. 100. — Anteflexion pessary.

Should the pains be so great that morphine must be given, the employment per rectum, in the form of suppositories, is to be most recommended: Morphiæ hydrochlor., Extr. belladon., ãã 0.01 to 0.015; Ol. theobrom., 2.0 gms. Fiant supposit. Two to three to be used daily.

B. ANTEVERSION.

Etiology and Anatomy.

If the uterus is so infiltrated, thickened, and hard that thereby the physiological angle of flexion at the internal os is obliterated, and if, moreover, some perimetritic process has fixed either the lower end of the uterus above, or the upper end of the uterus below, we have an *anteversion*. Accordingly, the uterine parenchyma, and its surroundings besides, must be changed pathologically. Therefore we shall encounter anteversions in perimetritis and metritis. But it is also possible that the metritis has run its course and almost healed, while the posterior fixation keeps the cervix permanently fastened high up. Thereafter the uterus may again have diminished. Usually, however, the uterus is found in a state of chronic metritis, *i. e.*, hypertrophic.

Not only a posterior, but also an anterior fixation causes anteversion. The tube and ovary of one side may be fixed laterally to the inner surface of the pelvis by peritonitic adhesions, so that the finger is unable to raise the fundus. However, a direct forward fixation cannot very well be assumed, owing to the interposition of the bladder between uterus and symphysis. With a sound bent at a right angle I have always been able, in front of the uterus, to get upward into the bladder. Should the posterior fixation crowd the vaginal portion upward and backward, and besides, should the full bladder, expanding between fundus and symphysis, push the uterus back, the vaginal portion will bend slightly downward, as represented in Fig. 101. At the internal os, however, corpus and cervix merge together in a straight line.

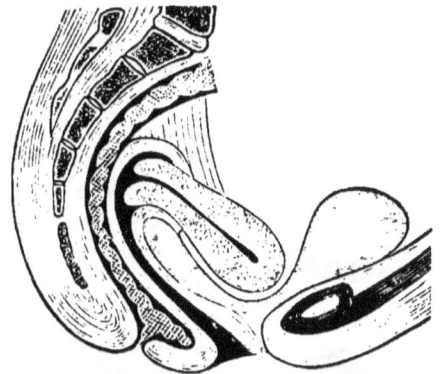

FIG. 101.—Anteversion, the result of abortion, metritis, and perimetritis. Chronic metritis. Adhesions in Douglas's fossa.

A fibroma may likewise cause a high degree of anteversion. One such case I examined post mortem (Fig. 102). During life the uterus was absolutely immovable and could not be sounded. At the autopsy no fossa of Douglas was found. The rectum, the sigmoid flexure, the atretic tube filled with blood, and the left ovary changed into a hæmatoma, together formed a coil cov-

FIG. 102.—Strong anteversio-flexion; myoma of the upper wall; total adherence of the posterior uterine wall. Nullipara.

ering the posterior wall of the uterus. Should, in such a case, the myoma be more prominent, the lower wall of the uterus has the same feel as in anteflexion.

Symptoms and Course.

As anteversion is always the sequel or accompaniment of metritis or perimetritis, the symptoms of these affections occupy the foreground. Peculiar to anteversion are vesical difficulties. It is readily understood that the heavy uterus, especially so during menstruation, prevents the expansion of the bladder. Aside from this, considerable menorrhagias occur. If the anteversion be recent and the uterus relaxed, the blood effused into the uterine cavity dilates the latter, and is expelled only after the uterus reacts on the expansion. Some long-continued hemorrhages are also observed; they are, like the other symptoms, to be referred to the metritis rather than the displacement. After the acute or subacute inflammations have run their course, the patient may feel absolutely well even during the persistence of the displacement.

Fig. 103. — Mayer's rubber ring.

Diagnosis and Prognosis.

The diagnosis is formed by combined examination. On feeling the flat uterus covering the pelvis like a roofing board, we ascertain whether the uterus is still movable, that is to say, whether the vaginal portion can be pushed downward and the body upward. In reference to the treatment it is necessary to carefully palpate also the surroundings to discover exudations. Should the complication be successfully removed, the dislocation *per se* does no harm. Hence the prognosis is favorable.

Treatment.

All the complications are to be treated. No special treatment is required for the anteversion. But should the uterus press heavily on the bladder, or should its weight drag upon perimetritic adhesions, a simple round rubber ring will often be of great service. The ring receives the vaginal portion in its lumen which is filled by it, and the vaginal portion is thus held captive. The ring is so placed that its surface lies parallel to the posterior vaginal wall or to the lowest part of the sacrum and coccyx. The inflexible uterine body, following the movements of the vaginal portion, must therefore be at a right angle to the plane of the ring, that is, be elevated. Fig. 103 represents such a ring; Fig. 104, forceps for the introduction of these rings, constructed by myself. The ring is first grasped in the forceps, then it is lubricated. Without uncovering the genitals of the patient, the ring is then introduced along the posterior vaginal wall until it can pass no further or causes pain. The forceps blades are then separated, the ring, immediately in front of the vaginal portion, opens, and draws the vaginal portion into its lumen by atmospheric pressure, forcing the forceps apart. The blades are removed singly.

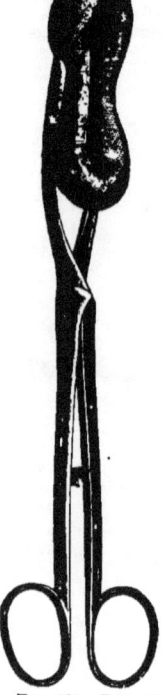

Fig. 104.—Fritsch's pessary forceps for the painless introduction of the rings.

DISLOCATIONS OF THE UTERUS. 191

If such a ring is to be painlessly removed, a curved sound is hooked *around* the ring from within, or a tenaculum *into* it. It is then slowly pulled out while a finger lies in the vagina.

Of course, rings can also be introduced and removed without these instruments. Only the method described is the most gentle.

We shall recur to the treatment of the hemorrhages under the head of retroflexion. Compare also p. 164.

C. Retroversion.

Etiology and Anatomy.

In retroversion, the uterine axis lies behind the axis of the pelvic inlet. Unfortunately, the importance of this anomaly of position is often undervalued. It is usually a transitional position, developing either into the normal anteversion, or into a retroflexion or a descensus of the uterus. The family physician will often have the opportunity of verifying these processes; while the specialist, who diagnosticates the fully developed displacement which causes many symptoms, more frequently finds an advanced stage, *i. e.*, a flexion.

In the puerperium, the uterus must lie in the same position as in the latter part of pregnancy. A dislocation of the muscular bundles, a stretching of the anterior, and flexion of the posterior wall, in the sense of causing a backward bending, is not imaginable post partum. The relaxed vaginal portion has no influence on the uterus. If it be too heavy, or if it be depressed, it may become somewhat retroposed in the *anteverted position* and descend. But it is different when the vaginal portion has become firm, in the late puerperium, after ten to fourteen days. Then the thickness of the uterus still permits no flexion at the internal os.

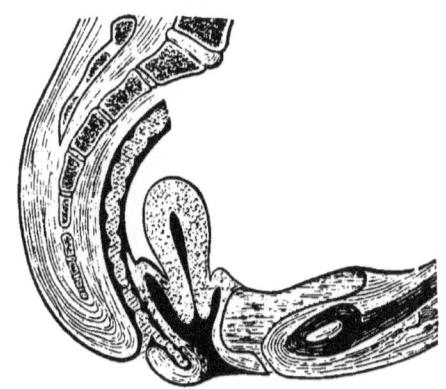

FIG. 105.—Late puerperal retroversion, incipient inversion of the vaginal vault, slight cystocele, perineal defect.

A movement of the cervix, therefore, must be communicated to the body. If the still voluminous portion glides backward, it will rest upon the inclined plane of the pelvic floor; there every pressure from above will force the vaginal portion downward and forward. In this way, the body moves upward and backward. But as there is no free space above, the resistance of the intestines depresses the fundus likewise. However, if the intestines rest upon the anterior surface of the uterus, they prevent the spontaneous replacement of the organ. During defecation, the vaginal portion is also pushed forward. But if the uterus has got into the pelvic axis, it has no support below to rest upon, hence it is only held by its superior fastenings; these—the peritoneal suspensory ligaments—are not firm enough, and the uterus sinks down.

This process is also set up after abortions, or in hypertrophy of the uterus with lax vagina, with increased intra-abdominal pressure.

Therefore the uterus lies as in Fig. 105.

Another form of retroversion was represented in Fig. 96. In that case, the uterus is fastened posteriorly by adhesions—retroposed. It has maintained its anteflexion. Not rarely, however, the body bends upward, becomes likewise adherent, and we find a retroposed, retroverted, elevated, completely fixed uterus.

Symptoms and Course.

The symptoms consist in bearing-down, a very annoying, though not painful, pressure on the rectum, a sensation of heaviness in the pelvis, incapacity for physical exertion, and constipation. In recent cases, hemorrhages are superadded. Often a very considerable hemorrhage occurs after the getting up of the puerpera, in consequence of the retroversion. More frequently the hemorrhage is slight, and is interpreted by the patients as commencing menstruation.

The hemorrhage continues often in a slight form for a very long time, simulating menstruation. Inasmuch as the retroversion cannot readily become spontaneously replaced in the recumbent position, and as the constipation has an injurious influence, the hemorrhage continues. Dysuria likewise ensues, brought about by the flexion of the urethra and incipient cystocele.

In the further course, anteversion often occurs spontaneously. Frequently it is quite sudden. But if the untoward influences persist, the retroversion is a preliminary stage of flexion or of prolapsus. If the fastenings are relatively firm and the body relaxed, the latter becomes flexed—retroflexion arises. On the contrary, if the fastenings are lax and the body firm, badly involuted, stiff, chronically inflamed, and heavy, the uterus descends, inverting the vaginal vault; the process of prolapsus begins.

Diagnosis and Prognosis.

Even the sensation of bearing-down and the pressure on the rectum point to the diagnosis. The same holds good for the hemorrhages. On exploration in the recumbent position, the soft, tumid vaginal portion is found covered with blood or mucus; the finger pushes the posterior vaginal vault high up on the posterior uterine wall. During combined pressure we notice the great mobility of the uterus up and down. The degree of descensus is diagnosed by examination in the erect posture. In the latter, the vaginal portion often lies close behind the introitus, surrounded by the folds of the relaxed vagina.

In retroposition with elevation, the direction of the vaginal portion and the absence of the uterus in the anterior vaginal vault permit of the formation of a diagnosis. With appropriate treatment, the prognosis is good.

Treatment.

In the way of prophylaxis, every lying-in should be carefully conducted; especially a woman who has previously suffered from retroversion should have daily astringent injections in the puerperium, should not always lie on her back, provide for daily evacuation of the rectum, and never retain her water too long. Furthermore, the puerpera must abstain from every exertion, and for six weeks she should, at least for one hour in the middle of the day, keep the horizontal position. Ergotin is to be also used, both subcutaneously and internally.

If the retroversion is present, a Mayer's rubber ring embracing the vaginal portion is to be inserted. Thereby the uterus is not only elevated, but also placed so that the intestines thereafter rest upon its posterior surface. This favors the regaining of the normal position. A small ring of from six to seven centimetres in diameter suffices. The patient need not remain abed; on the contrary, the hemorrhage ceases sooner during walking; nearly always the flow stops as soon as the uterus is in the normal position. A ring can be applied as early as two or three weeks post partum.

Later, when the vagina has become firmer, an oval Hodge pessary, bent according to the vagina, also does good service. These pessaries are made of tin tubing and hard-rubber, and owing to their smooth surface they cause but little irritation, and may remain in place for a long time. As a plaster dressing laid *around* a fractured limb replaces the *internal* support by an *external* one, so inversely the Hodge pessary lying *in* the vagina and stretching it, replaces the defective resistance of the vaginal wall. Strictly, therefore, the pessary is to assist the faulty function of the vagina. In a large number of cases, these easily applied and harmless pessaries stretch the vagina sufficiently, and raise the uterus so that it definitively resumes its normal position. The disappearance of the congestive hyperæmia proceeds *pari passu* with it.

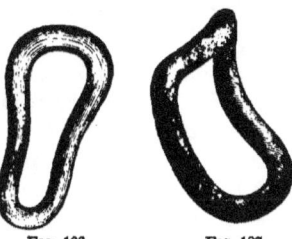

Fig. 106.　　　　Fig. 107.
Hodge pessaries.

Fig. 106 represents a pessary of tin tubing which may be given different forms by careful bending. Fig. 107 is a Hodge pessary of hard-rubber.

Hot injections may also be tried for the arrest of the hemorrhage. At times they are promptly effective. They should have a temperature of 40–43° C. (104–110° F.).

D. RETROFLEXION.
Etiology and Anatomy.

In the majority of cases, retroflexion arises from retroversion. The fastenings of the uterus do not yield *ad infinitum;* the intestines rest upon the anterior surface of the uterus; every increase of the intra-abdominal pressure adds to this weight. When the uterus involutes it becomes flexible, the upper half gradually bends more and more backward from the lower half. As there is still a certain amount of resistance at the angle of flexion, the vaginal portion follows the movement, that is to say, the more the fundus falls backward and downward, the more the vaginal portion glides forward and upward. Hence there is a stage of transition in which the uterus at times roofs over the pelvis as in anteversion, but in the opposite direction. The uterus at that time is often soft, so soft that it may be indented, that thin pessaries leave grooves within it, that the compression from behind—rectum—and from in front—bladder—renders the body shorter and broader. But gradually the fundus must sink more and more downward, the vaginal portion moving forward. Thereby the body, as it were, bores its way between the vaginal portion and the rectum, until the fundus rests upon the floor of the fossa of Douglas. The rectum is considerably compressed; the fundus can be felt from the rectum immediately above the anus. The vaginal portion is pressed closely against the symphysis, or is, as it were, pulled out of the vagina, quite short. If there be lacerations, the traction of the upper convex uterine wall draws the anterior lip of the os upward; the posterior lip lies in the direction of the short compressed vagina which now passes in front vertically downward (comp. Fig. 108). The vaginal portion may even be very short, while it appears extraordinarily long to the finger pushing the fundus of the vagina upward, after reposition of the uterus.

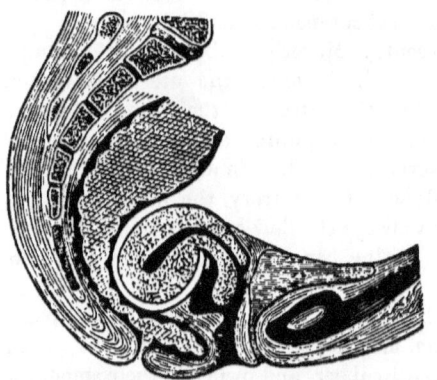

FIG. 108.—High degree of retroflexion; congestive hypertrophy of the body; long posterior, short anterior, lip of the os; compressed posterior, stretched anterior, vaginal wall; vertical direction of the vagina; compression of the rectum by the body of the uterus.

The ovaries usually lie laterally above the uterus. However, if they were previously descended and adherent behind, they may also lie between the uterus and the rectum, or in any other place.

In retroflexions, adhesions in the surroundings of the uterus are very frequent. Although there are not only recent but also very long-standing retroflexions in which the uterus is perfectly movable, still, in the majority of retroflexions, remnants of previous peritoneal inflammations are present in the surroundings of the uterus. These partly date from the puerperium, partly they are to be interpreted as the consequences of the chronic metritis, of the disturbances of circulation. But the flexion itself also causes adhesions of opposite peritoneal surfaces. The dislocation to which the hyperæmic body of the uterus is subjected by suddenly increasing intra-abdominal pressure, the pressure exerted by a large column of fæces, may lead to lesions of the peritoneum. Gonorrhœal perimetritides also play their part, as do the remnants of hæmatoceles and inflammations, the consequence of an over-vigorous treatment.

These adhesions at first only join peritoneum to peritoneum. Therefore, if the peritoneum is expansible, or if it can be displaced on its substratum, even a firm adhesion will permit a displacement of the uterus. Besides, the adhesions themselves are not always in planes or very firm, but they extend, as residues of former adhesions, like a spider's web from one organ to another. During the normal change of position of the pelvic organs, the pseudo-membranes stretch or tear as they do in pregnancy. On the other hand, as stated above on p. 181, the inflammation may propagate itself to the subperitoneal tissues; then there will be an absolute fixation.

Besides the retroflexions hitherto described, others, both congenital and acquired, occur in nulliparæ. Ruge found congenital retroflexion with abnormally thin uterine wall in a new-born child. We have stated above that the small infantile uterus often moves in the region of the internal os as on a hinge-joint. The excessively lax uterus, yielding to every external impulse, sometimes lies anteflexed, another time retroflexed. This extraordinary laxity of the uterus is said to arise also from masturbation or from continual attempts at coition by an impotent man.

In these cases, too, frequent sequels of the displacement are: congestion of the body of the uterus, perimetritis, ovarian inflammation, and fixation of the uterus in the fossa of Douglas.

Symptoms.

The symptoms depend, *first*, upon alterations in the uterus proper; *second*, they may be referred to the influence exerted by the dislocation upon surrounding organs; and, *third*, a number of consensual troubles and effects on the general health will have to be described.

The retroflexed uterus, previous to senile involution, is usually hyperæmic. Hence the menstruation will be copious. As the uterus is lax, the mucosa can swell without producing pain. A profuse, painless menstruation is the characteristic symptom of retroflexion. The hemorrhage has a variable character; sometimes it sets in with especial violence and lasts but few days; in other cases, it continues uninterrupted for twelve or fourteen days, so that the patient is free from hemorrhage for a short time only. The hemorrhage may even be permanent in a mild degree, and may increase during menstruation. After the hemorrhage, bloody mucus continues to escape for some time from the relaxed uterus. Should the angle of flexion lie on a higher level than the fundus, and should the parenchyma be extraordinarily yielding, the secretions even accumulate in the uterus, dilate it, and are expelled as offensive masses.

But in the case of an infantile retroflexed uterus of a nullipara, the symptoms of anteflexion, sterility and dysmenorrhœa, may continue and even become increased.

The symptoms from the side of the rectum consist in difficult and painful defecation. As in all female diseases in which perimetritis forms a complication, so in retroflexion the act of defecation will cause pain, owing to the displacement of the pelvic organs. The small amount of physical exertion, the dread of the painful defecation, the wilful retention of fæces due thereto, the loss of fluids during menstruation—all conspire to bring about constipation. The uterus is pressed against the rectum during increase of the intra-abdominal pressure. Many women distinctly state that they experience a disagreeable pressure upon the rectum after defecation.

Besides the symptoms which admit of explanation, there is a whole series of obscure phenomena occurring during retroflexion—symptoms hitherto comprised in the convenient term of consensual phenomena or hysterical symptoms. Although it has been repeatedly demonstrated that these symptoms disappear after replacing the retroflexion, that does not furnish the proof that they are to be referred solely to the retroflexion. The intimate connection of the ovaries with the symptoms which we call hysterical, gives rise to the conjecture that complicating ovarian diseases frequently furnish the cause. Even the mere pressure to which the dislocated, abnormally fixed ovaries are subjected, allows symptoms of irritation of the ovarian nerves to appear. And if we bear in mind that the altered nutrition, the hyperæmia of the ovaries, largely influences and modifies the processes during the maturing and rupturing of the follicle, we may be justified in referring the hysterical symptoms more to the ovarian troubles than to the retroflexion.

Such hysterical symptoms, with which, by the way, we shall become better acquainted in the description of hysteria, are: hiccough, vomiting, paroxysmal cough, facial and intercostal neuralgias, sciatica, hemicrania, pains in the abdomen and in the nerves of the extremities, trembling of the arms and legs, the latter symptom having been known to the ancients under the name of *anxietas tibiarum*. The dyspepsia and the chronic indefinable stomachic troubles (hystero-neurosis ventriculi) also belong to this category.

Finally it may be stated that the retroflexion may lead to the interruption of pregnancy. Should a deeply seated, retroflexed uterus be impregnated, it grows and often rises out of the pelvis, but, on the other hand, in its growth it also fills the lesser pelvis in such a manner that it cannot get above the promontory. Then the uterus is hindered in its growth, disturbances of circulation, congestions, and decidual hemorrhages occur which lead to abortion. The uterus also presses upon the pelvic organs. The urethra is crowded against the symphysis pubis, causing retention of urine. Often this is the first and most important symptom of incarceration of the retroflexed, gravid uterus. If replaced in time, the pregnancy continues. Otherwise, abortion occurs. Very grave vesical diseases may also form (comp. p. 107).

The general health suffers greatly from the hemorrhages, disturbances of digestion, and the hysterical symptoms. Not a few women having acquired a retroflexion after the first labor or an abortion, are invalids from that time forth. The weakened organism loses the power of resistance. Should additional bad influences be superadded, or should there be a disposition to phthisis, the retroflexion with its consequences may be dangerous to life.

Diagnosis.

We have mentioned already under the head of symptoms that the sensation of pressure in the pelvis, constipation, and hemorrhages point to retroflexion, especially in the case of women who have frequently borne children or have had abortions.

The direct examination demonstrates the displacement: the vaginal portion is felt quite in front at the symphysis, and from it we can follow the uterus backward. The latter is usually somewhat enlarged, and has so characteristic a form that a mistake is not easily made. On either side above, we not rarely feel one or the other ovary. As most patients are women who have borne children, I urgently advise to examine always with two fingers; if the fingers are spread above, the vulva of course is not injured, and during the spreading of the fingers the margins of the uterus can be best palpated, and the organ pressed against one finger with the other. Combined examination is always necessary. Although the abdominal coverings are often so firm or thick that the uterus cannot be felt from the outside, still the approximation of the finger tips in the anterior halves of the pelvis proves the absence of the uterus from its normal position.

Of course, any tumor filling the fossa of Douglas may be mistaken for the normal or the enlarged uterus. Should a tumor of that nature be very painful, the minute diagnosis should be desisted from, if there be no symptoms which urgently demand treatment. The treatment should be expectant, while a result is sought to be obtained by repeated careful examinations. Nothing is more erroneous in gynecology than the attempt to exhaust the diagnosis in all its details *on principle* during the first examination. We might do much injury by too protracted reckless examinations. For instance, should the uterus be anteposed by an exudation posterior to it as in Fig. 109, but a diagnosis be made of retroflexion of the uterus, and a reposition perhaps even be forced owing to an assumed gravidity, the greatest dangers may result. Adhesions may be separated, encysted exudations may penetrate into the abdominal cavity, or at least an inflammation or a relapse may be caused.

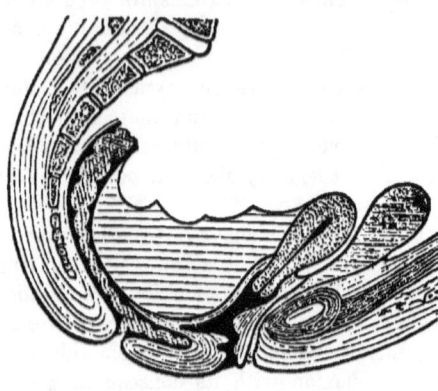

FIG. 109.—Intraperitoneal exudation anteposing the uterus.

Hence caution and care are demanded. Should there be sensibility, we must cautiously try to ascertain *what* is sensitive. Gently and by turns the fingers press in every possible direction, palpate every attainable point in the pelvis, push the uterus hither and thither, search for tumors and resistances in the immediate neighborhood of the uterus and in the entire pelvis. In this way it is possible to find where there is any inflammation.

But if the examination is not painful, we test if the uterus is adherent, that is to say, whether it is replaceable. For this purpose quite a number of different manipulations are possible and necessary. At first two fingers are forced so high up behind the uterus that, stretching the latter, it and the whole pelvic organs are, as it were, pushed and lifted out of the pelvis into the abdominal cavity. Now one and then the other finger passes from the posterior vaginal vault into the anterior, pressing the vaginal portion backward and upward. At the same time the extended fingers of the other hand, pressing from the umbilicus toward the posterior pelvic wall, crowd the intestines between the uterus and the posterior abdominal wall, in that way displace the intestines and thus reach the posterior wall of the uterus. Then the hands below and above lie immediately at the uterus and hold it in almost normal position, as represented in Fig. 110.

Should this replacement succeed easily and painlessly, should the position of the uterus remain normal after removal of the hands, as shown by repeated examination with the patient in the recumbent and the erect posture, no adhesions are present.

But if the entire manipulation causes pain, and the former position returns immediately after reposition, the uterus is held posteriorly by peritoneal adhesions. It may happen, too, that the replacement of the uterus fails altogether. Then the knee-elbow or Sims's lateral position may be employed. The former has many advantages over the latter. In the knee-elbow position the reposition is attempted; should the uterus, after removal of the finger, be again drawn up, spontaneously retroposed, posterior adhesions are present.

The sound may also be used to advantage. At first it is inserted with a curvature corresponding to the retroflexion, that is to say, the sound is governed by the position of the uterus, not the reverse. Then the hand raises the slightly stretched uterus and places it retroverted or cen-

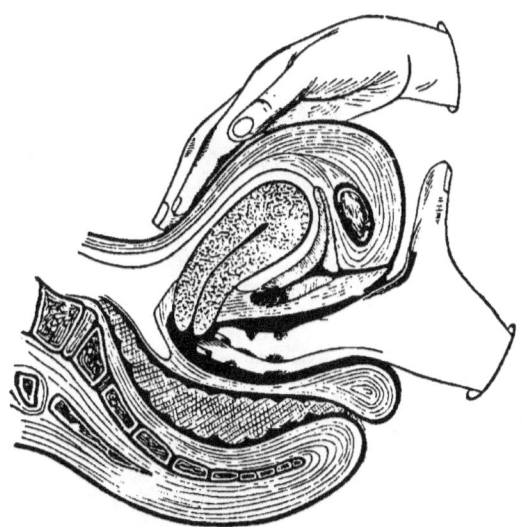

Fig. 110—Successful reposition; also position of the hands in combined examination.

tral; should during the slow withdrawal of the sound the uterus immediately drop again upon the tip of the finger held in the posterior vaginal vault, the uterus is likewise adherent.

Finally, total adherence exists if any movement of the uterus is absolutely impossible.

In retroflexion the prognosis depends upon whether the patient is treated in time and correctly. It is entirely beyond doubt that by timely replacement it is possible to restore the normal position, that the folds of Douglas become shortened, and that the uterus permanently remains in an anteverted position. In general, therefore, the retroflexion is curable. But if the folds of Douglas and the peritoneal fastenings of the suture are relaxed by long standing of the flexion, it is possible to hold the uterus in its normal position only by vaginal supports. Even then the treatment will mitigate the suffering. Therefore, improvement may be attained in every case.

Treatment.

Should a retroflexion be found with symptoms which may be interpreted directly or indirectly by the retroflexion, the indication is to restore the normal position of the uterus. Should this fail at first, we attempt as far as possible to remove the obstacles to the replacement. Having become convinced that the normal position produced is permanent, the uterus is artificially fixed in its normal position. If the normal position cannot be restored, or should the symptoms remain the same in the normal position, or should they even become worse, we, for the time

FIG. 111.　　　　　　FIG. 112.　　　　　　FIG. 113.

FIG. 111.—Thomas's pessary. The pointed lower end unfortunately allows the pessary to glide easily into the vulva, where it incommodes.
FIGS. 112, 113.—Figure-of-eight shaped pessaries, after Schultze.

being, desist from rational or mechanical therapeutics and treat symptomatically.

The method of replacement is described on p. 198. From time to time we come across a fortunate, generally quite recent case, in which the uterus remains anteverted after a single reposition. In order to test this, the patient is made to rise after successful replacement, to strain, walk about, and is then ex-examined. This is repeated after a few days, weeks, and months. In most cases the retroflexion returns already when the replacing fingers leave the vaginal portion, or when the patient descends from the examining table.

Then the uterus must be fixed in the normal position. In flexions of not very long standing, this is possible by the mere stretching of the vagina. Therefore all pessaries which answer this purpose in any manner are appropriate, such as those of Hodge (p. 193), and that of Thomas (Fig. 111). With these pessaries, often *the fundus of the vagina alone* stretches so greatly that the vaginal portion again glides forward, the uterus again becomes retroflexed, and despite the stretching of the vagina the old pathological condition is once more present. It is necessary, therefore, especially if the symptoms recur, frequently to control the position of the pessary and the uterus.

In dealing with recent cases, often one or two weeks suffice to reaccustom the uterus to its normal position. In the majority of cases, however, we do not succeed in this simple manner. *It will be necessary, not only to stretch the vagina, but also to fasten the vaginal portion posteriorly above.* For this purpose we employ Schultze's figure-of-eight pessaries, of which two forms are represented in Figs. 112 and 113.

These pessaries, first recommended by Sims, consist of very flexible copper wire coated with rubber. They are marked with numbers designating the width of the ring in centimetres. Nos. seven to twelve are the numbers required.

When such an instrument is to be introduced, we first ascertain how much the vagina can be stretched, how large the vaginal portion is, whether and where there are in the vaginal vault any resistances, tumors, and painful spots, and the width of the pubic arch. The length of the pessary and the size of the ring are measured accordingly. The upper ring must be bent to fit the vaginal portion, that is to say, it must be large enough to admit that part. The lower ring must be broad if the pubic arch is wide; if narrow, the diameter may be less. In nulliparæ, indeed, the figure-of-eight pessary maintains itself in the vagina, nowhere resting on the bone, only stretching the vagina and on all sides fixed or borne and retained by it. For this reason in nulliparæ the pessary may be much smaller (No. 7 or 8).

But in women who have borne many children and have a relaxed vagina, if we wish to be sure that the pessary will permanently answer its purpose, a considerable stretching of the vagina is certainly necessary. In that event the pessary, crowding the relaxed levator against the bone, rests upon the pubic arch.

The pessary having been bent appropriately as in Fig. 112, and the uterus replaced, the instrument is inserted into the vagina obliquely, so as to avoid the urethral eminence. The upper ring is kept as close as possible to the posterior vaginal wall. After the pessary has passed into the vagina, it is moved toward the centre, and the vaginal portion is caught in the lower ring by lever motions. Then the touch is practised below the pessary to ascertain whether the vaginal portion is in the upper ring, and by rectal examination we make sure that the fundus of the uterus cannot be felt, *i. e.*, that the uterus has not perhaps remained retroflexed and only been elevated *in toto*.

Rectal exploration is especially important when the narrowness of the vagina makes a vaginal examination impossible by the side of or below the applied pessary. We then test whether pains are caused by pressing the lower ring upward; whether the patient, without being incommoded by the pessary, can sit down, stoop, rise, and walk about. If these various movements can be performed without difficulty and if the pessary has not been displaced even by a defecation, it may be allowed to remain —"it fits." Still the examination must be repeated after a few days.

If the pessary causes pain, it is removed and a most careful exploration made in order to find the sensitive spot. Then a smaller instrument is chosen, or the former instrument is bent so that no pressure is exerted on the sensitive spot. Often we may succeed by bending the pessary according to the vagina, *i. e.*, concave above; or by twisting the upper ring laterally, or by pressing the instrument somewhat shorter, or making the upper ring smaller—in short, we must try many times in order to find a fitting form.

Frequently the retroflexion returns immediately after removal of the replacing finger; the anteversion failing to remain even when the body of the uterus is fixed from the outside. In that case the introduction of the pessary is to be attempted in the knee-elbow or the lateral position. Should that method, too, fail, I have several times employed the following procedure: The uterus is replaced *manually* while a flexible copper sound is within it. The sound is gradually bent at a right angle *during the reposition* of the uterus, that is to say, the index and middle finger fix, *in the vagina*, that part of the sound close beneath the vaginal portion, and the external hand lifts the handle strongly upward. Thus the sound is bent at a right angle. Then the uterus is completely replaced *manually*, and the smaller ring is pushed over the handle of the sound, so that, in gliding upward, it must receive the vaginal portion. Not until we have thoroughly convinced ourselves of the correct position of the ring and the uterus, is the sound carefully removed. This procedure is not quite easy, but it is certain and free from danger, according to my experience. Slight lesions of the mucosa are possible, but are harmless when antisepsis is employed.

After menstruation, disinfecting irrigations are given for two or three days in order to remove the blood at and behind the ring. Should leucorrhœa arise, a *new* ring is inserted after the vagina is cleansed.

If the pelvic floor is very lax, and the correct application of the figure-of-eight pessary does not succeed, the sleigh pessary represented in Fig. 114 will be of service.

In this pessary the vaginal portion lies between the two arches projecting upward. The uterus is elevated in toto, the flexion is obliterated, so that the uterus, as it were, stands in the pessary.

In recent cases, the pessary is removed on trial after the lapse of a single menstruation. In long-standing cases, the pessary remains until an obvious improvement of all symptoms or pregnancy has occurred.

Not rarely a cure, *i. e.*, a recontraction of the uterine ligaments, is impossible, and the retroflexion immediately returns even if the instrument has been worn for years. Nevertheless the pessary may often be definitely removed even in these cases, because in the course of treatment the uterus has become so small and all complications have been so far removed, that thereafter the retroflexion continues *without symptoms*. Should a relapse of the complications ensue, a renewed treatment will be necessary.

Should there be an active perimetritis, of course any orthopædic treatment is to be abandoned for the time being, the inflammation of the pelvic peritoneum being first attended to.

Should the replacement be impossible, owing to tenderness on pressure of the *body of the uterus alone*, we make a *very cautious* attempt at reposition with the sound. There are not a few cases in which the reposition with the sound is easy and quite painless, while pressure on the posterior vaginal vault at once provokes paroxysms of pain and straining. Of course this attempt has to be at once abandoned if even the sounding of the retroflexed uterus is in any way painful. I have performed reposition in this manner a countless number of times without injurious consequences. The dangers of elevation with the sound have been exaggerated.

Fig. 114.—Schultze's sleigh pessary.

If the reposition fail, it will be advantageous to insert a soft Mayer's ring pessary. Although it cannot keep the uterus in place, it lifts up the entire pelvic contents without exerting a one-sided pressure at any spot. Thereby the traction is diminished which the uterus exerts on its ligaments, as well as the pressure on adjoining organs. I have repeatedly witnessed a contraction of the relaxed folds of Douglas, and the occurrence of anteversion, by this simple treatment. Even if this be not the case, the vagina and the uterus are prepared for subsequent orthopædic treatment, and the pains are diminished for the time being.

After a former perimetritis, firm adhesions, and consequent inability of replacement, Schultze has proposed and performed a new procedure. The uterus is dilated, one finger is inserted up to the fundus, the finger is bent forward, the other hand loosens the adhesions by stroking down the posterior part of the uterus which is then fixed in its normal position by an appropriate pessary. This method is feasible only with very lax abdominal walls. The greatest caution is required, for undue force may lacerate the rectum or cause severe internal hemorrhage.

If it be absolutely impossible to place the uterus correctly, and if there be symptoms calling for interference, *symptomatic treatment* is to be instituted. Its object will be the amelioration of perimetritis and metritis. The treatment, however, will be more particularly directed against the symptoms of pressure and the hemorrhages. Long-continued use of vaginal injections, baths, laxatives, and aperient waters (p. 138) are required. For the hemorrhage I employ two methods with good results. *First*, ergot is used for two or three months, five grammes made into sixty pills, five to be taken morning and night. During the hemorrhage the pills are discontinued. After prolonged use, menstruation usually occurs one or two days later and is less prolonged. *Second,* laminaria tents are introduced into the uterus from time to time, liquor ferri is injected after twelve hours, and at the same time ergotin (1 : 8), one or two syringefuls, given subcutaneously. The marked decrease in size thereby produced in a relaxed hypertrophic and hyperæmic uterus often has a diminishing influence on the hemorrhage for two or three menstrual periods. This treatment may also be instituted when the uterus remains large in its correct position. Then we really have a chronic metritis, the treatment of which has been discussed on page 158.

E. Prolapsus of the Uterus.

In this section we shall consider descensus of the vagina together with that of the uterus, dividing the various forms as follows:

1. Isolated descensus of the vaginal wall,
  "   "   "   anterior wall—cystocele;
  "   "   "   posterior " —rectocele;
  "   "   of both walls.

2. Primary descensus of the vaginal wall with descensus of the uterus.

3. Primary descensus of the uterus with inversion of the vagina.

4. Prolapse of the uterus by increased pressure from above, traction from below, or lack of the physiological supports.

1. *Isolated Descensus of the Vaginal Wall.*

During ever labor the advancing head presses the anterior lip of the os downward, compresses and shortens the anterior vaginal wall. This is particularly the case with rigidity of the os uteri, premature escape of the liquor amnii, and swelling of the anterior lip of the os. This shortening can not rarely be demonstrated—before and a short time after labor we see immediately behind the vulva the bluish-red lip of the os when separating the labia. Of course the anterior depressed vaginal wall is loosened from all its attachments. As the vagina is considerably shortened, as retention of urine is of frequent occurrence in the puerperium, and as the heavy puerperal uterus likewise inserts a large segment into the pelvic inlet, this descent is not obliterated. It remains, not only in the first few days of the lying-in, but forever. Only when the involution of the uterus, and that of the peritoneum which lifts and fastens it, progress favorably, when at the same time the vagina again becomes rigid and firm, contracts yet elongates, the vaginal portion rises up again and the anterior vaginal wall once more becomes tense. But nearly always a slight descensus of the anterior vaginal wall remains behind.

Rupture of the perineum is also of etiological importance. If the vagina be severed behind, the front wall is crowded together. Then there is too much material in front, as it were; the anterior wall relaxes and may descend. But, behind, the normal support for the anterior wall is lacking. While normally (comp. Fig. 6, p. 12) the anterior rests upon the horizontal posterior wall, in perineal rupture the anterior wall may glide down the posterior wall which now runs vertically, and may descend unhindered into the introitus (comp. Fig. 115).

The involution of the vagina is also of importance. Especially the rugous column and the urethral eminence often remain extraordinarily thick and hyperæmic. The median portion of the vagina depends isolated, like a tag, into the introitus, while immediately alongside the fastenings are intact. Of course, the bladder closely adjoins the descended vagina.

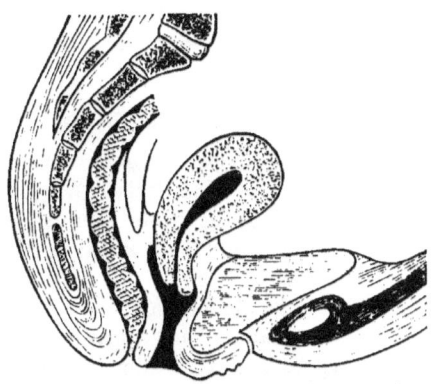

FIG. 115.—The uterus at the end of the puerperium. The posterior vaginal wall has been destroyed by a perineal rupture and runs vertically. The anterior vaginal wall, deprived of its support, descends, hypertrophies, and drags the uterus after it; beginning prolapse.

In most cases, congestion, hyperæmia, and hypertrophy occur in the descending vaginal wall. The latter becomes thick and heavy, hence sinks down and drags the uterus after it. However, though rarely, some isolated descensus of the anterior vaginal wall occurs—prolapse of a cystocele, in which the extraordinarily thin, dry, anterior vaginal wall projects even in front of the vulva without alteration of the position of the uterus. *This would be an isolated prolapse of the anterior vaginal wall with cystocele.*

The lower part of the posterior vaginal wall may also project into the vagina. Not rarely a perineal rupture lacerates the perineum in such a manner that the fissure extends up the vagina on one side of the rugous column. The latter may hang like a tag into the introitus, in the same way as the urethral eminence in front. Sometimes, though by no means always, the rectum participates in the descensus. Then we have *vaginal rectocele*. In that form, too, a prolapse of the vagina and rectum in front of the external genitals may occur without any participation of the uterus. Of course, this is much rarer than the formation of a cystocele. In the formation of a rectocele, a gradual expansion and protrusion of the rectum by retention of fæces is of etiological importance.

Thus far we have considered the *lower half* of the vagina. The upper walls, too, descend primarily in very rare cases. For instance, it has happened that, under pathological conditions, the bond between bladder and uterus becomes dissolved, and the vesico-uterine excavation prolapses into the vagina. The fossa of Douglas may also be abnormally dilated; then the interspace between the two folds of Douglas forms, as it were, the hernial opening through which intestines or pathological contents (ascites, thin-walled ovarian tumors) enter the fossa of Douglas, expand it, and force it into and in front of the vagina. If intestines are present in the descended vesico-uterine excavation or in Douglas's fossa which are to be diagnosticated from the vagina, we have an *enterocele vaginalis anterior* or *posterior*. Both are extremely rare, perhaps even doubtful cases.

It is obvious that prolapse of the anterior may combine with that of the posterior wall. In that case, however, the lateral connection of the vagina with the underlying tissue is generally intact. Only when the uterus is likewise prolapsed, will there be a total detachment of the vagina with complete prolapse of all its walls.

2. *Descensus of the Vaginal Wall with Descensus of the Uterus.*

In the great majority of cases, the vaginal prolapse is merely a stage of the combined uterine *and* vaginal descensus. The displaced vagina drags the uterus after it, or prevents the reascent of the uterus or its elevation by the involution of the peritoneum. The vaginal wall remains extraordinarily thick; perhaps because the descensus leads to congestion by traction on the vessels and impeded return-flow; perhaps because the slight pressure in the roomy, relaxed, gaping vagina leads to hyperæmia in the descended vagina, according to the same law in which, for instance, a swelling of the head occurs. In this hyperæmia the vaginal portion or the lower segment of the uterus must participate. The cervix is hypertrophied both circularly and longitudinally, that it is to say, the vaginal portion becomes extraordinarily thick, the cervix surprisingly long.

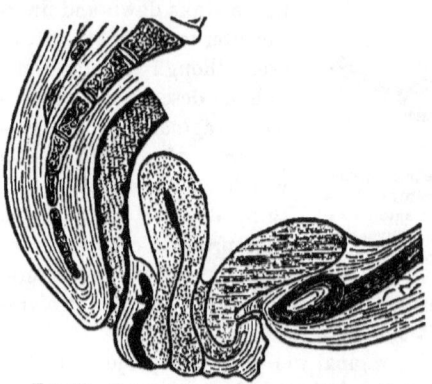

Fig. 116.—Prolapse of the anterior vaginal wall; descensus of the uterus; posterior vaginal vault still present. (Hypertrophy of the median portion of the cervix.)

Gradually the anterior vaginal wall sinks so low that, completely everted, it lies below the external urethral orifice. The posterior vaginal fornix may be still present. In that condition the uterus must be retroverted, lying in the direction of the pelvic axis, as represented in Fig. 116.

If the process advances unchecked, the anterior vaginal wall will naturally make traction on the posterior. Thus the posterior vaginal wall is dragged down and by degrees sinks lower and lower. Once in front of the vulva, the effect of the congestion on the entire mass outside of the pelvis must become much greater than before. Especially the circular hypertrophy and the thickening of the vaginal walls will be considerable. This condition we have illustrated in Fig. 117.

3. *Primary Descensus of the Uterus with Inversion of the Vagina.*

Finally, if we ask ourselves whether there are, besides, any etiological factors attributable to the uterus which favor or impede the occurrence of prolapse, we must first enumerate the general loosening of the fastenings of the uterus and the position of the uterus.

After the puerperium, the peritoneal attachments must involute in the same way as the uterus. Should this involution fail to occur, should the peritoneum remain hyperæmic, relaxed, and its superficial extent too large—which again is connected with the involution of the uterus — the peritoneal ligaments will be unable to keep the uterus elevated, it will sag down, descend. Or if the uterus is abnormally heavy and large, the normal supports will not suffice whenever the position of the uterus is such that it can glide downward. In the normal position at most an increased anteversion can result (comp. p. 191). But when there is retroversion, the uterus easily slips downward, as has been explained on p. 191, Fig. 105.

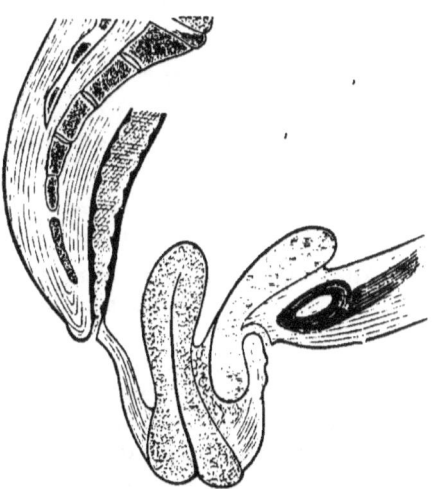

Fig. 117.—Total prolapse of the uterus and of both vaginal walls; great hypertrophy of the vaginal portion and of the vagina at its point of attachment to the uterus; circular hypertrophy.

It is even possible that the last factor, the retroversion, is the most important. Then the descending uterus first inverts the fundus of the vagina, unrolls the vagina, as it were, from the underlying tissue, and sinks in front of the external genitals without the occurrence of any descensus of the anterior vaginal wall with consecutive hypertrophy of the vagina and the cervix. That would constitute a primary prolapse of the uterus with secondary prolapse of the vagina. The result of this process is depicted in Fig. 118 on the following page.

Of course, there are cases—and they constitute the majority—in which all unfavorable factors (perineal rupture, lax vagina, primary formation of cystocele, retroversion, and relaxed peritoneal attachments) concur in bringing about the unfortunate result—the genesis of prolapse.

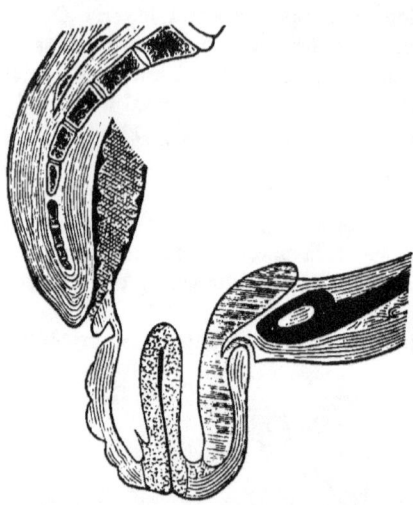

More rare is the opposite condition — that, despite all favorable elements, the fixation of the uterus in an anteverted position alone prevents the occurrence of prolapse. In a similar way, the persisting normal position of the uterus may retard the process of prolapsus, or a possible peritoneal inflammation during that time occasionally fixes the uterus so that the process remains interrupted and a *stage* merely continues permanently.

FIG. 118.—Primary Prolapse of the Uterus after retroversion, without hypertrophy of the cervix.

4. *Prolapse of the Uterus by Increased Pressure from above, Traction from Below, or Absence of the Physiological Supports.*

Although that may have been the etiology in the majority of cases, the occurrence of prolapse in nulliparæ proves that other causes may exist.

Thus, though rarely, prolapses occur quite suddenly in lying-in women and convalescents from grave diseases. During pregnancy, as in protracted diseases, the fat in the neighborhood of the uterus is absorbed; in consequence thereof the internal genitals are loosened at their attachments, and should a favorable position of the uterus be superadded, together with great sudden straining, the uterus may at one bound glide in front of the external genitals.

Excessive dilatation of the vagina by too large pessaries, likewise, stretches and loosens the vagina to a degree that from a slight descensus of the vagina prolapsus may arise. Furthermore, a growing abdominal tumor crowds the uterus down, or a swelling springing from the uterus drags it in front of the vulva. In fissured pelvis with gaping symphysis, the uterus finds no support, it descends when the soft parts have become loosened by a labor, and finally prolapses.

Anatomy.

In describing the anatomy, we must consider the uterus, the vagina, the peritoneum, the rectum, and the bladder.

The *uterus* is usually hypertrophied, especially in its lower, extra-peritoneal part, that is to say, from the internal os down; there is hypertrophy of the cervix. The cervix, moreover, is often so greatly elongated that it lies far in front of the external genitals, although the fundus of the uterus is at its normal height. The uterus, as it were, hangs on the anterior vaginal wall which reaches from the pubic arch and the symphysis as far as the vaginal portion. While formerly the vagina dragged down the uterus, the latter is now held by it and by its slight dilatability further descent of the uterus is prevented.

These conditions are best understoood by a schematic drawing made by Schroeder.

If *a*, the vaginal portion, is hypertrophic, we have a probosciform portion (comp. p. 139, Fig. 78, and p. 180, Fig. 94), while the vaginal vault, of course, remains in its normal position in front and behind—the condition having nothing in common with the cases of prolapse here under consideration. If *b*, the median portion, is hypertrophic, we have the condition represented in Fig. 116, p. 206, a stage in the occurrence of prolapse. If *c*, the supravaginal portion, is hypertrophic, we have the state shown in Fig. 117, p. 207—total prolapse.

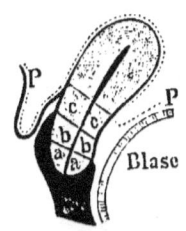

Fig. 119.—Schroeder's division of the cervix for the elucidation of the anatomy of prolapsus. *a*, Vaginal portion; *b*, median portion; *c*, supravaginal portion; *p*, peritoneum.

This anatomical division is not to be understood as meaning that the cervix hypertrophies *primarily* and displaces the attached parts *secondarily*. Should the supravaginal part of the uterus hypertrophy primarily, it would have to grow into the abdominal cavity above, as in chronic metritis, the development of tumors, or in pregnancy. On the other hand, should the uterus descend primarily, there being an absence of the etiologically important transition stage of descensus of the vagina, congestion, and hypertrophy, then there is no cervical hypertrophy (comp. Fig. 118, p. 208).

Subsequently the uterus may undergo senile involution, but usually the influence of the hypertrophy present during the genesis of a prolapse is so persistent that even in the sixtieth and seventieth year of life the uterus is still extraordinarily large.

If the cervix is considerably hypertrophied, we always find the adjoining portions of tissue of increased dimensions. Thus the mass of connective tissue fastening the bladder to the uterus is considerably elongated in Figs. 116 and 117—a condition favorable to the extirpation of the vaginal portion, because the point of reduplication of the peritoneum is at a much greater distance from the vagina than usual. On the other hand, in cases in which the non-enlarged uterus is totally prolapsed, the distance from the vagina to the vesico-uterine excavation is very short (Figs. 118, 120).

If the vaginal portion was irregular before the prolapse occurred, deeply lacerated for instance, the prolapse will likewise have an irregular form. The cervical canal may be exposed completely separated, gaping, *i. e.*, everted.

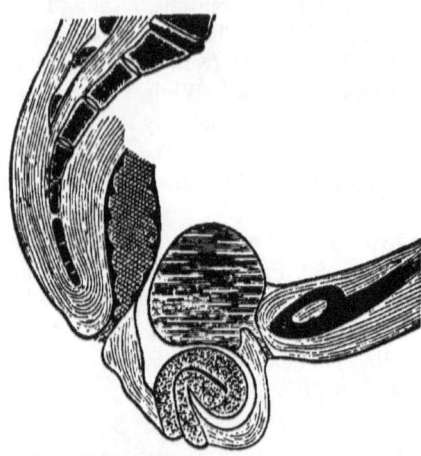

FIG. 120.—Anteflexion of the uterus with prolapse, complete separation of the bladder from the uterus. Bladder and rectum lie above the posterior surface of the uterus where they meet.

At times the uterus is flexed. Either it is so relaxed that the kind of flexion is accidental, or else a flexion existed before the occurrence of prolapse and continued subsequently. In anteflexion the fundus of the uterus penetrates between bladder and cervix, thus completely severing the bladder from the uterus (comp. Fig. 120).

Retroflexion is also possible. Fig. 121 represents that condition, together with considerable rectocele, after Freund.

Finally the uterus may be inverted, that is to say, there is a prolapse of the inverted uterus.

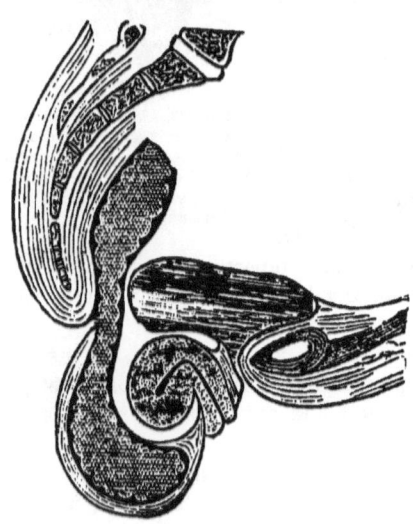

FIG. 121.—Retroflexion of the uterus with prolapse, considerable rectocele.

During sounding we not rarely find an atresia in the cervical canal, usually in the region of the internal os, but I have found it also much more deeply seated. For the occurrence of the atresia an endometritic process is probably necessary, see p. 142.

The *peritoneum*, which is very elastic, is dragged along by the uterus. While the uterine ends of the folds of Douglas are usually distinctly present (Fig. 118), they cannot be followed toward the parietal peritoneum.

The broad ligaments are also stretched on both sides. The peritoneum is not loosened beyond the innominate line upward. The round ligaments play no part. But while behind, generally close above the vaginal attachment, the peritoneum, the fossa of Douglas, is present, the vesico-uterine excavation is not deepened. But, as has been described, as the hypertrophic part of the cervix inserts itself between the vaginal portion or the vagina and the body of the uterus, the peritoneum must remain at its old place, that is to say, turn back on the bladder at the internal os above the cervix. Assuming the length of the normal uterus at seven centimetres, and that of the prolapsed at twelve centimetres, we shall not err if we count at least four and a half of the five centimetres on the cervical hypertrophy. *Therefore, the vaginal attachment would be four and a half centimetres distant from the vesico-uterine excavation.* But if the uterus prolapses primarily, the peritoneum is close above the vaginal attachment, in the absence of any cervical hypertrophy. In that case the sound would also demonstrate but a slight length of the uterus.

The parts of the prolapse exposed externally swell œdematously in acute cases. In prolapses of gradual occurrence the lower surface likewise often increases materially, so that the vaginal portion, measured from the anterior to the posterior vaginal attachment, may be from six to eight centimetres in diameter. This is termed circular hypertrophy.

Generally the parts of the vagina in front of the vulva lose their characteristic surface, the mucosa becomes dry, leathery, hard. As the uterine secretions no longer moisten the vagina, this is quite explicable. At the swollen, somewhat œdematous portion, ulcers, solutions of continuity, and losses of substance in the skin not rarely arise. Even a fistula has been produced by such a rodent ulcer at the vesico-vaginal wall.

The entire vagina is considerably thickened, especially at the uterine attachment (comp. p. 207, Fig. 117).

The *rectum* is not so intimately connected with the vagina as to be forced to follow the descending posterior vaginal wall. On the contrary, even in total prolapses, the rectum is usually found in the correct position, the vagina being, as it were, unrolled from the rectum. Should the rectum sink in front of the external genitals with the vagina, as in Fig. 121, a part of the rectum will be below the anus; there the fæces may stagnate, the liquid portions are absorbed, and firm scybala remain. In moderate rectocele, however, the rectum contracts during defecation and expels the contents, although with difficulty.

The *bladder* follows the anterior vaginal wall, not because the connection between them is very intimate, but because the intra-abdominal pressure keeps the bladder close to the vaginal wall, and because there are no ligaments above holding the bladder up. But as the lower half of the urethra *is* firmly attached to the bone, the urethra cannot follow this motion. Therefore the bladder and the upper half of the urethra flexes downward, a cystocele forms (comp. Figs. 116-118). The internal orifice of the urethra lies about in the middle of the hour glass shaped bladder. The lower part of the bladder can evacuate the urine only with difficulty. The bladder never separates from the cervix except in anteflexion, or in enormous ascitic expansion of the abdomen. The sound always demonstrates the lower sacculation, even if but little urine is apparently retained.

Symptoms and Course.

During the genesis of a uterine prolapse the women complain both of pain in the abdomen—traction upon the peritoneal attachments—and of pressure and bearing-down. Conjoined with this there is usually dysuria due to the cystocele, and constipation with rectocele. If the prolapse lies in front of the vulva, it prevents the patient from walking and working, especially if the prolapse is inflamed in consequence of mechanical injuries and uncleanliness. Erysipelas, extensive degeneration of the skin, pain with great swelling, force the patient to lie down. Should such inflammations recur frequently; should medical aid, rest, and good nourishment be impossible, the condition of the patient becomes deplorable. On the other hand, not a few cases occur in which the patient has nothing to complain of and believes medical assistance to be uncalled for.

Menstruation shows no abnormalities characteristic of prolapse. Pregnancy may occur. Then the growing uterus remains in the abdomen and the patient is temporarily cured.

If the prolapse occurs suddenly (compare p. 208), the pain is usually excruciating, so that shock and syncope ensue. Peritonitis may succeed. At times, however, sudden prolapses also develop without symptoms.

As will be gathered from the description on the preceding pages, the *course* is chronic. It may extent over many years, and may be retarded or interrupted by treatment. Even complete spontaneous cure, though rarely, has occurred. Namely, if peritonitis ensued, and the uterus was replaced by the mere rest in bed, peritonitic pseudo-membranes may fasten the uterus above and hold it permanently.

Diagnosis.

The differential diagnosis between the prolapse of a tumor and prolapse of the uterus is easily made by inspection, palpation, and exploration.

It is especially requisite to determine the *degree* or the *form* of the prolapse. Therefore it is necessary to proceed quite systematically.

In incomplete prolapse we examine in the erect position, causing the woman to bear down, while we pull on the vaginal portion to see how far the uterus descends. The pulling is interrupted at the slightest indication of pain. Should the prolapse have been replaced by the patient, she is directed to strain so that the prolapse recurs.

Then, with the patient in the recumbent position, we examine *first the vagina* to ascertain whether the vaginal vault is still present in front or behind; and how deep the vagina is. *Second*, we examine with the uterine sound from the *bladder*, with the finger in the *rectum*, for cystocele and rectocele. Then we determine by the sound the length of the uterus, and by the combined examination the size, mobility, form, and the possibility of reposition of the uterus.

Treatment.

The *prophylaxis* of prolapse is covered by the rules for the correct management of labor and the puerperium. During labor, incisions must be made into the strongly depressed, non-dilating os. Perineal lacerations are to be accurately united. The vagina is to be cleansed by irrigations, lest its walls remain inflamed, hyperæmic, and heavy. The bladder is to be emptied; the importance of this fact has been previously pointed out. In the same way, easy defecation must be provided for, as much straining and a thick column of fæces produce the etiologically important and dangerous, late puerperal retroversion. All inflammations disturbing the involution of the uterus are to be carefully treated. The puerpera must not be allowed to get up too early and exert herself. If bearing-down occurs, searching examination must be made and if necessary the proper treatment instituted. In the case of very lax vaginal walls, strongly astringent injections must be used several times daily.

To begin with, if we find a *rectocele* or *cystocele* in a normal or normally situated uterus, positive cure may be attained by the excision of a flap of mucous membrane appropriate to the degree of descensus, and union of the margins. This harmless operation should be performed much more frequently; by the removal of cystocele, especially, we might prophylactically prevent many a subsequently developed uterine prolapse.

In this operation, *elytrorrhaphy*, we freshen an oval of from five to eight centimetres long and three to six centimetres broad, according to the size of the cystocele. If the anterior vaginal wall lies *in front* of the vulva or can be drawn down, the freshening is to be performed in front of the external genitals, where also the sutures are inserted, and tied as far as feasible. The remainder of the sutures are united after the reposition.

For this operation I employ the leg braces depicted in Fig. 122. A strong silk thread is passed through the vaginal portion. By means of this thread the vaginal portion is drawn downward and backward and, if no assistant be present, fastened to the perineal holder (Fig. 122, centre). A Simon's speculum may also be introduced, and the silken thread wrapped around the clamp fastening the speculum to the handle.

Then the self-retaining speculum hangs down, depressing the uterus by its weight. But if an assistant be at hand, the vaginal portion had better be seized with the forceps illustrated in Fig. 15, p. 36. As we freshen at the right or left side, the assistant renders the surface tense by corresponding traction or twisting of the vaginal portion with the forceps.

For the mode of freshening, the kind of suture, and the after-treatment comp. pp. 89-91.

Recently Hegar has devised a clamping pincette which greatly facilitates the excision of an oval. Fig. 123 represents the instrument, previously constructed by Sims for the same purpose. With sharp tenacula or American bullet forceps we first seize the fold of mucosa to be removed. This fold is clamped in the pincette, the sutures are passed beneath it,

FIG. 122.   FIG. 123.

FIG. 122.—Leg-braces screwed to an ordinary table. Between them the perineal holder.
FIG. 123.—Pincette of Sims and Hegar for grasping folds of vaginal mucosa during excision.

and the fold cut away close to the pincette and above the ligatures. I did not succeed in this; I always passed the needles close under the pincette and abscised the fold of mucous membrane close above it. The threads are then knotted, and the wound thus rapidly closed, almost without any loss of blood. Should the operator fear to have included the vesical mucosa in the clamp, he introduces a sound into the bladder while a needle remains inserted in the middle of the fold. The sound is passed from above down in the groove between the two arms of the pincette, and we convince ourselves whether the needle is to be felt there, that is to say, projects free into the bladder or not. In the former case, a smaller fold must be seized and the needle passed more superficially.

We proceed in a similar manner at the posterior wall in the rare isolated rectocele. In excessive dilatation of the rectum it is certainly permissible to remove a piece of the rectal mucosa.

But the cases are very rare in which there exists a quite isolated descensus of the vagina. Usually the uterus participates in the dislocation and the treatment is hence to be directed against the latter likewise. We have seen above that the retroversion and the secondary cervical hypertrophy are the most essential complications. For the retroversion conjoined with descensus, we shall have good results from a pessary which stretches the vagina and thus keeps the vaginal portion away from the vulva and holds it in the position of anteversion into which it has been brought by the hand. Pessaries appropriate for this condition are the longitudinally oval ones of Hodge, bent according to the direction of the vagina (comp. p. 193, Figs. 106 and 107). A Schultze's figure-of-eight pessary, in the upper ring of which the vaginal portion is maintained behind and above, is often employed with advantage. In fact, we properly have to deal with the treatment of a retroversion described on p. 193.

Mayer's rubber rings are not to be recommended in incipient prolapse. They operate by stretching the vagina also to the right and left, opposing the desired extension. In general, they, moreover, occupy too much space, as their volume is large. Cases are not rare in which the soft rubber ring *helps palliatively, but is positively injurious*. For the dilatation of the vagina increases more and more, and thus also the size of the rings required. Finally the vagina is so stretched, relaxed, and dilated that is easily becomes inverted and prolapses with any sudden exertion. In this way prolapses have occurred even in nulliparæ. On the other hand, the Hodge pessary requires but little room. It *stretches* the vagina, but does not *dilate* it. While it is in place, the peritoneal fastenings as well as the vagina may gain strength and a positive cure is possible.

If the uterus, too, is hyperæmic and badly involuted, it will, when correctly placed, soon become smaller and normal, provided the case be not one of chronic metritis and of hypertrophy of many years' standing.

In the latter case, the cervix is hypertrophic, and an amputation or excision of the vaginal portion is indicated: *first*, because the vaginal portion constitutes a large part of the prolapse; *second*, because after amputation the uterus, or the cervix, does not shorten merely by the amputated piece, but in its entirety (comp. p. 159). The operation is not dangerous, because the anterior peritoneal pocket (comp. pp. 206, 207, Figs. 116 and 117) is at a distance from the vaginal attachment.

The operation is performed in the manner described (p. 139, Figs. 78–80). The operation by the knife and the suture is to be preferred to all methods formerly in use—the galvano-caustic ablation (comp. p. 66, Fig. 34) and the écrasement.

As in the present instance the object is not to produce a *normal* vaginal portion, but to remove as much tissue as possible; moreover, as in great circular hypertrophy the wedge-shaped excision is very slow and difficult, I have operated according to the following method (Fig. 124). After the lateral division of the vaginal portion, the posterior lip was first drawn down and severed in such a manner as to make the surface of the incision form half of a right angle to the uterine axis. The anterior lip was removed in the same way. Then two deep sutures closed the external halves of the wound. Now stretching the specula upward and downward, the wound surface bulges strongly above and below, as represented in Fig. 124. The wound is then united so as to form a cross. In the centre a large os remains which does not easily become atretic.

In many cases in which the low position of the uterus is due rather to the size of the vaginal portion than to the prolapse of the vagina, amputation of the vaginal portion is positively sufficient. The *small* uterus is kept in place above and assumes the normal anteversion. Should this not be the case, a pessary will help, or elytrorrhaphy, above described.

Fig. 124.—Amputation of the vaginal portion with great circular hypertrophy; cruciform union. Below and laterally the walls are united, above the sutures are inserted, the os remains in the centre.

But if the support of the posterior vaginal wall be lacking, if the perineum behind be considerably lacerated or the vagina greatly relaxed, the descensus may gradually reform. It is necessary, therefore, to furnish also a posterior support, to add a perineoplasty. The posterior support is so important and effective even, that during its integrity or artificial reinforcement the descensus of the anterior wall can proceed only to a non-dangerous degree, that the anterior vaginal prolapse is kept replaced by the restored posterior wall! And hence many operators, neglecting the nearer *anterior* elytrorrhaphy with excision of the vaginal portion, recommend and perform only *colpoperineoplasty*, which is to extend as high up as possible, as treatment for prolapse.

While hitherto we have especially had reference to cases which are designated hypertrophy of the median portion of the cervix (comp. Fig. 119), those, therefore, in which the posterior vaginal vault is still present; we now come to the subject of total prolapsus (Figs. 117 and 118), in which the vaginal attachment to the uterus no longer forms the upper end of the vagina, but in which the uterus is suspended from the pelvis by the vagina, and the vaginal attachment forms the *lowest* part of the vagina.

Until recently, even in these cases, help was sought to be given exclusively by pessaries. A pessary may be of use in three ways: *first*, it is so voluminous as to press the vagina against the pelvic wall, then the vagina cannot become inverted. *Second*, a pessary is laid like a bar in front of the lower opening of the pelvic cavity, then the pessary prevents the prolapse passing the vulva, but it lies immediately over and upon the pessary or behind the vulva. *Third*, a pessary, by being fastened by *external* bandages, may keep the uterus in place.

The first series of pessaries is represented chiefly by the larger and largest numbers of Mayer's rings. Besides, there were used—large wooden balls covered with wax; rings of tow, rubber, and wax covered with leather and thickly varnished; pyramidal and tubular instruments, inflatable rubber pessaries, etc. Although all pessaries which dilate the vagina excessively are irrational, it cannot be denied that actually a large number of patients are perfectly contented with these instruments. To be sure, the majority of these pessaries are not applied by physicians, but by midwives, trussmakers, or the patients themselves.

The most important of the instruments of the *second series* is Zwanck's wing pessary, Fig. 125.

It is introduced into the vagina with closed, folded wings; then the lower screw is turned until the wings are completely separated (see Fig. 125). Thus expanded, the wings hold themselves right and left at the inner surface of the ischium, or at the anterior pelvic wall, especially the lower part of the pubic arch. If the pessary is well placed, it forms a transverse bar occluding the lower opening. In front there remains only the angle of the pubic arch, and behind a narrow opening to the coccyx. The uterus and the vaginal folds rest upon the pessary and can reach the outside neither in front nor behind.

But the sharp margins of the wings, pressed strongly against the bone by the superincumbent prolapse, gradually crush the soft parts, bore their way in, as it were, so that the instrument becomes covered with granulations and immovable. This is a serious matter when the instrument

Fig. 125.—Zwanck's pessary.

places itself with its longest diameter in the straight diameter of the pelvis. Should this rotation occur, owing probably in the first place to the column of fæces, the wings may perforate the bladder and the rectum. These cases of fistulæ due to Zwanck pessaries are extremely frequent, so that the total banishment of the instrument is demanded nowadays. However, there are careful patients who remove the instrument every night and feel very well with it. Accordingly it is not justifiable to give this dangerous instrument to unknown women who are lost sight of. On the other hand, the Zwanck pessary may be used when we are convinced of the conscientiousness of the wearer and are in the position to prevent mischief by accurate observation.

I have also employed a "transverse bar pessary" in the following manner: A small flexible rubber ring of six to seven centimetres diameter is bent like the contour of the wings of Zwanck's pessary. The size of the instrument depends on the width of the pubic arch; if the latter be very wide, we need a large instrument. The centre of the anterior bow is marked with a thread. Then the pessary is passed into the vagina, narrow side first. The pessary is now moved to and fro until the previously marked centre lies exactly in the middle when the vulva is separated, after the pessary is no longer held. With the finger hooked over, we try by strong traction whether the pessary sits firmly; the woman is made to get up, walk about, strain, and the pessary is again examined to see if it remains unchanged.

In this way the prolapse is often kept back by a non-voluminous, non-compressive, soft pessary.

But if the vagina is completely separated from the underlying tissue, folds will crowd past any pessary and displace it.

If in such cases we cannot remedy the fault by placing several pessaries one above the other, or if the voluminous apparatus cannot be borne, we have recourse to the third series of pessaries.

These apparatus, "hysterophores," a much used form of which is represented in Fig. 126, consist of a cup a, situated in the vagina, which bears the uterus; an abdominal belt b, from which four elastic tubes $c\,c$ extend downward to unite in a leather plate to which the cup on its stem is fastened by a screw d.

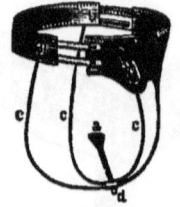

Fig. 126.—Hysterophore.

The apparatus here illustrated, an American invention, possesses the advantage that the tubes c are readily adjustable on the abdominal belt and easily fastened. As the tubes $c\,c$ cross between the legs, the anus behind and the urethra in front are left free. Such apparatus, as a rule, are not continually worn by the patients, as they are troublesome when changing from sitting to standing, walking, or lying.

Although there is any quantity of pessaries besides those described, constructed partly by physicians, partly by laymen, we content ourselves with what has been stated. The other apparatus are either but slight modifications of those here mentioned, or so unpractical as to be worthless to the practitioner.

Inasmuch as all pessaries have their inconviences and disadvantages and are worthless in many cases, it was natural that cure was sought to be attained by operative means. This was done in various ways.

In the first place, in cases in which a pessary would not remain or could not be borne, the vulva was sought to be partially occluded—Fricke's episiorrhaphy. Soon it was found that the stitching together of the labia majora was insufficient, that the uterus gradually tore the membranous bridge and again prolapsed.

The obstacle was then placed higher up. The posterior vaginal wall was united and a contracted portion of the vagina there formed, narrow

DISLOCATIONS OF THE UTERUS.

enough to prevent the passage of the uterus. However, even this method proved insufficient. Therefore, flaps of skin were also excised laterally, the vagina was burned and cauterized, or contracted circularly by ligatures.

This method likewise failed.

A great advance was Simon's posterior colporrhaphy. Simon freshened, in the posterior vaginal wall, a surface represented in the adjoining Fig. 127. The base of the figure *c d c* is above the labia minora within the vagina. The size is adapted to the proportions of the vagina, usually *b a b* was made equal to five centimetres; *b c* equal to six centimetres; *c d c* equal to seven centimetres in length. *a d* lies exactly in the middle behind. The surface is approximated, as indicated in Fig. 127, in such a manner that *a b* falls on *a b*, *b c* on *b c*, *d c* on *d c*. When united, a cicatrix the form of Figure 128 will be exactly in the median line. The uterus is elevated the height of this cicatrix. The vaginal portion rests

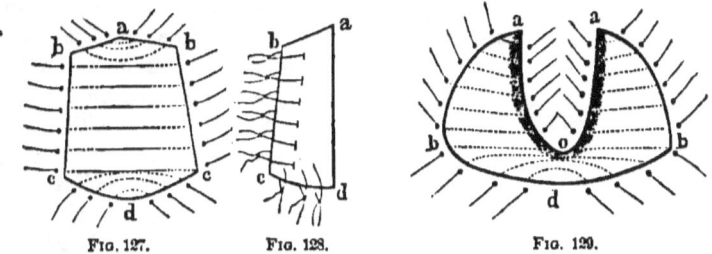

Fig. 127. Fig. 128. Fig. 129.

Figs. 127 and 128.—Diagram of the surface freshened by Simon; the sutures marked in the former, united in the latter.
Fig. 129.—Diagram of the surface freshened by Bischoff.

upon the line *a b* (Fig. 128), the cicatrix forming, as it were, a firm pedestal for the uterus. But if the uterus does not abut against this line; if, retroverted, it crowds between *b c* and the anterior vaginal wall, the cicatrix and the vagina gradually yield, and the uterus begins to prolapse anew. However, there are quite a number of cases, by Simon and other operators, in which the method described was followed by permanent results.

Bischoff, of Basle, had another idea. He placed the chief value on the direction of the vagina and the uterus. He justly emphasized that a prolapse is impossible whenever the angle formed by the uterus with the vagina is acute. Accordingly, Bischoff seeks the importance of his operation in the *forward flexion of the vagina*. He freshens the above Figure 129. First the flap *a o a* is dissected off with the handle of the scalpel, finger, and knife. This flap, then, lies exactly in the median line and is formed by the posterior rugous column. Then the mucous membrane is removed from the surface *a b d b a*. Now *a b* on both sides is joined to *a o*, and thereafter *b d* to *b d*. In this manner, from a vertical posterior vaginal wall is formed an almost horizontally extending portion, at the upper end of which the vagina is acutely flexed.

Winckel again operates similarly to Simon. His freshening is much lower, but he obtains a better barrier by not removing the mucous membrane, but dissecting from the middle toward the sides and uniting the flaps folded toward the lumen of the vagina. Thus, if Simon's figure is *higher*, Winckel's freshened surface is *broader*, encircling the Vagina in form of a ring.

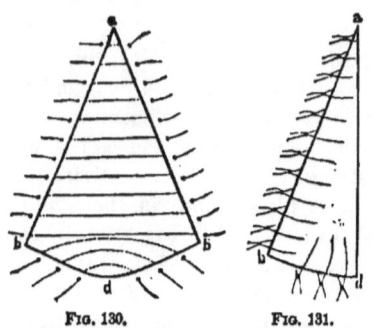

FIG. 130.   FIG. 131.

Diagram of the surface freshened by Hegar, the sutures indicated in the former, united in the latter. According to the size of the vagina, $a\,d$ is from 7 to 10 cm. long ; $a\,b$, 8 to 9 cm. ; $b\,d\,d$, 7 to 8 cm.

Neugebauer took up an older proposition, and operated by freshening an oval at the anterior and posterior vaginal walls. These surfaces were united, the anterior wall adhering to the posterior.

Simultaneously with Simon, Hegar had performed many operations and indicated a method having the advantage of being simple. Hegar set himself the task of restoring the natural conditions as nearly as possible. To this end he freshens a triangle, Fig. 130, $a\,b\,d\,b$. $b\,d\,b$ lies more deeply than in Simon's operation (Fig. 127, $c\,d$), so that a new perineum is formed. The point a is close to the os uteri. After this triangle is united, there is in the posterior vaginal wall a firm triangular cicatrix, Fig. 131, $a\,b\,d$. This cicatrix has the purpose of preventing an inversion of the vagina. Although it is impossible to restore the integrity of the *upper* fastenings of the uterus, the narrowness and rigidity of the vagina can be artificially reproduced. This is done by the formation of a firm, unyielding cicatrix which, re-creating the perineum, extends up to the vaginal portion. Thereby the vaginal portion is, as it were, kept above and the uterus elevated. Gradually it diminishes and by the

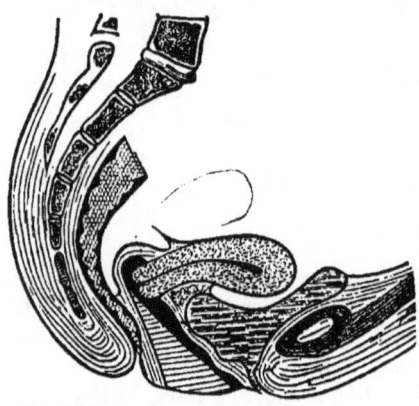

FIG. 132.—Complete effect of Hegar's prolapsus operation; the united surface indicated by transverse shading. Of course, it is impossible to obtain this form of the cicatrix. As is shown by the examination immediately after the operation, the cicatrix contracts from above downward, thus slightly folding the entire vaginal surface.

weight of the intestine is pressed upon the bladder; thus restoring the normal position in which prolapse does not recur. In Fig. 132 we see the strong cicatrix, marked by transverse shading, keeping the still hypertrophied uterus elevated.

If the prolapse is very large, extensive pieces of skin are removed from the anterior and lateral vaginal wall in the manner described on p. 213. The cicatrices resulting therefrom, likewise render the vagina narrow and rigid. A hypertrophied vaginal portion may be diminished by the various methods, but the correct position of the uterus, the hemorrhage during the operation, and the rest in bed subsequent to it, have an atrophying influence on the cervix even without excision of the vaginal portion.

Inasmuch as in prolapsus operations very large wound surfaces have frequently to be united, the surface corrugates and folds too much if it be brought together by a single loop of thread. Therefore the deeper half of the freshened surface may be united by deep catgut sutures, superficial silk sutures being used above. In this way the wound surface is well coapted. The technique of the freshening and approximation, as well as the after-treatment, are the same in general as those of perineoplasty (comp. p. 89 and Chapter IV.). It should be remarked, however, that the patient must remain abed at least two weeks after a prolapsus operation. The sutures may be allowed to remain and to suppurate out gradually. The suppuration in a suture track renders the cicatrix but the firmer.

As to the value of the different methods, that of Hegar stands at the head, both as regards the facility of execution and result. The objection is is made to it that by it the vagina is rendered too narrow; that not only is healing retarded by the great traction during the stitching, but a serious obstacle may exist at a subsequent labor. For this reason A. Martin has devised a method in which two longitudinal strips instead of a triangle are denuded and united. Whether thereby as firm a cicatrix will be secured as would appear desirable according to Hegar, will have to be settled. However, subsequent labors enter into the consideration in the least number of patients. Minor details of the operation, especially as regards the further excision of ovals, will have to be modified to suit individual cases.

Bischoff's operation is certainly appropriate to cases in which the posterior rugous column depends somewhat hypertrophied, tag-like; but in prolapses of very long standing, in which the posterior vaginal wall is quite thin and lax, it might be impossible to secure a thick flap. In Bischoff's and in Winckel's operations, the danger in subsequent labors is slight, as incisions soon make room. Hence it is possible that in future the latter operations may be chosen for younger women; for the older, the more certain and complete operation of Hegar.

The only danger from prolapsus operations would be by sepsis, but this may be avoided with certainty. Thus far about eighty per cent were cured by the operation.

F. INVERSION.

Etiology.

After full-time labors, exceedingly rarely after abortions, the fundus of the uterus, for reasons that do not belong here, occasionally becomes depressed; gradually it descends, passes through the cervical canal, and finally the inverted uterus lies in the vagina or prolapses even in front of the vulva. Should the woman survive the consequences of the acute inversion—hemorrhage or progressive septic inflammation—the affection remains in a chronic form.

Besides, inversion arises from tumors. Should there be a tumor ex-

FIG. 183.—Inversion of the uterus, with a myoma in the fundus.

actly in the fundus, should the surrounding uterine tissue atrophy or undergo fatty degeneration, the tumor glides into the uterine cavity. Partly by the weight of the tumor, partly by the uterine contractions, the tumor—dragging the uterus along—is forced deeper. Finally the tumor passes through the os uteri, even in front of the vulva. Both benign and malignant tumors may lead to inversion.

Anatomy.

In the majority of the carefully observed cases, the cervical canal was still present, so that the inverted uterus, and above it the os uteri, could be felt in the vagina. The bladder is not dislocated. In recent puerperal inversions the ovaries will probably lie in the funnel; but, at least according to my observations, there are no intestines in the funnel. Subsequently the ovaries are drawn up, so that the uterus becomes small and the funnel narrow, so much so that not even the finger can be pressed into it. The size of the uterus is very variable. While in one case it was felt as a small, firm, round, spherical tumor, in other cases it appeared large, soft, compressible, flattened, almost œdematous. In Fig. 133 is represented an inverted uterus, in the fundus of which is a large myoma surrounded but by little uterine tissue above and below.

Symptoms and Course.

The most important symptom of acute inversion, the hemorrhage continues during the chronic inversion, partly as menorrhagia, partly as metrorrhagia, for instance, during coition. Pus, shreds of membrane, and mucus are also passed. The traction on the peritoneal ligaments causes pains, increasing especially during the development of a prolapse. The compression exercised by the contracting os upon the cervix may lead to congestion and even gangrene of the body of the uterus. However, cases have also been observed in which the inversion has lasted for decades without symptoms. Just because every symptom was absent, the uterus was taken for a tumor and removed.

In the further course, all the consequences of repeated great losses of blood ensue. The patients may be bedridden for years, as every movement provokes the most violent hemorrhages.

Diagnosis and Prognosis.

In order to make the diagnosis, we must *first* demonstrate that the uterus is absent from the place where it ought to be, and *second*, that the tumor felt in the vagina is really the uterus.

When the anamnesis has made inversion appear probable, combined exploration is performed. If the abdominal coverings are too resistant to allow of thoroughly palpating the pelvic inlet, examination is made from the rectum and vagina, alone and combined, from the vagina and bladder, and from the bladder and rectum combined. Narcosis is often required. Owing to the importance of a correct diagnosis, no pains must be spared.

The tumor is then circumscribed at its upper end while traction is made upon it. Should the cervix become inverted, so that the groove at the pedicle disappears by flattening, an inversion is present. If the vaginal portion remains unchanged, while the tumor descends low, it is a case of pediculated polypus. If it be impossible to penetrate with the finger between the tumor and the cervix, the sound is pushed up round about the pedicle. Should it penetrate deeply, the knob of the sound is searched for externally. Thus guided, the uterus can be clearly felt and inversion excluded.

But should the sound be arrested all around close above the external os, that will be the point of inversion. Finally we try to feel the funnel distinctly by combining downward traction on the uterus and upward pressure, with the controlling hand placed externally.

Even if the lowest part of the prolapse is certainly formed by a tumor, *the pedicle* of the tumor may, at least in part, be formed by the inverted uterus. This will always be suspected when the pedicle is extraordinarily thick. Therefore the most careful examination is necessary to find the limit between tumor and uterus. The uterus is softer than the tumor, the latter usually much paler.

The *prognosis* is unfavorable without operative interference, because the hemorrhages continue, and the degeneration of the tumor may lead to sepsis. The reposition succeeded after thirteen and more years; then not only perfect health results, but pregnancy may even occur. Spontaneous reinversion has been once observed by Spiegelberg.

Treatment.

The treatment consists in enucleation of eventual tumors, reposition of the uterus, and, if that fail, amputation of the organ. The latter operation is unnecessary when the inversion no longer causes any symptoms.

The reposition is performed in narcosis. Under external control, lest the forced-up organ be torn off the vagina, the uterus is pushed in and through the cervix. The uterus must be grasped with the whole hand, and by this pressure made softer and less plethoric; the spreading fingers bore into the grooves of the cervix and crowd the margin of the os outward. It has also been recommended first to reinvert one tubal region, to make longitudinal incisions into the body, instead of pressure from without to draw the vaginal portion down with tenacula, Muzeux's forceps, or by silk ligatures passed through it. Furthermore, wire sutures have been placed below the fundus from one lip to the other, so as to make continuous pressure against the fundus.

Should we fail in the above-described manner, a colpeurynter filled with water or air is placed in front of the uterus. The permanent pressure softens the uterus, diminishes it, and forces it through the cervix. This method succeeds in cases in which the employment of the greatest force and repeated attempts have led to nothing. The reinversion ensues both slowly and imperceptibly and with violent contractions of the uterus.

If this manner likewise fail, firm adhesions in and above the funnel probably prevent the reduction of the organ. Then, if the symptoms be inconsiderable, the uterine mucosa will be sought to be altered by cauterizations or strong astringents so that the discharge of blood and mucus will almost cease.

If the hemorrhages continue to be very profuse, nothing remains but to amputate the uterus. This must not be done by a simple abscission, because the bleeding and subsequently suppurating wound surface will at once invert upward into the abdominal cavity. It is necessary, therefore, to place a ligature above the place of amputation, or to put several needles through the uterus. It might also be feasible to join the wound surface accurately all around with catgut sutures, and then to favor the spontaneous reinversion of the remnant of the uterus.

Antiseptic precautions evidently are here pre-eminently required.

Should the inversion have been caused by a tumor, the latter is to be carefully removed or enucleated. After that, the reinversion is usually easy. The separation by the ligature, the écraseur, and the galvano-cautery has been recently abandoned.

G. The Rarer Displacements of the Uterus.

The extra-median position of the uterus is without practical importance. A deviation of the vaginal portion to one or the other side may be congenital or aquired. In the latter case, exudations, pseudo-membranes or tumors pushed or pulled the uterus to the right or left. In the same way a latero-version occurs. The latter, besides, may be the consequence of defective development of one half of the uterus. Thus a unicorn uterus always inclines strongly toward the defective side.

The elevation of the uterus likewise is merely a complication of other pathological conditions. By parametritic exudations at the anterior abdominal wall the uterus is often lifted very high, or fibromata and ovarian tumors pull the vaginal portion high out of the pelvis, so that it can barely be reached.

Very rarely the uterus gets into a hernial sac. If we bear in mind that in foetal life the uterus lies above the pelvis, it will be seen that it is possible for an ovary or the tube to be inclosed in a hernial sac, dragging the uterus after it. Then we have an inguinal *hernia of the uterus*. Such a uterus may become impregnated. If symptoms be present, extirpation of the uterus would have to be performed.

CHAPTER XIII.

NEW-FORMATIONS OF THE UTERUS.

A. MYOMATA.

MYOMA of the uterus is a partial hyperplasia of the uterine tissue; when the hyperplasia of the connective tissue preponderates, the tumor is called a fibro-myoma. The etiology of these tumors is unknown. Nearly every tenth woman has one or more myomata in the uterus. Myomata occur from puberty on, but usually they arise later.

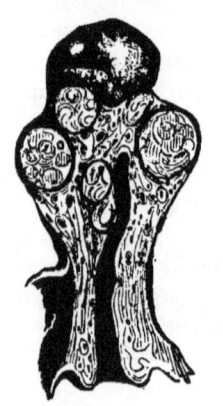

FIG. 134.—Several myomata in a single uterus; above a subperitoneal, at the right and left a firm myoma surrounded by a connective tissue capsule; near the uterine cavity, two submucous a myomata. Vascular dilatation of the uterus.

In a single uterus may be as many as fifty tumors which mutually hinder their growth, producing flattened, hemispherical, or quite irregular bodies. If there be many tumors, they are usually small and hard; while with the more diffuse, large, vascular, interstitial myomata smaller tumors are but rarely demonstrable.

Myomata occur in all regions of the uterus; most frequently above, rarely below. Cervical myomata belong among the rarer cases.

The *point of origin* is of importance for a myoma; three different types have been distinguished accordingly: *subperitoneal, interstitial*, and *submucous myomata*.

Should a tumor develop in the external layers of the uterus, it will of course grow toward the abdominal cavity and not into the resistant uterus (Fig. 134). The tumor is covered by but a thin layer of uterine tissue or is close under the peritoneum, hence "*subperitoneal myoma*." As long as the tumor is small, it may be sessile on the uterus in the shape of a small, hard swelling the size of a nut; but if it grow more and more, the myoma, following the law of gravity, will glide down the external wall of the uterus, descend, and possibly drag the uterus after it and flex it, or the myoma, as it were, detaches itself from the uterus, forming a pedicle.

When the subperitoneal myomata become larger, they may also pull the uterus upward, so that the cervix becomes quite thin and even ruptures. As in ovarian tumors, axial rotations occur. Thereby the cervical canal may be so compressed that the imprisoned blood may form a hæmatometra. Much more rarely the myoma will depress the uterus. A descensus of the uterus results more readily from the traction of a polypus from below than from the pressure of a myoma from above. Should subperitoneal myomata develop nearer the lateral margin of the uterus, they unfold the broad ligament and lie completely between its layers, in the parametrium, that is, extraperitoneally.

The myomata being hard, they irritate the peritoneum mechanically, causing adhesions and agglutinations. If a woman with such affections becomes pregnant, the adhesions usually yield without symptoms and are stretched to cords—pseudo-ligaments. The latter, after the delivery and involution of the uterus, cross the abdominal cavity, so that the intestine may be flexed around the ligaments and the puerpera perish from ileus. By such adhesions further nutrition of a subperitoneal myoma is also brought about. The growth, for instance, may adhere to the parietal peritoneum. I have seen a case in which a small myoma was firmly adherent to the ovary, making the impression at first sight of a double ovary. Some very rare cases are also known in which, after stretching, thinning, and rupture of the pedicle, the myoma was found free in the peritoneal cavity.

If the subperitoneal myomata are sessile on a broad base, they obtain sufficient nutriment and often attain an enormous size. Myomata have been found at post-mortems and removed by laparatomy which distended the abdomen to and beyond the size of a uterus near term. In very rare cases, gangrene of the distended abdominal wall ensued, and the myoma was exposed openly. I know of a case terminating favorably, in which an abscess opened at the point of rupture for parametritic exudations, above Poupart's ligament, from which suppurating myomatous masses the size of a man's head were removed in several sittings.

If a myoma arises about in the middle of the uterine wall, it is an *interstitial* myoma. These tumors especially, being surrounded on all sides by vascular portions, find the most favorable conditions for further growth. We shall have to look for the diffuse, soft, cavernous myomata almost exclusively among the interstitial. It is obvious that such myoma, too, in its further growth may partly reach closely to the peritoneum, or may, and possibly must, get into the uterus (Fig. 135). These myomata are comparatively the largest. The *opposite* uterine wall may become quite thin, membranous, so as to rupture during labor. The uterine cavity is considerably distorted and enlarged.

If a myoma springs from the muscular layers above the mucosa, it will not develop into the firm uterine parenchyma, but into the uterine cavity. A *submucous* myoma forms. Such a one can only bulge the uterine mucosa into the cavity in a segment of a circle. Or else it may to a great extent lie within the cavity, in the form of a *polypus sessile on a broad base* (Fig. 136). Finally it is possible that the myoma prolifer-

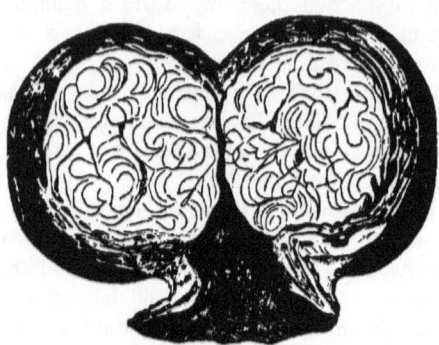

FIG. 135.—Large myoma of the anterior uterine wall, of interstitial origin; no clear demarcation of the uterine basis; enormously dilated vessels at the border of the myoma.

ates into the uterine cavity in such a manner as to be united to the parenchyma merely by a pedicle, that is to say, so as to give rise to a polypus (Figs. 137 and 138). The latter again may remain in the uterus or be delivered into the vagina. If it be very large, it irritates the vagina by pressure and often becomes adherent to it. Then the diagnosis will be doubtful; a vaginal origin may be suspected (comp. p. 100). Although the polypi have usually a round form, they may be grooved if partly in the uterus, partly in the vagina. Hour-glass shaped tumors arise in this way.

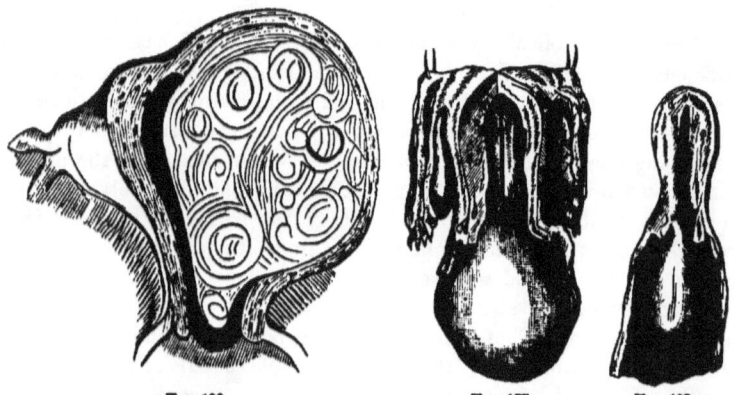

FIG. 136.   FIG. 137.   FIG. 138.

FIG. 136.—Large submucous myoma sessile on a broad base.
FIGS. 137 and 138.—Pediculated myomata; fibrous polypi; the former springing from the fundus, the latter from the cervical canal.

The pedicle, of course, leads the nutrient vessels to the polypus; thick, profusely bleeding vessels occasionally coursing through the pedicle. Usually but a single polypus is present, yet cases occur in which a second polypus is palpable after removal of the first. All forms of myoma may combine, so that within or upon one uterus quite a number of tumors are found. By pressure, traction, and adhesion of the myomata to the different uterine walls the cavity becomes so irregular that sounding or intra-uterine treatment is impossible. Even in the case of a single myoma sounding may fail, as in that illustrated in Fig. 102, p. 189.

A peculiar form of myoma is the so-called *cysto-myoma*—a class of tumors which only recently has awakened the interest of gynecologists, especially on account of the difficulty of the differential diagnosis from other abdominal tumors—ovarian cysts. Perhaps it would be more correct to treat of these tumors by themselves, not as a subdivision of myomata. They reach a colossal size by cyst-formation; they are not specially confined to the external or internal surface of the uterus, but are met with both close beneath the mucous membrane and subperitoneally. The origin and significance of the cyst-formation are not as yet quite clear. It is certain, however, that these cysts in some tumors formed teleangiectatic lymph sinuses containing lymph. Others interpret the cysts as the result of an œdema, of a fatty degeneration, of various apoplexies, of myxomatous degeneration, or of a parenchymatous softening generally. At all events, the cysts have no definite form, are not invested with epithelium. The spontaneous coagulation of the fluid obtained by puncture pointed to the connection with the lymphatic vessels.

Various *regressive metamorphoses*, both chronic and acute in character, occur in uterine myomata. In the first place, small myomata, especially, become very hard, firm, and tough, so as to creak when cut. Then, lime salts may be deposited in the connective-tissue septa, the myomata calcify. The muscular tissue at the same time seems to soften and melt down; at least when sawing open a calcified myoma we usually find between the calcareous masses softened patches filled with detritus, reddish shreds, and thick liquid. Quite large myomata may also calcify. Thus at an autopsy I saw a myoma the size of a child's head, fastened to the uterus merely by a peritoneal fold, which had been diagnosticated for years as a lithopædion. The degree of calcification, however, is very variable. Defective nutrition must always be accepted as the preliminary condition to the calcification. In this connection we must mention the so-called uteroliths—completely calcified myomata separated from their place of growth, found accidentally post mortem in the uterus, or expelled during life. In the centre of a myoma we not rarely find a small cavity with fluid or softened contents.

Another regressive process is that of fatty metamorphosis. Quite spontaneous fatty degeneration has not been observed, but there are—though very few—cases in which the fatty involution of the puerperal uterus likewise affected the myoma which disappeared completely.

The transition to the acute metamorphosis is caused by *œdema of the myoma.*

The myoma becomes œdematus for two reasons. *First*, the pedicle of a polypus delivered into the vagina may be compressed by the rapidly or gradually contracting os. Then we have a congestive œdema which is occasionally followed by gangrene. But the myoma may also remain in the condition of œdema for a longer time. In that case the tumor feels soft, almost fluctuating, as if there were a cyst within it.

Second, a myoma becomes a œdematous after injury and infection. In that case there is not only infiltration, but also very rapid occurrence of thromboses in the wide, sinuous vessels of the tumor. The consequence of this process frequently is suppuration with the same final result as in congestive œdema—gangrene, sloughing of the whole tumor. The suppuration seems to form a demarcation toward the healthy tissue. Often during the removal of sloughing myomata, when reaching healthy tissue while tearing out particles, I have found large portions intersected by purulent trabeculæ. This demarcating process it not unfavorable, for in many cases it seems to protect the organism against acute sepsis.

The least incentive suffices for the production of sloughing; an erysipelatous process, an acutely purulent œdema rapidly occupies the whole tumor. By enormous purulent or rather sanious discharge, the entire tumor may be gradually eliminated, positive cure resulting.

Symptoms and Course.

The symptoms of myomata vary according to their seat. If a myoma arises *subperitoneally*, if it is attached to the uterus in the shape of a small tumor, all symptoms may be absent. Even during its growth it causes no more symptoms than the enlarging gravid uterus. Even when the myoma reaches a considerable size, there will be no symptoms characteristic of *the myoma*, but only those found with all larger abdominal tumors—a sense of fulness of the abdomen, constipation, dysuria, bearing-down, impaired nutrition, dyspnœa, etc.

If a subperitoneal myoma is very hard, pediculated, or very prominent, it mechanically irritates those portions of the peritoneum against which it is occasionally pressed; there occur circumscribed peritonitides with their characteristic symptoms, adhesions, agglutinations, formation of false ligaments. This is especially the case when the myoma, by being situated in the fossa of Douglas, is moved about much in the varying repletion of the rectum and bladder. In rare cases ascites is also found.

The symptoms become more dangerous when the myoma, as more rarely happens, does not spring from the fundus, but from the cervix or at least a lower part of the uterus. Then in its growth the myoma, situated below the promontory, will be incarcerated in the lesser pelvis. Both dysuria and retention of fæces are the consequences. A communication between the inner part of the myoma and the rectum has also been brought about by adhesion to the rectum and partial necrosis of the myoma. In that case the myoma may slough rapidly. In other cases, peritonitides occur only around the myoma, numerous membranes surround the entire lower portion of the pelvis. Recrudescences of the peritonitis, circulatory disturbances in the tube and ovary lead to secondary ovarian affections, for instance, the formation of hæmatoma. Thereby the symptoms of a considerable, incurable pelvic peritonitis come into the foreground.

Growing myomata not rarely lead to compression of the veins and œdema of the lower extremities. Conditions similar to phlegmasia dolens occur. True pyæmia also ensues. Whether these affections have their first cause in mechanical (compression) or in thrombotic infectious conditions, by suppuration of the myoma, must be decided by observation in the special case.

Furthermore, nerves may be pressed upon by myomata, thus occasioning neuralgias in the lower extremities.

Of *interstitial myomata* we must distinguish between the small encapsulated and the large diffuse myomata.

The former assuredly not rarely cause dysmenorrhœa, because they render the expansion of the uterus during menstruation difficult or impossible. The main effect of these small tumors is that the upper part of the uterus in which they are usually situated becomes heavier and hence the normal position more pronounced. Almost invariably we find an increased anteflexion and anteversion (comp. Fig. 102, p. 189). Hence arises also the strangury present in every anteflexion, which increases especially during the enlargement of the uterus (comp. p. 182).

In the case of the large, soft, interstitial myomata all symptoms may be absent. With these tumors the symptom of swelling of the uterus during the period and diminution during the pause is observed with especial frequency. The great plethora of the tumor and consequently of the uterus leads to considerable menorrhagias. An additional cause for these is the fact that the uterine cavity and hence the mucosa has greatly increased in extent. The hemorrhages at times are so colossal that the patients are nearly exsanguinated. Hardly have they recuperated a little in the intermenstrual period, when the loss of blood recommences and continues violent for many days. Or else the menstrual type disappears completely, the menorrhagia changes into permanent metrorrhagia. This is variable only in that, at times, dark, decomposed, almost tarry blood, and again bright and fresh blood or else sanguineous serous fluid is passed. Such patients may appear very anæmic ; the anæmia is distinguished from that of carcinoma by the absence of the waxy pallor of the complexion and the disappearance of the subcutaneous fat. Should the bleeding become at once profuse, there may be somnolence, amaurosis, and anæmic convulsions as in death from hemorrhage.

In the case of *submucous myomata*, all those symptoms become prominent which are connected with pathological conditions in the mucous membrane. This is pre-eminently the hemorrhage. But while in the case of interstitial, large, soft, diffuse myomata the cavity of the uterus, enlarged but regular in form, readily allows the blood to escape ; a submucous myoma may obstruct the cavity, or several tumors pressing one against the other may completely occlude the efferent channel. Should there be, besides, some axial rotation of the uterus, this will likewise lead to retardation of the flow. The uterine cavity becomes dilated, the uterus expels its contents under paroxysms of pain, frequent attacks of uterine colic occur. The spermatozooids, too, have a long sinuous road and do not get into the tubes. Still more frequently occlusions of the tubes, compression of the interstitial part or of the uterine ostium, and perimetritic processes generally, are causes of sterility. Even when the fecundated ovum gets into the uterus, the alteration in the mucosa will render its implantation impossible. But if implantation of the ovum take place, the expansibility of the uterus is insufficient, and abortion occurs sooner or later. Yet should the ovum attain its normal development in spite of all obstacles, the delivery may be mechanically obstructed. And if the child has been delivered in some manner, after-hemorrhages and sloughing of the myoma in the puerperium will endanger the life of the mother.

The fibrous or myomatous "*polypi*" are submucous myomata which have formed a pedicle. It is evident, therefore, that the symptoms of polypi coincide to a great extent with those of submucous myomata. On the other hand, the possibility of displacement in the uterine mucous membrane causes special symptoms. Not rarely the uterine mucous membrane is irritated by such polypi so that hypersecretion results. Often the general hyperæmia of the entire organ leads to a more active growth of the mucosa, and *mucous polypi* spring up by the side of the myoma. This complication is very frequent. Then particularly a profuse serous discharge may be present which may lead us to suspect a large mucous polypus alone.

As the uterus softens during menstruation, the os opens slightly, and the conjointly swelling myoma, not finding sufficient room in the uterine cavity, crowds into the cervical canal. Thus it happens that a polypus can be readily felt in the external os during menstruation, while it is retracted subsequently. The polypus may even be partly or wholly delivered into the vagina during every menstruation and later again disappear. When the polypus becomes œdematous by compression of the pedicle, it cannot at times slip back again and must remain in the vagina. Then, there being the possibility of coming in contact with infectious matters, and being no longer or but imperfectly nourished through the compressed pedicle, the danger of rapid gangrenous degeneration is especially great. But the polypus may also become habituated to the new relations, remain sufficiently nourished in the vagina, and continue to grow. Then it may fill the vagina completely, the tumor cannot be grasped, and thus we may be in ignorance of what is beyond the greatest circumference of the swelling felt.

In rare cases the pedicle becomes more and more elongated, the polypus drops out of the vagina and changes to a myoma pendulum. The end of the pedicle is usually at the fundus, but a polypus may also spring from the cervix (comp. Fig. 138). It is not improbable that the point of attachment may migrate downward by the permanent traction.

When a polypus is removed, at times a second one is felt above it which now commences its descent. More frequently, however, but a single polypus is in the uterus.

As regards the *course* in general, it is characterized by the fact that uterine myomata are thoroughly benign tumors. Unless the hemorrhages or other complications alter the general condition, the myomata do not directly injure the health. The growth is usually limited by the regressive process, so that we may speak of a kind of spontaneous cure. On the other hand, the growth of an interstitial soft myoma by no means ceases with the menopause. Myomata may reach quite an enormous size even after the climacteric. Cysto-fibromata become especially large. But the course is slow in all varieties. Years are required for the development of a really large tumor.

Spontaneous cure may ensue in various ways. First, a myoma may undergo fatty degeneration in the puerperium and disappear. Second, a myoma situated close beneath the mucous membrane may erode and perforate the latter, so that it gradually parts from its bed and is expelled; this process has occurred in the above-mentioned uteroliths. Finally, a myoma may degenerate by sloughing and be cast off.

Diagnosis.

Small myomata situated on the outside of the uterus are easily felt on combined examination. Not infrequently such myomata, characterized by their harder consistence, are accidentally discovered during a gynecological examination.

The conditions are more difficult when the myoma is larger. Here the aid of the sound will be required to decide which of the tumors felt is the uterus, which the myoma. Or possibly the myoma may have grown between the layers of the broad ligament; then the differentiation from an adjoining parametritic exudation is important. For the differential diagnosis we must bear in mind that myomata are usually round, easily palpated, uniformly hard tumors, while the exudations have an irregular form and surface, and variable consistence in different portions. Of course, the history will elucidate many a case even before the detailed examination, and the further course soon dispels every doubt. When the myoma has dropped into the fossa of Douglas, or when it springs from a neighboring portion of the uterus, great differential diagnostic difficulties at times arise. In that case it is often necessary to arrive at the diagnosis by exclusion, every possible tumor being taken into consideration and the pro and contra being duly weighed. Especially when the myoma, imbedded in pseudo-ligaments and surrounded by small adjoining exudations, has displaced the uterus from its normal position, it is often impossible to arrive at a correct diagnosis except by prolonged observation. If we were to form the diagnosis at the first, often very painful examination, we should injure the patient more than we could benefit her, even if certainty were to have been secured. Careful sounding, painfully thorough combined examination, often even only the treatment, settle the diagnosis. It is quite impossible to form a correct diagnosis when the subperitoneal myoma has sprung from the region of the ovarian ligament and hangs freely movable on a long pedicle.

Once I have mistaken an ovarian sarcoma, firmly adherent to the posterior surface of the uterus, for a myoma, until undeceived by the autopsy.

If the myoma be very large, the tumor filling the entire abdomen, perhaps a cysto-fibroma, we must have a clear idea that the physical relations of a very large abdominal tumor must be the same, *no matter what its source*. Whether the tumor has sprung from above or below, right or left, in front or behind, if its size hinders the mobility, it must follow the law of gravity and adapt itself to the room naturally given. Hence from the result of the examination *alone* it is often impossible to decide whence the tumor arose.

Nevertheless the diagnosis of the large, subperitoneal myomata is not difficult. The history, perhaps also the diagnoses of formerly attending physicians, demonstrate a very gradual enlargement dating back for years. The general condition need not be disturbed in the least. The myomatous consistence, the equably round form of the entire tumor or of its several sections, the connection between uterus and tumor as ascertained by combined examination, the result of the sounding—all point to a myoma. Should there be doubt despite all this, akidopeirastic or an exploratory puncture with the Pravaz syringe may be made. By the latter, as by a wet cup, pure blood is drawn from the myoma.

In the case of large, soft, interstitial myomata the diagnosis is still easier, often an interminably long uterine cavity being sounded. The form of the tumor, too, which ordinarily reproduces the shape of the uterus enlarged, is in favor of a myoma. Besides, the tumor lies clearly in the midst of the abdomen, the right and left side being free. Not rarely, while moving and displacing the tumor, its connection with the segment of the uterus felt per vaginam can be rendered evident.

That cases occur in which mistakes are made despite these landmarks is proved by the literature of every year. Thus larger cysto-myomata, especially, were often diagnosticated as ovarian tumors. Here, too, the contents of the cysts are important. The fluid removed from a cysto-myoma is coagulable lymph, solidifying on exposure to the air. No morphological elements besides lymph-corpuscles are found. In ovarian fluids, however, are found their characteristic chemical substances (see below) and cylindrical epithelial cells, etc.

Smaller *interstitial* myomata, *i.e.*, hard, solitary encapsulated tumors, are often difficult to recognize when they are completely imbedded in the muscular tissue. The diagnosis is indeed easy when the myoma projects somewhat or at least lies so close beneath the peritoneum that its harder consistence can be clearly demonstrated during palpation from the softer one of the uterus. But if the tumor is smaller and more deeply situated, we must take into consideration all conditions which in any way lead to enlargement of the uterus. Chronic metritis and pregnancy should be named pre-eminently, particularly because obscure symptoms likewise exist with both. The gravid uterus, to be sure, is softer; that enlarged by metritis more uniform than with myoma; still the diagnoses here are often "authoritative diagnoses." It is evident that the possibility of small interstitial myomata should alway be borne in mind, for ten per cent of all women have myomata.

The *submucous myomata*, respectively *the polypi*, are often more easily diagnosticated, by being, at least partially, accessible to the exploring finger. In the first place it should be remarked that the body of the uterus in these cases at times possesses a peculiar form. While the cervix, extraordinarily slender and thin, extends deeply into the vagina, the uterus is attached above like a round, smooth ball—a proof that a round tumor contained within it expands all its walls equally. On sounding it is often possible to pass the knob of the sound clearly toward the fundus in various directions along the tumor.

When other symptoms point to a myoma, it is particularly important to explore during menstruation. Not infrequently the uterus opens just at that time and the spherical portion of a myoma is distinctly felt in the internal os or even in the cervix. But this is not meant to imply that all tumors to be felt in the cervix are polypi. Myomata quite interstitially situated may grow into the cavity and extend a part of their surface downward (comp. Fig. 136, p. 228).

An excellent method in such cases is to grasp the myoma directly with forceps (Fig. 15, p. 36), and to twist the tumor during combined examination. If the tumor can be moved only with difficulty and the uterus follows these motions, it is a non-pediculated myoma or one with a very broad pedicle; if the tumor can be easily turned around its longitudinal axis in the uterus, the pedicle is thin. Positive antisepsis is absolutely necessary during this attempt, as a tumor of this class not rarely sloughs after the slightest injuries.

Without operative interference it is often quite impossible to satisfy ourselves as to the upper terminus or the origin of the myoma. On the other hand, there are also fortunate cases in which the finger distinctly passes the greatest periphery of a polypus and is enabled to circumscribe the pedicle. Special difficulties are caused by large tumors situated in the vagina. Partly owing to their size, it is impossible to palpate them completely; partly there are adhesions to the vaginal walls which prevent upward penetration. A doubtful case is to be examined in narcosis by all possible combinations from the rectum and bladder; all anamnestic data, results of preceding personal and foreign examinations are to be taken into consideration; but then hardly a single case will perpetually remain obscure.

Where the tumor is sloughing when the case is first seen, we might think of carcinoma; then microscopic examination will be the most important, showing in the one case smooth muscular fibres, in the other characteristic carcinoma cells. A gangrenous myoma is shreddy, brownish-red, feeling like the placenta of a dead fœtus; a carcinoma is friable, lighter-

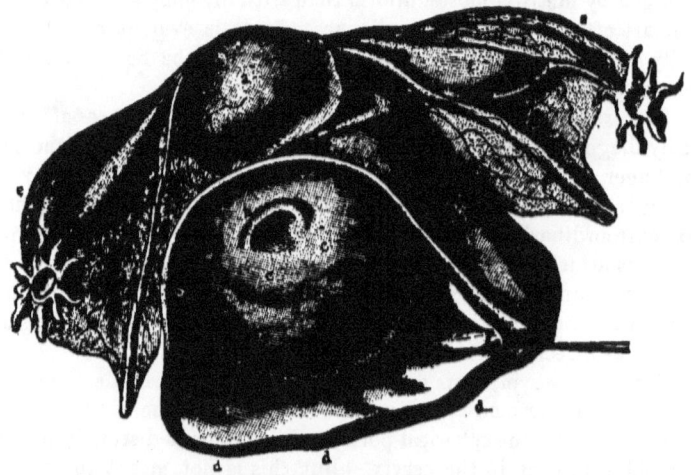

Fig. 139.—Large cervical myoma, after Boivin and Dugès. a, fundus of the uterus; b, os uteri; $c\,c\,c$, posterior lip of the os, greatly expanded by the tumor seated in the cervical parenchyma; $d\,d\,d$, vaginal wall; $e\,e$, tube; $f\,f$, ovary; h, round ligament.

colored, softer. Moreover, a sloughing carcinoma bleeds when the finger detaches particles, while there is no or only very slight hemorrhage when tearing shreds off a sloughing myoma.

Sometimes tumors spring directly from the vaginal portion. Here, equably round carcinomata arise, ordinarily in front of the lower surface of the vaginal portion, while myomata come from the cervical canal. Besides, the latter are usually much harder than the carcinomata. However, even if in doubt as to the diagnosis, such tumors would always be cut off.

In the case of an interstitial myoma of the cervix, a myomatous hypertrophy of one lip or of one half of the vaginal portion (comp. Fig. 139) the os is displaced laterally; it may be scythe-shaped, adjoining the myoma with a sharp margin, so that a hæmatometra may suggest itself. Such cases are often obscure until exploratory puncture is made, because with strong internal pressure in hæmatometra the consistence, likewise, appears firm. An exploratory puncture is certain to clear up these cases. Only when the blood is tarry will there be a hæmatometra. In myoma, too, a great deal of blood often escapes when a large blood-vessel has been accidentally pierced, but of course the blood is quite fresh.

Treatment.

Formerly the bath treatment, salt-water baths, and all iodine preparations were employed internally and externally. It was hoped that, being "resorbents," they would prove their virtues also on myomata. It cannot be denied that the family physician has the right to paint the abdomen of an inconsolable patient with tincture of iodine, or to send her to a spring of iodine, bromine or salt water; but the reports of the results thus obtained, of course, are not to be taken in good earnest.

The only drug having any influence on uterine myomata is ergotin. It is Hildebrant's great merit to have insisted on the renewed employment of this remedy. The following solution is the best for this: ℞ Ergotini bisdepurati, 2.0 gms.; Aquæ dest.; 8.0 gms.; Acidi carbolici fluidi, gtt. i. S. Inject once a day, subcutaneously, one Pravaz syringeful (one gramme). Before and after the application, the syringe is to be carefully cleansed in cold water. The spot where the injection is given must be washed off and disinfected. The injection is best made into the abdomen; there is room enough for a number of injections. The point must penetrate some distance, the ergotin should be injected more deeply than morphine. With this method no abscesses come under observation. But the injections are painful and the punctures remain for a long time as sensitive hard nodules.

The ergotin injections are to be discontinued as soon as symptoms of ergotism appear. These symptoms consist in a sense of formication in the tips of the fingers and the often distinct bluish tinge perceptible through the nails. Different individuals react variously an ergotin. Ergotism sometimes appears after fifteen to twenty injections, while other patients readily bear one hundred injections.

The theory of its action is, that the intact layers of uterine parenchyma contract; these contractions cause both shrinkage of the vessels with consequent defective nutrition of the myoma, and direct pressure upon the myoma. A proof of the correctness of this view is furnished by the not infrequent cases in which myomata necrose during the ergotin treatment, and the every-day observation that painful uterine contractions follow on injections of ergotin.

Success is to be expected only when an intact muscular layer surrounds the myoma. Subperitoneal myomata with thin pedicles are not influenced by ergotin. Interstitial or submucous myomata experience both effects of ergotin, while polypi may be pressed out of the uterus by ergotin, but of course cannot be directly diminished.

Should a polypus be observed to descend after ergotin injections, the process is sought to be hastened. In such cases I have several times given a double dose of ergotin. When the ergotin presumably can be used for but a few days longer, it is permissible to increase the doses rapidly in strength.

The ergotin treament must always be continued for some time, and *daily* injections conscientiously given. An incomplete, interrupted course is a needless torture of the patient. On the other hand, if absolutely no influence has been gained during two menstrual epochs, or after about forty injections, the case must be treated differently.

As to the result of the ergotin treatment, cases undoubtedly occur in which myomata *disappear completely*. I myself have observed the disappearance of a myoma the size of a child's head and one the size of a fist, with alarming hemorrhages, after three months' injections of ergotin. However, it is a *very small* percentage in which radical results are to be noted. Still, even without radical cure, improvement of the symptoms, especially of the hemorrhages, is very frequent. On the other hand, cases likewise occur in which ergotin is devoid of any effect.

After all, a trial of this disagreeable but not dangerous method is always to be recommended, for experience teaches that, even without radical cure, the good influence of an ergotin treatment on the hemorrhages lasts for years, and that the growth of the myoma proceeds with extraordinary slowness or ceases entirely.

For the hemorrhages *tincture of iodine* has also been injected both into the uterine cavity and into the tumor. The former is quite rational, but useless in severe cases and impossible of execution if the uterine cavity is blocked. Even *injections of liquor ferri* have not the slightest effect, because the bleeding surface is altogether too large, too uneven, and insufficiently accessible to enable any injected remedy to come in contact with the mucous membrane *all over*.

The injections of tincture of iodine into the tumor are free from danger, as I have repeatedly convinced myself; but I have never seen any perceptible effect from them. The same be said of *electrolysis*, with this difference, that it is not harmless.

Another method was based on the view that the hemorrhage was the consequence of stretching of the mucous membrane. This stretching was to be diminished by longitudinal incisions over the tumor, thereby arresting the hemorrhage. Cases in which effects were obtained in this manner have been repeatedly described.

Sewing up of the os uteri with lead wire, of course without freshening it, has been proposed by Freund and carried out with good effect.

Of old, injections of *ice water* have also been given, and pellets of ice placed in the vagina, both loosely and inclosed in a rubber bag. The cold injections sometimes provoke vesical tenesmus, are very disagreeable to the woman, and are effective only when very long continued.

Quite recently, *hot injections* have also been proposed—four to six litres of water at 32–34° R. (104–109° F.) injected into the vagina. Indubitable results have been obtained from this method. But in really dangerous hemorrhages all methods fail.

Precisely the circumstance that these hemorrhages continue for days forces us eventually to the *tamponade*. Even during its continuance the ergotin treatment should be begun. Although it cannot be denied that the manipulation of tamponing and the frequent change of the tampons favors the hemorrhage, there are cases in which the vital indication leaves no other choice.

Attention must be paid also to the couch on which the patient is placed. If the patient lies in a warm feather bed, perhaps even on feather pillows; if the pelvis, in a depressed portion of the bed, is the lowest part of the body, the couch will have to be changed. The patient must lie quite horizontally ; the pelvis, if possible, must be on a higher plane than the trunk and the legs. Absolute rest is necessary.

All heating beverages—coffee, tea, beer, and wine—are to be avoided. Acid lemonade, milk, and bland diet are to be advised. As the hemorrhage is often more profuse during constipation, some bitter-water should be given before the onset of menstruation, so as to keep the bowels rather empty during the subsequently necessary repose.

The favorite internal medication of acid tinctures of iron, liquor ferri in solution, etc., injures the teeth and the stomach, but is of no use. However, narcotics are required for pain, eccoprotics for constipation, analeptics during convalescence.

Should the symptomatic treatment bring about no improvement in the symptoms ; should, on the contrary, the tumor grow very rapidly ; should the injurious consequences increase, *operative removal of the tumor* will have to be taken into consideration. The myoma *per se* does not indicate myomotomy, but the impossibility to remove the dangers of the myoma by symptomatic treatment and to arrest the growth of the tumor. *Myomotomy*, therefore, is always the last resort, and in every single case the danger of the operation and that of the continuance of the symptoms must be most conscientiously balanced.

Therefore, the conditions are different here from what they are, for instance, in ovarian tumors which must be removed on principle, because they always grow and certainly lead to a painful death.

We distinguish between *colpo-myomotomy* in which the operator attacks the tumor from below, and *laparo-myomotomy* by which the tumor is removed through the abdominal cavity. Of course, in most cases it will be clear from which point the attack is to be made. In the case of polypi and tumors generally, the limitation of which can be distinctly felt from below, colpo-myomotomy is to be performed. On the other hand, there is no doubt, in the case of subperitoneal myomata, that laparo-myomotomy should be done. However, we shall see that there are doubtful cases.

To begin with the *operation for polypi*. Should the tumor lie in the vagina and the finger feel the pedicle reaching into the uterus (comp. Fig. 137, p. 228), two fingers are placed against the pedicle, Siebold's scissors are carried up along the palmar surface of the fingers, and the pedicle is severed under the control of the finger-tips. Two curved Siebold polypus scissors are represented in Figs. 140 and 141.

These two scissors, especially the one shown in Fig. 140, are quite excellent instruments, indispensable to every gynecologist. The masses of tissue to be divided are firm and resistant so that very strong scissors are needed. None of the multifariously curved and complicated instruments answer as well as Siebold's scissors. In myoma operations weak scissors cannot be used at all.

Often the cutting may be facilitated by grasping the polypus with Muzeux's forceps (Fig. 15, p. 36) and thus drawing the pedicle taut. Twisting of the pedicle also renders the division easier. Such polypi have been also simply twisted off. When the pedicle is very thin, this can be easily done, but in the case of pedicles of some thickness cutting is easier and less dangerous.

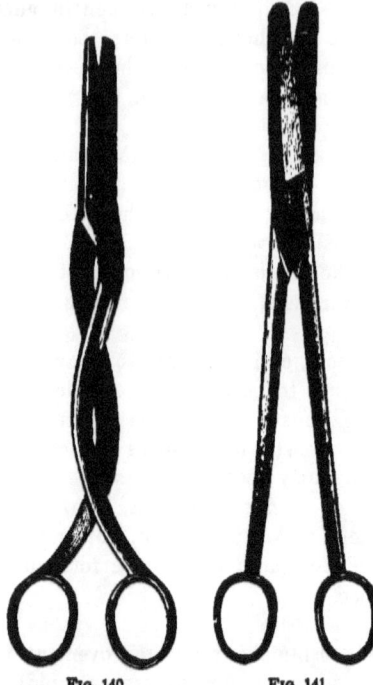

FIG. 140.   FIG. 141.

FIG. 140.—Siebold's polypus scissors; the double crossing of the handles renders it possible to cut at a great depth, for the scissors can be opened widely without requiring much room for the handles.

FIG. 141.—Superficially curved scissors capable of reaching behind and around the tumor to the pedicle.

But in many cases the operation is not so easy. The difficulties consist *in the size, the unfavorable accessibility, and the attachment of the tumor on a broad base.*

When a myoma lying in the vagina is so large that its greatest periphery cannot be felt, the tumor must be treated similar to a child's head, that is to say, the forceps are applied and we endeavor to draw out the tumor, or at least to get around it with the fingers during traction. If this fails, twisting around the longitudinal axis is to be attempted; if the myoma readily yields to the torsion, it is movable and pediculated; if it cannot be twisted at all, it is either attached to the uterus on a broad base, diffusely blending with its tissue, or is adherent to the vagina. The former is ascertained by combined palpation of the uterus during the torsion, the latter by circumscription with a forceps blade or a sound. Having diagnosticated a pediculated myoma, a large polypus, we endeavor to pass the finger around the greatest periphery, if necessary under chloroform. If it be possible to reach the pedicle with curved scissors, the pedicle is to be cut off. With every cut the tumor descends a little more; the fingers feel the firm, tightening cords of the pedicle; many slight incisions sever one cord after another; the assistant pulls vigorously on the tumor, and suddenly a fresh incision divides the rest of the pedicle or it tears under the downward traction. The hemorrhage usually is inconsiderable, but the greater it is the more rapidly must the operation be done. After incomplete operations the danger to life is imminent, for the tumor almost invariably sloughs (comp. p. 230, supra).

But if it be impossible to reach the pedicle with scissors, we may succeed in crowding a wire snare with the finger tips so high all around that it will adjoin the pedicle. Levret has already employed two tubes through which a snare was run. With these tubes he pushed up the snare and carried it around the tumor. However, in very difficult cases I have been able to do more with my fingers than with such instruments. By means of the snare the tumor is now pulled strongly downward, and the scissors are again sought to be passed up.

A direct division of the pedicle with the wire snare or the wire écraseur will succeed only with a very thin pedicle. No reliance should be placed on the use of the wire écraseur so common nowadays. There is no wire which will not break or tear whenever it is intended to sever a pedicle of a myoma thicker than a finger. Even the écraseur is unreliable. The most superior method of operation, whenever an application of the scissors is impossible, is the *galvano-caustic ablation*. The practitioner had better send such, certainly rare, cases to an institution where the galvano-caustic removal can be performed than to risk the operation with imperfect procedures.

If écrasement can be with justice termed an antiquated procedure, then, of course, ligation with subsequent decomposition may be considered a thing of the past.

If the pedicle has been severed in some manner, the tumor must be removed. The vulva not being distensible as in pregnancy, quite extraordinary exertions are sometimes required for the extraction of a large tumor. The most appropriate instrument for this purpose is the obstetric forceps. There is also a special instrument, made like a miniature forceps. The latter instrument, too, has been provided with barbs directed downward, so that it has a very firm grip. Muzeux's forceps are also employed, but they tear out when using some degree of force. Obstetrical extraction instruments, such as the cranioclast or an excerebration forceps, and the Stein-Mesnard bone-forceps have also been used in lieu of a better. Covered hooks and corkscrew-like instruments have also been specially devised for this purpose. But in general it is more correct to complete these operations with the simplest auxiliaries than to add to the gynecologist's stock of instruments.

If it be absolutely impossible to reach the pedicle, a passage must be made by *diminishing the tumor*. This may be done by either of two methods—by excising wedge-shaped pieces from the tumor and quickly sewing up the profusely bleeding wound, or by employing the so-called *allongement opératoire*. In the latter, the tumor is to be surrounded by spiral incisions and then drawn down.

In cases requiring an excessive amount of force for the withdrawal of the tumor, or in which the hand with the instrument can be only with difficulty insinuated between the vagina and the tumor, a fatal injury might easily be caused by rupture of the vagina or detachment of the uterus from the vaginal vault. It is decidedly less dangerous to create a passage by the diminution of the tumor. Certainly the hemorrhage is quite enormous; therefore, we must not proceed haphazard, but everything must have been previously well considered and arranged, lest the operation might have to be interrupted. Only too readily, with imminent danger to life, the plan is finally abandoned, and the tumor removed in any manner, not to say torn out. It is one of the most difficult tasks to finish such a case correctly, according to the preconceived plan.

Another difficulty arises when the myoma has caused an *inversion* of the uterus, or when a differential diagnosis must be formed between inversion of the uterus and polypus (comp. p. 223). If a myoma projects, we cut down upon it and try to isolate and separate the tumor from the uterus with the finger. When certain of proceeding in the limiting layer, we operate partly without, partly with cutting, very firm cords being divided with the scissors.

If the tumor lies partly in the vagina and partly in the uterus, the constricted part in the internal or external os often simulates a pedicle. Particularly when the vaginal part has become œdematous, such an error may easily be made. In such a case the doubt is dispelled, especially, by the roomy, long uterine cavity demonstrated by the sound, as well as by the size of the uterus ascertained by the combined examination. Although in such cases it may not be possible to reach the pedicle with the finger, the galvano-caustic loop may be guided upward. It is not absolutely necessary to exterminate every part and parcel of the tumor in order to obtain a positive result. Even when a long pedicle has remained behind, it is retracted and, as it were, merges in the uterine parenchyma.

The second difficulty consists in the inconvenient accessibility of the tumor. Here we have to deal mostly with intra-uterine polypi in nulliparæ. In cases of that nature, dilatation with laminaria tents often fails. To repeat it for days and follow it by a laborious operation is much more dangerous than to gain access by *deep, lateral incisions* and forcible procedure. Even then the separation is still very laborious, owing to the firmness and narrowness of the uterus. In these cases it may happen that barely one finger with the scissors can be forced through the narrow, tubular cervical canal and that an hour or more is required for cutting through a pedicle the thickness of a finger. A very good instrument is the spoon forceps represented in Fig. 142. It consists of two oval spoons with sharp and dull margins at the extremity. The tumor is held very firmly in the concavity, especially on account of the fenestræ. The entire tumor or a part of it is grasped and twisted off by vigorous rotation. We must have gained the conviction that when a myoma extends into the uterine cavity on a *distinct pedicle*, it is of submucous origin, or at least has a sufficiently thick layer of uterine tissue above it to render perforation of the body during the torsion an impossibility. It is certainly of importance that whatever be grasped should not be drawn down during the torsion, but *be twisted off as close as possible to the basis.*

If but *one* finger and a *single* instrument can be got into the uterine cavity, we can hardly warn enough against any sharp instrument. With the latter we lacerate the finger and the uterus without being able to avoid it. Myomata are too tough for the sharp spoon with which nothing can be done.

As the last difficulty we enumerate *the attachment on a broad base* or the *interstitial seat of the tumor.* The decision whether operation is to be done at all and if so, in what manner, is difficult in those myomata which clearly project into the uterine cavity or even into the os uteri without being pediculated. Thus, for instance, the case illustrated in Fig. 136, p. 228, simulates a polypus, while the portion felt is merely the small segment of an enormous myoma pressed into the cervix. We encounter both encapsulated and soft, diffuse myomata gradually blending with the surroundings. Many an operator has learned to his great sorrow that it is absolutely impossible to know beforehand whether a

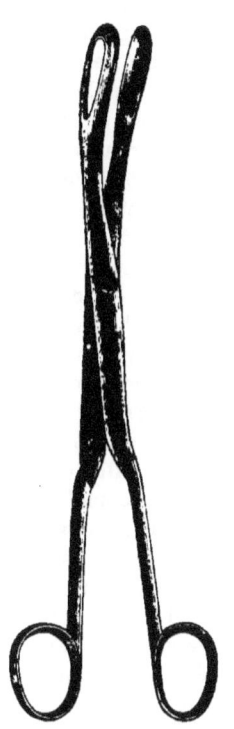

FIG. 142.—Spoon-forceps for grasping, twisting, and tearing off small myomata or their particles.

myoma lies loosely in a capsule or whether it merges gradually into the surroundings.

We often read of polypi "with broad base" which are nothing but myomata which, having originated near the internal surface, expanded the uterine cavity, projected into it, and gradually became submucus tumors.

Here especially the principle is valid—not the presence of a myoma indicates the operation, but the fact that the tumor causes symptoms the removal of which is *absolutely necessary*. The dangers of these operations are incalculable and no conscientious physician will resolve upon it with alacrity. On the other hand, if the hemorrhages are dangerous to life, if the myoma grows so rapidly that life will presumably be put in jeopardy, and if no other treatment is effective, we must shoulder the responsibilities of the operation which is then justified.

The procedure is governed by the condition. We must examine, if necessary under chloroform, precisely whence the myoma springs and to how great an extent it is connected with the uterus. If the combined examination shows that the myoma is large and virtually represents an enlargement of one of the uterine walls; that the cervix is intact, perfectly preserved, and very hard and firm, laparo-myomotomy must be chosen, the dangers of which can be calculated and met. But if convinced that the myoma springs from a comparatively small space, that it is a *polypus* with broad base and short pedicle, that the os is dilatable and the tumor not particularly large, colpo-myomotomy may be selected.

The method practised in the latter has received the euphemistic name of "enucleation." It cannot be denied that there are fortunate cases in which, after division of the mucosa or a thin muscular layer we encounter a hard myoma that can be isolated. In such a case the tumor is enucleated with the finger, raspatory, elevatorium, spoon-forceps, or other blunt instrument, as a firm atheromatous cyst is removed from the scalp. The remaining cavity is irrigated.

The hemorrhage usually ceases after a few hot injections. Otherwise we may try to approximate the walls of the cavity with deep sutures. If this fails, owing to the high seat of the cavity, and the hemorrhage is strong, a tampon dipped in liquor ferri is placed in the cavity. Of course, the removal of the tampon is difficult; often small, very gradually detached particles of cotton and crusts remain behind for a long time, maintaining suppuration and sloughing. Then perhaps resorptive fever ensues and urgently demands the disinfection of the sloughing cavity The procedure to be employed here is described on p. 50—permanent irrigation.

If it can possibly be avoided we save ourselves and the patient the inconveniences of the styptic tamponade.

Judging from my experience, the myomata which can be enucleated are decidedly the rarest. Much more frequently the interstitial myomata gradually blend with the uterine tissue, and after incision of the supposed capsule we are in the disagreeable plight of being forced to work on in the dark, barely aided by the touch. Blood streams over the hand; some few though shreds are torn off with the fingers. One instrument after another is introduced—none suffices. Finally the operation must be interrupted. The operation can hardly be called a "curative" procedure. But the patient has in her uterus a large part of the tumor, bruised, suggillated, and severed from its nutrient vessels. Of course, a process of demarcation may develop. Then the remnants of the tumor slough and are cast off, but the dangers thereof are very serious for the patient. In the first place, the hemorrhage is often enormous. Then, acute sepsis sometimes sets in; peritonitis, without as well as after perforation of the uterus. Especially frequent in these cases is a phlebitis proceeding with thrombosis and embolism and finally as pyæmia. The thrombosis begins in the wide vessels of the tumor; owing to the great anæmia they continue in the shape of marantic thrombi into the wide efferent vessels lacking the *vis a tergo*, the sloughing remnants of the tumor furnish the required germs of decomposition, and the pyæmia is established.

Experience, however, teaches that many cases terminate favorably. The chief essential will be that as much as possible is removed. Hesitating operators who work slowly and cease too soon, alarmed by the hemorrhage, will have less favorable results than those who remove at least the *greater part* of the mass as rapidly as possible, always guided by the combined manipulation.

But in every case, whether the object has been attained more or less completely, the antiseptic apparatus must be employed in the most careful manner. *The most important antiseptic instrument is the thermometer.* After the operation the temperature must be taken every two hours. Immediately after every rise the wound must be irrigated with two to five per cent carbolic acid solution. During every irrigation we feel for shreds which have been subsequently detached or are but slightly adherent, and remove them. After forty-eight hours, when the danger of fatal hemorrhage and of *primary* sepsis is past, I prefer to take less energetic disinfectants—salicylic acid, chlorine water, sodium sulphite, alcoholized water, or only salt-water. In the many recesses of the ragged tumor, too much carbolic acid remains behind. And a carbolic intoxication has its great dangers in very anemic persons.

Should the fever, *which must be exactly controlled even in the most favorable case,* not have subsided after the irrigation, we would have to resort to permanent irrigation, as described on p. 50, the same as after the styptic tamponade.

Colpo-myotomy is modified not a little if the myoma should have become necrotic already before the operation. In such cases, the tumor is often found quite relaxed; it opens like an unrolled curl, and drops in front of the vulva in the shape of a red, tough, villous mass. This condition lasts at most twenty-four hours, then the mass grows softer, more smeary; it putrefies, becomes offensive, and falls to pieces.

Under these circumstances, the treatment is very simple. There can can no longer be any doubt whether the procedure is permitted; the operation must be done at all hazards.

The hemorrhage is usually not very considerable; the greater the hemorrhage the more should the operation be hastened. But we must beware of tearing or cutting the tumor *below*. The shreddy mass leads to its point of origin as the funis does to the placenta. Having found it or palpated it to some extent, all that can be should be removed. For this purpose, too, Siebold's scissors (Fig. 140) are the best *sharp*, the spoon-forceps (Fig. 142) the best *blunt* instrument. Precise directions for the procedure cannot be given. Only the operation must be done rapidly; often the os closes as soon as the tumor which dilates it is removed.

In the case of hemorrhage after the operation, the simplest procedure is the styptic tamponade.

In these cases especially, permanent irrigation is in place, for we must not only produce a *prophylactic* effect, but *disinfect a wound already infected*. In every case, unless several days' sloughing had effaced every distinct outline, I found, in the remnants ultimately removed, some purulent tracks, there being a quite distinct line of demarcation separating the sequestrum, the necrotic myoma, from the healthy or still nourished tissue.

The observation that a myoma becomes necrotic after minor surgical procedures has also found therapeutic application. Some operators made an incision above the myoma, that is to say, through the covering muscular layer; thereby the myoma sequestrated itself, and was removed in a second session. It may also be possible by these incisions to open the capsule of a firm myoma and thereby to promote its spontaneous expulsion.

But reports of such operations with good results should by no means induce any one to believe that every myoma projecting into the uterine cavity may be caused to be expelled in a harmless way by longitudinal incisions across the tumor. I have repeatedly pointed out the danger of incompleted myoma operations.

I hope my demonstration may have rendered the dangers evident, and, in conclusion, I repeat the warning not to attempt colpo-myomotomy in interstitial myoma or in submucous polypi with broad base, without the most urgent indications.

Large uterine myomata of rapid growth, especially the cystic, must be removed by *laparo-myomotomy*. Interstitial myomata, too, unless they are very small and the vagina and cervix very roomy, should be operated on from the abdominal cavity.

To be sure, in the case of laparo-myomotics, all the dangers may be calculated beforehand, the tumor can usually be completely removed with certainty, the hemorrhage arrested, and we may rely on having no subsequent disease unless there be sepsis. *However, despite all asepsis, every laparo-myomotomy is an operation endangering life.* Therefore, only when the indication is urgent, should the operation be done. Apart from the cases in which symptoms of incarceration furnish the vital indication, laparo-myomotomy is in place generally when the tumor is of rapid growth, with reflex effect on the general condition, and where there are hemorrhages which resist every measure for their arrest. It is clear that, in proportion as laparotomies in general become less dangerous, the circle including the indications for laparo-myomotomy may be extended.

The operative procedures are still in the course of development, and it will suffice, therefore, to describe the operation briefly. The general principles are: rigorous antisepsis, covering the incised wound in the uterus with peritoneum, prompt arrest of the hemorrhage, complete closure of the abdominal wound, *i. e.*, dropping of the pedicle.

The size of the abdominal incision will vary according to that of the tumor. In very large tumors the abdomen has been opened from the xiphoid process to the symphysis. As the tumor ordinarily is not spherical, it is brought through the wound with its narrow side. If there is a distinct pedicle, in the case of subperitoneal polypi, the pedicle, if thin, is ligated with a strong silk thread. Then the tumor is cut off, the pedicle is watched for some time to see if any blood oozes through it, perhaps the ligatures are tightened, a second thread is passed around the pedicle, vessels which look suspicious are ligated, the peritoneum is isolated, and, if at all possible, drawn over the incision by fine sutures.

In the case of thicker, shorter pedicles, that is to say, where the tumor is immediately attached to the uterus, a preventive ligature is laid around the pedicle and subsequently tightened. Later the thread is laid in the grooves of the snare. The wire écraseur serves for the constriction of the pedicle. These instruments have been constructed in very various modifications. A simple instrument is Koeberlé's wire écraseur (Fig. 143). The free end of the wire is passed around the pedicle of the tumor, then through the upper opening, and fastened around the lower button, like the other end of the wire. Turning the screw then will diminish the loop.

If the tumor has been separated, after the application of the snare, the field of operation is accessible. If the cut surface be broad, the incisions are made wedge-shaped. Then very deep, strong sutures will approximate the wound surfaces (Schroeder). The snare is relaxed cautiously, so as to tighten it again at once in the case of hemorrhage and apply additional sutures. .When the hemorrhage has been definitely arrested, the snare is removed. This is effected most rapidly by dividing the wire with cutting pliers.

If there be no pedicle, the ligature is applied immediately below the tumor. Therefore, in the case of tumors extending into the cervix, the ligature may have to be placed around the latter. For this purpose it is advantageous to use a thin rubber tube, which is placed in several coils below the tumor after the latter has been greatly elevated. It will then be possible to enucleate the tumor above. Schroeder has shown that the uterus may also be ligated directly without being followed by sloughing of the ligated portion. The latter author introduced a needle with double thread into the uterus, ligated toward both sides, and excised the tumor like a slice from a melon. The wedge-shaped wound was stitched together. In tumors extending far down, Schroeder has applied the ligature below the tube and ovaries, and thus has removed a portion of the broad ligament, tube and ovary. The removal of the latter is always necessary whenever the entire uterus is amputated.

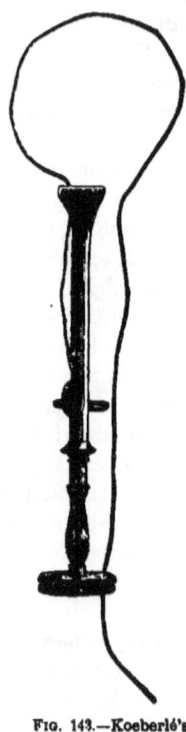

FIG. 143.—Koeberlé's wire écraseur.

But if the myoma lies deeper than the internal os, the removal will be impossible; it is also questionable whether myomata can be extirpated which have proliferated between the layers of the broad ligament and below the pelvic peritoneum.

In the case of very large tumors which he could not lift out of the abdomen, Péan has passed *needles through the tumor*, and thus ligated or constricted it in portions. Then the tumor was cut off piecemeal. Péan did not venture to drop the large wound-surface, and fastened it in the abdominal wound. This method permitted that the écraseur could be allowed to remain for the time being, and therefore seemed to offer greater security against after-hemorrhage.

The larger the portions of tumor left behind and the more extensive the cut surface uncovered by peritoneum the greater will be the dangers to life.

Under the name of "anticipated climax," Hegar, after Trenholme's procedure, has proposed and performed an operation which is rational in theory and salutary in practice. It is certain that, after cessation of menstruation, the most dangerous symptom, the hemorrhage, often disappears completely, and that, if no retrogression, at least arrested growth of the tumor is frequently observed. Accordingly he hoped to anticipate the favorable influence of the menopause by *castration*, the removal of both ovaries, whether they were healthy or not.

Experience thus far has not permitted a positive decision. Indubitable cases occur in which the hemorrhage ceases definitely, the tumor becomes smaller and even disappears, that is, cases of relative cure. On the other hand, Schroeder reports that the "cure" had lasted but a short time, and that the tumor had grown after castration. The last-named author attributes the favorable influence of castration partly to the necessary ligation of large nutrient vessels and even saw a case of shrinkage of a myoma after ligation of a very vascular omental adhesion.

Therefore, if laparotomy be done at all and the tumor is to be operated on in some manner, it is better to remove the whole tumor. The result of castration is uncertain and the dangers are great. Owing to the doubtful result, laparotomy should not be risked unless the indication is vital.

More than all, however, castration is often very difficult, as during the growth of the tumor the ovaries may be dragged toward the tumor, or displaced by it, or be quite inaccessible. We, therefore, side with Schroeder in that he would see castration resorted to only as a forced substitute when the removal of the myoma is impossible.

Finally, we would urgently advise, in every case, very carefully to consider the chances of the operation. For nothing is more embarrassing than to fail to find the ovaries, to be unable to reach them, or to fail to isolate them when the tumor cannot be removed, and the abdomen must be closed again without having accomplished anything.

B. Carcinoma of the Uterus.

Anatomy.

As carcinomata generally evince a preference for places where two different species of epithelium adjoin, so uterine carcinomata are found most frequently at the vaginal portion, both relatively and absolutely.

The etiology is unknown. Carcinoma occurs most frequently between the fortieth and fiftieth years of life, but cases have also been described affecting girls from sixteen to seventeen years and old women of seventy to eighty.

Carcinoma of the uterus appears in different forms. In the first place, with intact mucous surface of the vaginal portion, we find carcinomatous polypi sessile on a broad base, of regular, round shape. Also beneath the mucous membrane one or several nodes form, slightly prominent at first, which gradually render the vaginal portion more and more irregular in shape. Other nodes form adjoining to the former ones, so that, while at first but one lip was affected, eventually the whole segment of the uterus palpable per vaginam is changed into a knotty mass.

The carcinoma proliferates downward in the vaginal walls. Carcinomatous nodes form in it, down to the introitus vaginæ, which are directly encountered by the examining finger, so that the touch is often altogether impossible. In the same way, this utero-vaginal carcinoma forms regionary metastases in the parametria. These tumors enlarge so greatly that they fix the uterus completely and frequently extend to the pelvic wall. Thereby the ureters coursing along the parametrium become dislocated, compressed, and stenosed. Thus hydronephrosis and dilatation of the ureter arises—an almost constant condition in uterine carcinoma. The neighboring organs, too, bladder and rectum, are implicated. The latter is dislocated, flexures and fistulæ form. The intestinal gases penetrate into the carcinomatous masses; the latter disintegrate, become inflamed, abscesses break downward and upward. Thus I made the autopsy on a case in which an abscess opened above the symphysis and discharged ichor, fæces, and intestinal gases. The *bladder* is also attacked by the carcinoma, perhaps perforated, leading to the formation of fistulæ, atrophy of the bladder, and catarrh of the vesical mucosa.

Above, the body of the uterus is usually preserved; the pale carcinomatous nodes are clearly demarcated from the bright-red uterus. On the other hand, we find occasionally a carcinomatous infiltration of the entire organ, so that the malacic uterus, the size of the puerperal one, consists throughout of masses of carcinoma only.

The carcinomatous nodes extending below the *peritoneum*, dislocate the latter and lead to its inflammation. The peritoneum of the fossa of Douglas is pushed upward. Numerous peritoneal agglutinations arise; thereby the uterus becomes imbedded up to the fundus behind and in front in pseudo-membranes and membranous inflammatory products. In rare cases, carcinomatous nodes grow all over the peritoneum—*carcinosis of the peritoneum*.

The characteristic feature of all carcinomata is the rapid growth, and the speedy disintegration, in some places, connected with it. Thus also uterine carcinoma disintegrates in portions, by the pressure of its growth, through fatty degeneration. The cancer melts down in some places, ulcerations result which discharge the necrotic stroma of the carcinoma —fatty epithelia, shreds of connective tissue, serum, and blood. Later on the secretion becomes offensive, sanious.

Should the primary cancerous node arise somewhat higher up, it will also proliferate from the orifice of the otherwise intact vaginal portion. The os is crowded apart, the pale vaginal portion surrounds the bright-red, small tumor. In its further development the forms described occur here too.

Thus far we have spoken of a deep *parenchymatous* carcinoma, but the superficial *epithelioma* also occurs upon and in the uterus—first in the form of *papillomatous carcinoma* or cauliflower excrescence. This carcinoma, springing from the lower surface of the vaginal portion, forms a dendritic tumor of rapid development which, beginning at a circumscribed spot, has the tendency to fill the vagina in its growth. At the same time processes, or glandular tubes in carcinomatous degeneration, proliferate upward, so that even after operative removal of the tumor a relapse occurs in the cut surface. The fact that almost invariably after ablation of such a tumor the same final forms as in parenchymatous cancer eventually arise, shows most clearly its carcinomatous character.

Furthermore, real *epitheliomata* are found in the cervix—superficial cancerous ulcers. They change the cervix into a deep wide crater. But the os may also be preserved for a long time, only the cervical canal being dilated by the carcinoma. Subsequently, however, the cancerous ulcer extends to the vaginal portion and the vagina, eats into the bladder and rectum, and forms secondary cancerous nodes in the neighborhood. Here, too, the final result may exactly correspond to that of the parenchymatous carcinoma.

The last and rarest form is carcinoma of the body, really a malignant adenoma. It is a cylinder-cell carcinoma, destroys the body of the uterus, but leaves the vaginal portion intact.

The metastases of cancer of the uterus have nothing characteristic. In general, they are rare. Usually there is no mention of the fact that carcinomatous nodes occur in the spongy portion of the bones of the pelvis and thigh. I have also seen a metastasis in the clavicle, and once I removed from the external skin between vagina and anus a pendulous carcinoma the size of the fist which had appeared simultaneously with a relapse on the vaginal portion.

Symptoms and Course.

The *objective symptoms*, aside from the effects on the general condition, consist of *discharges from the vagina*. Hemorrhage and sloughing are characteristic. The former occurs both early and late, the latter at a later period.

Not rarely the patients report that for some time past menstruation was surprisingly profuse and prolonged, that blood also escaped often in the intermenstrual period, or else that the type of menstruation had entirely disappeared. As the carcinomata usually occur in the climacteric period, the patients are not at first much alarmed about the irregularities.

Frequently they report, too, that menstruation had ceased one, two, or more years, then it had recurred in an intensified degree, in quite irregular manner, and with particularly weakening effect. In other cases again, the statement is made that every trauma produces hemorrhages, for instance, much straining at stool, coition, and physical exertion.

Quantitatively the hemorrhage is very variable; in some cases, but little blood may pass continually, while in others so large an amount of blood may be lost that the patients faint and report flooding.

Hardly ever shall we receive a negative reply to the question whether bloody water has escaped. This discharge, described also as "meat-water like," has often a purely serous character in cauliflower growths, and at first is not at all offensive. Careful inspection always shows the admixture of small white flocculi—lumps of epithelia and shreds of the cancer stroma. In parenchymatous carcinoma, secretion may be altogether absent for a long time, only the menstrual hemorrhage is very profuse, owing to the consecutive metritis. On the other hand, the secretion of a carcinoma of the cervical mucosa is very profuse from the beginning. Here the discharge at first resembles a catarrh dating back to the puerperium, in which mucus, pus, and blood escape.

Of course, in all three cases, decomposition of the secretions will readily succeed the invasion of bacteria. Examinations and vaginal injections without antiseptic precautions, traumatic and spontaneous destruction of particles of the tumor, retained coagula and secretions, manipulations during the arrest of hemorrhage, uncleanliness of the external genitals, etc., with the roomy vulva of multiparæ, will transplant the agents of decomposition to the soil most favorably prepared for them. If decomposition of the secretions has been caused in this way, it can hardly be arrested again, and thenceforth the characteristic *cancer-juice* escapes. The latter has a most disagreeable, sweetish, nauseating odor which becomes especially penetrating when mixed with urine, owing to the presence of a fistula. Among the lower classes devoid of the facilities for thoroughly cleansing the patient, where the care of the relatives relaxes but too soon when certain of the incurability, where all the members of the family are crowded into one narrow room, and where economy preserves the expensive stove-heat as long as possible, an atmosphere sometimes forms which, being almost unbearable, restrains the entry of the physician.

Constipation nearly always occurs. It may be mechanical—the carcinoma may constrict the passage—and of more dynamic origin; the fæces are voluntarily retained for fear of the pain and hemorrhage connected with defecation. The loss of fluids, too, renders the fæces harder.

In other cases, the direct proliferation of the carcinoma into the lowest part of the large intestine seems to put the whole of the latter in a state of irritation, causing colliquative diarrhœas. In such cases, too, amyloid degeneration of the intestinal mucosa may be suspected.

The symptoms associated with the extension of the carcinoma to the *bladder* consist, in the first place, of those of a vesical catarrh. Frequent strangury, burning sensation during micturition, and a purulent sediment in the urine occur. Subsequently the lumen of one or both ureters may become diminished. With the gradual growth of the tumor, the other kidney will act vicariously without symptoms. But if there be bilateral compression of the ureters, they may dilate to the volume of the small intestine, hydronephrosis, and even pyelonephritis and uræmia occur, with their sequelæ to be discussed below. If a fistula have formed, the urine escapes per vaginam. Then the vulva and its surroundings become erythematous and painful.

If the tumor extend to the neighborhood of the large vessels, compressing or even projecting into them, œdema of one or both lower extremities occurs. The œdema soon extends upward, the vulva and the lower abdomen becoming swelled. The vulva may be so obstructed by this œdematous swelling that the opening can hardly be found. The tumefaction extending beyond the anus, and the obstacles encountered by the blood flowing from the hemorrhoidal veins, lead to the formation of large nodes at the anus and, in connection with uncleanliness, facilitate the development of bedsores.

Toward the end of life, universal œdema often occurs in consequence of renal disease.

The *subjective symptom* of carcinoma—*the pain*—is by no means constant; it is certain that more carcinomata run their course without than with pain. Especially when the vaginal portion is alone affected, it is not at all clear whence the pain should come, for the vaginal portion is not sensitive. But if the new-formation implicates the uterus, pains must occur. They are often described as radiating into the legs, boring, lancinating, or labor-like. The latter character, even to paroxysms of uterine colic, will be assumed by the pains when the secretions are retained within the uterine cavity by carcinomatous masses. Not rarely the patients distinctly state that fluid is expelled with labor-like pains. Also the pressure exerted by the growth of a carcinoma, *e. g.*, within the body of the uterus, the direct destruction and irritation of the nerves are causes of pain.

The bladder likewise reacts by tenesmus upon the irritation of the intruding neoplasm.

The palpation of the lower abdomen, as well as digital movements of the uterus, are almost invariably painful. The above-mentioned participation of the peritoneum makes this appear not at all surprising. Not rarely the pains increase to those of general peritonitis. In such cases, we must assume the rupture of an ichorous patch—certainly a rare circumstance—or an acute general carcinosis of the peritoneum. The latter at times runs a febrile course.

Should the tumor disintegrate more and more, should the profuse hemorrhages weaken the organism, should there be the first indication of coma—incipient uræmia—the pains often cease *pari passu* with the deterioration of the condition.

Especially annoying, but fortunately very rare, is a hyperæsthesia of the vulva occurring in the shape of pruritus or vaginismus.

As regards the effects on the general condition, it is remarkable, in the first place, that septic processes are rare despite the sloughing. But the ichor can usually escape freely; there are within the tumor no healthy lymphatics capable of absorption, and should a blood-vessel be opened, the escaping blood carries away the infectious matter. Nor must we forget that the body, as it were, is callous. In the same way as many an anatomist with wounds makes dissections with impunity, while others would at once suffer from lymphangitis, so here, too, the body has become thoroughly impregnated. It is not impossible that a part of the symptoms relating to the general condition are due to a gradual reception of septic matters.

In a carcinoma, *anorexia* nearly always exists in consequence of the constipation. This lack of appetite, the sensation of nausea, and the recurring vomiting are to be referred to the continuance in the often horrible atmosphere and to uræmic intoxication and hydræmia. As has been repeatedly stated, symptoms pointing to *renal disease* are particularly frequent. Continual lassitude, actual somnolence, headache, and visual disturbances are to be referred to uræmia. Eclamptic convulsions occur, as a rule, only immediately before death. Albuminuria is not infrequent.

The aspect of the patient is designated as the *cancerous cachexia*, including the fawn-colored complexion; the slightly œdematous swelling of the face; the sunken dull eyes, seemingly larger on account of the absorption of fat; the slow motions of the anæmic lips which are laboriously drawn over the teeth in order to close the mouth; the suffering expression; the entire demeanor, indicative of profound physical suffering and resigned hopelessness.

Death ensues in various ways. The most frequent symptom, the *hemorrhage*, causes death in the minority of cases. Twice, however, I have witnessed death in consequence of spontaneous hemorrhage. Almost equally rare is death from acute peritonitis, at least when not due to an operative procedure. Acute carcinosis of the peritoneum, too, presenting almost the same symptoms as peritonitis, is but rarely fatal.

After surgical manipulations, of course, the accidental traumatic diseases, pyæmia and septicæmia, are likewise possible. By far the majority of the patients die of *uræmia*. A large proportion of those usually reported as having "died of exhaustion" belong here. The fact will not escape from the careful observer that, in the case of long-standing œdema, comatose conditions, mental hebetude, vomiting, and most violent headache temporarily appear and again disappear. Adding to this the use of narcotics, the weakness of the heart, the faulty quality of the blood, the disturbed assimilation connected with anorexia and constipation, the mental depression, and the respiration of tainted air, we should not be surprised that the machinery of the human body, even without any direct cause, finally cases to functionate. Often, however, the uræmia leads to notable phenomena, especially convulsions. They introduce the final stage, and the patient scarcely regains full consciousness. In every such case, I found at the autopsy œdema of the pia mater, and greatly distended, bluish-black veins.

As the beginning of a carcinoma does not come under observation, it is not possible to know exactly the duration of the affection. I have seen cases having a very rapid course, abscision of an isolated node at the vaginal portion being followed by death within three months, after reproduction of numerous cancerous nodes. On the other hand, I have observed a case for three and a half years. It was an uncommonly hard cervical carcinoma which caused no symptoms beyond profuse hemorrhages. In general the duration is assumed to be one and a half to two years.

Diagnosis.

The diagnosis of a carcinoma of some considerable duration is very easy. An error is almost impossible in such a case. The hemorrhages in the climacteric period, the offensive odor of the discharge, the cachectic appearance of the patient often permit, without examination, the formation of the diagnosis with the greatest probability. During exploration the finger easily discovers the variously hard, friable, or knotty tumor, either rendering the vaginal portion quite irregular or extending deeply into the vagina. A carcinoma very readily bleeds during the examination. For this reason care and gentleness are necessary when there is a suspicion of carcinoma.

Easy as is the diagnosis of a fully developed carcinoma, it is difficult, even impossible to recognize a carcinoma as such in the beginning. Here the differential diagnosis between carcinoma and the adenoid erosions of the vaginal portion described on p. 170 is of a paramount importance. The following may serve as landmarks: On the whole, benign erosions the consequences of the puerperium are more frequenly found in young women; only exceptionally does the general condition suffer as greatly as in carcinoma. In benign erosions, glistening ovula Nabothi are often seen. Within the erosion, there is no intact portion of mucosa; larger, gaping fissures lead into the cervical canal, the whole mass appears deep-red and bleeds readily. Carcinoma, however, in the beginning is mostly confined to a *single* spot, where it forms a distinct, small tumor with even surface, and at first leaves the other, larger portion of the cervix intact. But if the carcinoma has already attacked the entire cervix, the formation of nodes is distinct. In connection therewith, the vaginal portion of the cervix is usually pale. If disintegration has commenced here and there, the correspondingly varying condition of the vaginal portion is obvious to touch as well as to sight. Should a cancerous node proliferate in the cervix, as above described, the diagnosis will be very difficult at the first examination. Mistakes are here easily committed. I know of several cases, diagnosticated as cervical carcinoma and deeply cauterized, which completely recovered. They were certainly merely cases of cervical catarrh with proliferated mucous membrane.

Should it be impossible, therefore, to form a diagnosis *immediately*, a piece must be excised for microscopic examination. This excision will be preceded by the treatment for erosion. Should the erosions disappear under the treatment described on p. 175, it cannot be a case of carcinoma. Gentle treatment, even if it leave the diagnosis doubtful for three weeks, is absolutely required. There is no need for haste.

Should the erosions fail to heal, a wedge-shaped piece is excised with tenaculum and knife, in such a manner as to include *the margin of the cervix*. The base of the wedge at the external os should measure at least one centimetre. This wedge is inclosed in liver tissue, cemented in, and hardened. Sections are then cut and examined. A teased preparation does not furnish sufficient information. After excision of the wedge a suture is inserted through the wound surface. In the case of erosions, the wound closes by first intention, and the loss of blood and substance has a decidedly curative influence. In carcinoma, after-hemorrhages easily occur. The suture cuts through, the wound is soon again gaping. A tampon should always be placed on the vaginal portion after the excision.

A similar procedure is indicated for the diagnosis of carcinoma of the body of the uterus. With a sharp spoon, particles are scraped off and examined. The differential diagnosis must distinguish between carcinoma of the body, sarcoma, endometritis fungosa, and various adenomata.

In the case of *cancroid polypi* described on p. 249, the following circumstance will serve as a landmark: perfectly *smooth* tumors seldom spring from the lower surface of the vaginal portion. The rapid growth, the profuse hemorrhage during the removal, and a glance at the medulla-like, pale-red section of the tumor, permit of the formation of the diagnosis. At most a fibromatous sarcoma may be suspected. Then the microscopic examination will decide.

Cauliflower excrescences are probably always carcinomatous in character. Benign papillomata hardly ever occur quite *isolated* at the vaginal portion. Papillomata the size of the fist are now and then found in the vagina, and, associated with them, others at the vaginal portion, but in these cases pointed condylomata are also discoverable at the vulva. The age of the patient, the gonorrhoic origin, the rapid appearance, and the positive cure after a single removal obviate any difficulty in the differential diagnosis.

Some uncertainty in the diagnosis may exist, at least after the first examination, in the case of *sloughing myomata*. The history presents similar facts, and during the examination the finger feels shreds and soft masses as in disintegrating carcinoma. But in such cases, fever will always be found with fibromata, and a general participation of the body which permits of the diagnosis of a more acute affection. Then the examination of the discharged shreds will at once render the diagnosis clear. In fibroma—connective tissue, muscular fibres; in carcinoma—masses of large pavement epithelia in disintegrating, fatty, or intact condition. The differential diagnosis becomes clear also through the treatment—copious irrigations of the vagina.

With apathetic patients of inferior intellect the cases may be doubtful at first.

Furthermore, glandular polypi cause similar symptoms. Catarrhs also occur which have been protracted beyond the climacteric, which may be supposed to be carcinoma previous to examination.

After what has been said during the description of the symptoms, it is self-evident that the *prognosis* is most unfavorable.

Treatment.

The treatment is symptomatic or radical.

Unfortunately the women usually seek the aid of the physician when the carcinoma has so far advanced that a radical removal is no longer to be thought of. Then the symptomatic treatment is directed against the hemorrhage and the sloughing.

If a hemorrhage is to be arrested, the vaginal portion is exposed with a Sims speculum, and the bleeding spot is looked for while injecting cold water. To be sure, the hemorrhage will often have to be arrested at once, without speculum, on account of the vital indication. If not even a tubular speculum is at hand, the patient is placed on her side, two fingers strongly depress the perineum, and some tampons dipped in liquor ferri are applied to the vaginal portion. Some other tampons are placed over them, for the protection of the vagina. If not even liquor ferri can be speedily obtained, vinegar may be used to moisten the cotton or lint.

Villous, semi-putrid, *soft* masses are to be rapidly and completely removed with the *sharp spoon.* Thereupon the hemorrhage is best staunched by the styptic tamponade, the insertion into the wound of a tampon dipped in liquor ferri. The wound-surface may also be seared with Paquelin's thermo-cautery or the hot iron. If the hemorrhage is not too profuse during the scraping, the boro-glycerin (boracic acid, 1 ; glycerin, 10 parts) tamponade may be applied, or simply cotton moistened with two to five per cent carbolic solution. By the removal or scraping away of large villous, disintegrating tumors, the case is many a time improved for a number of weeks. This operation, too, must be done aseptically, as perimetritis and parametritis may easily follow it. The latter, however, cause pain and deterioration of the general condition, so that many a patient feels constrained to date a turn for the worse in her affection from this manipulation. Then, as usual, the physician is to blame for all the evil !

When the carcinoma is hard, it would be altogether wrong to injure the mucous membrane; that would only hasten the ulceration and hence the hemorrhage and sloughing. Now and then we see a small softened patch between two cancerous nodes which can be removed with a *very small* spoon. Or else the hot iron or a stick of caustic is introduced into this depression to destroy the bleeding spot. So, too, when the hemorrhage comes from the cervix or the uterus, scraping is to be done with small sharp spoons and liquor ferri applied either by a tampon or by Braun's syringe.

I emphasize particularly that the cauterizations are intended merely for the arrest of the hemorrhage. The hope of destroying, after scraping out the carcinomatous neoplasm, all the remnants of the tumor in the wound by caustics, is illusory. And as the strong cauterizations are not rarely followed by vaginal inflammation and perimetritis, they are urgently cautioned against.

The *softened* parts of the tumor must be removed also in order to prevent sloughing. In parenchymatous carcinomata, however, there is often no softened portion to be felt anywhere. Here, and in carcinoma generally, the most ample use is made of vaginal irrigation. Potassium hypermanganate, carbolic and salicylic acids, in short all disinfectants in existence can be employed. If it be impossible to introduce a tube, large disinfected compresses are laid in front of the vagina, to intercept the offensive odor, as it were. But in spite of all care and the most scrupulous cleanliness, it is impossible to control the fetor.

To the physician it is a hard task to treat a cancer patient until death, and yet it is a noble aim to secure to some unfortunate creature, by faithful attendance, daily some little relief, daily some new measure to meet one or other troublesome symptom. The different symptoms, present also in other uterine affections, are treated on general principles already repeatedly discussed.

The first modern attempts at the *radical removal of a uterine carcinoma* consisted in the galvano-caustic ablation of the vaginal portion, in the funnel-shaped excision, that is to say, the deepest possible excision of the vaginal portion, and in amputation with the écraseur.

Unfortunately, the locality is but little suitable for radical operations. In the case of an epithelioma, at least one centimetre of macroscopically healthy tissue must be removed conjointly. This requirement, however, meets great obstacles, in that the uterus is not an organ like the mammary gland or the ovary which may be totally extirpated. The intimate connection of the uterus with its surroundings renders it especially difficult to recognize the limit of the new-formation. By no means will everything morbid have been removed with certainty, when even the entire womb has been extirpated for uterine carcinoma. Besides, as it requires the statistics of at least a decade to judge of the results, no decisive opinion can be formed to-day of the methods of radical carcinoma operation which are only a few years old. Therefore we restrict ourselves to the brief description of the methods according to which the operation is performed.

When the new-formation affects the lower surface of the vaginal portion and perhaps also a part of the vaginal vault, Schroeder's *supravaginal amputation* is indicated. In this case the carcinoma generally does not extend to the internal os, hence the removal of the entire uterus would serve no purpose.

According to Schroeder, the procedure is the following: The vaginal portion is drawn down considerably; a strong loop of thread is passed through each vaginal vault. The vaginal portion is now separated from the vagina, or, if the vagina is likewise affected, the latter is severed at the distance of one centimetre from the carcinoma. The circumcised portion is then dissected off toward the centre, and the cervix is strongly pulled down. As the peritoneum invests the posterior vagina vault to some extent, a direct incised or lacerated wound of the peritoneum is here easily inflicted. This wound, as well as the lateral wound in the vagina, are to be united and spouting vessels ligated. The deeply protracted cervix is now separated obtusely, right and left, as much as feasible. Not till then is the anterior uterine wall severed and the anterior vaginal wall stitched to the uterine stump, so that, therefore, the vagina in front is fastened to the uterus and the wound closed there. The ligatures are not cut off, but employed for holding or drawing down the uterus. Now the posterior half of the cervix is divided and there, too, the vaginal wound is united to the uterus. The lateral portions, still gaping, are closed by deep sutures. The sutures are intended to diminish the opening in the pelvic connective tissue and to arrest the hemorrhage. These operations do not run quite typically. The mere circumstance that the vaginal portion is of varying size and that the neoplasm has attacked the vagina in different form and extent, necessitates variable incisions and different closures of the wound. Thus, in one case, the opening in the vaginal vault may be only three centimetres in diameter, while in another case the operation would have to be called rather a resection of the upper half of the vagina. In one case I have caused a loss of substance five and a half centimetres in diameter. In such a case the peritoneum will almost invariably be opened at the fossa of Douglas. When convinced that this injury will be unavoidable, it will facilitate the operation to make the wound according to a preconceived plan, and to precede it by the more difficult part, the stitching of the posterior vaginal wound to the remnant of the posterior lip.

For those cases in which either the body of the uterus is carcinomatous or a cervical carcinoma proliferates upward *in the mucous membrane,* the extirpation of the entire uterus would be the appropriate operation. This operation was believed to be impossible until W. A. Freund's labors inaugurated a new era. Freund has invented a method of extirpation of the uterus which he carefully studied out on the cadaver. This operation has been frequently performed. It is true, the results are not good, inasmuch as about eighty per cent of those operated upon have died either of the consequences of the operation or of relapses. Still the operation is a new one which must be tested in various ways, tried, and perfected.

A large abdominal incision having been made, the intestines are pushed into the vault of the diaphragm, or if they nevertheless prevent access to the uterus, they are placed externally, wrapped in a warm disinfected cloth. Then the operator separates the abdominal recti muscles from the os pubis, so that the pelvis can be conveniently inspected. The parietal peritoneum, which is but very loosely attached in front, is fastened by a few stitches to the abdominal coverings. Now the uterus is grasped with forceps devised by Freund or any suitable instrument and drawn upward. Then ligatures are applied in order to separate the uterus from the broad ligaments without hemorrhage. Freund ligated first on each side in three portions; the two upper ligatures passed through the broad ligament, while the lowest grasped the parametrium laterally, and with it the uterine arteries and the vaginal vault. For the latter suture Freund had constructed a special needle. After the ligatures were tied, the dissection of the uterus from the vagina began; for this purpose Freund likewise invented special instruments—a knife with covered blade and very short scissors bent at a right angle. During and after the dissection, often many additional vessels must be ligated, perhaps even the accidentally missed uterine artery. This hemorrhage may be particularly severe. Thereupon Freund closed the vaginal wound from the peritoneal cavity by stitching the anterior surface of the peritoneum to the posterior. The union of the vaginal wound was omitted.

The dangers of this operation consist in hemorrhage, shock, and inclusion of the ureter. Inasmuch as the operation is a typical one and unforeseen incidents can hardly occur, the hope is justified that operators will gradually learn to master the technique more and more, and secure correspondingly more favorable results.

Various operators, particularly Freund himself, Bardenheuer, and Schroeder have improved the method.

The preliminary ligatures through the vaginal vault have been abandoned, the procedure is more preparatory and, as in other operations, bleeding vessels are compressed digitally and ligated. The closure of the peritoneal cavity is dispensed with. On the contrary, leaving the peritoneal cavity open is believed to be more correct, as the secretions are thus able to escape and a disinfecting stream of water can irrigate the wounds.

An innovation of especial importance was introduced by Schroeder. According to him, the uterus should be removed from the vagina. After circumcising the vaginal portion, he penetrates in front and behind into the peritoneal cavity, flexes the uterus backward, draws it through the posterior vaginal wound *into* or *in front* of the vagina, ligates the broad ligaments *en masse*, and then separates the uterus. In this way the real danger of laparo-hysterotomy—shock—is avoided. It is certain that the displacement of the intestines and the manipulations in the abdominal cavity are very dangerous in this respect.

Although *colpo-hysterotomy*, therefore, is assuredly less dangerous, it will be impossible in all cases in which the uterus is enlarged or the vagina narrow.

Total extirpations are contra-indicated whenever the examining finger demonstrates nodes in the parametria. Therefore, a very thorough examination of the patient under chloroform is always necessary before deciding on the operation.

C. Sarcoma of the Uterus.

Sarcoma occurs in two forms in the uterus. First, as fibro-sarcoma, a new-formation related to fibroma, perhaps to be interpreted as a sarcomatous degeneration of a fibroma. The tumor forms in the body of the uterus, more rarely in the cervix; it may have a submucous situation, but may also be pediculated like a polypus, and depend from the fundus, the cervix, or the vaginal portion. The consistence is more encephaloid than fibromatous.

These tumors cause the same *symptoms* as myomata. But they relapse, eventually reach a considerable bulk, and change the whole form of the uterus to a quite irregular tumor the size of a man's head. After operation, nodular neoplasms form in the vagina and spring from the cut surface; finally the entire pelvic connective tissue and the neighboring lymphatic glands are filled with sarcomata.

Previous to extirpation, the *diagnosis* of fibroma is often made; but the examination of the remarkably soft tumor shows the small round or spindle cells, and hence its malignant character.

As neither the total extirpation of the uterus nor the completest possible removal from the vagina could prevent a relapse, the *prognosis* is absolutely unfavorable.

There occurs, besides, a *sarcoma of the mucosa;* this again is found in two forms—as a soft sarcoma-node of the mucous membrane, and as a diffuse, sarcomatous degeneration of the mucosa, which latter becomes completely disintegrated.

Destruction of the uterus also occurs here; in fact that organ may be perforated, the sarcoma, infecting the peritoneum, growing into the peritoneal cavity. Gusserow has described a case in which the sarcoma even perforated the abdominal coverings and proliferated outward.

In sarcoma of the mucosa, the *symptoms* of disease of the mucous membrane come into the foreground. Copious, watery, slightly sanguinolent, rarely (and then only in advanced stages) ichorous discharges escape. Particles of the tumor are expelled more frequently than in carcinoma. In one case I witnessed the expulsion, from time to time, of soft, friable lumps the size of a walnut, accompanied by pains. If the uterus enlarges, it reacts by labor-like pains. If the peritoneum becomes implicated, peritoneal sensitiveness to pressure arises.

The *diagnosis* is based on the examination of particles of the tumor (comp. p. 256). The *prognosis* is absolutely unfavorable; the *treatment* aims at the removal of the soft, necrotic masses, and the alleviation of the general symptoms.

Apart from the small-celled sarcomata, species of tumor also occur which must be classed as mixed forms, and be designated as *carcinosarcomata*.

D. ADENOMA.

We have already become acquainted with *adenoma* of the uterus in the shape of adenoid erosion, ovulum Nabothi, mucous and glandular

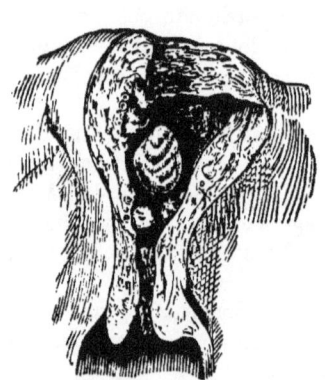

FIG. 144.—Several adenomata (mucous polypi) of the uterus.

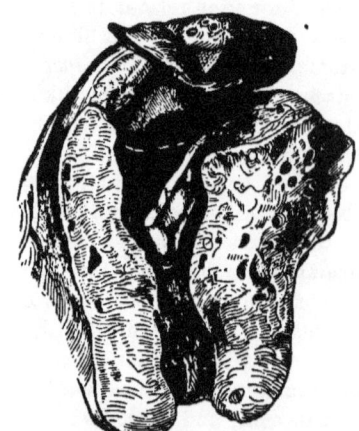

FIG. 145.—A mucous polypus extending into the uterine parenchyma; *a*, cut surface of the uterus; *b*, the polypus cut in half.

polypus of the cervix, and follicular hypertrophy of the vaginal portion. Glandular polypi occur also in the cavity of the uterus. Not rarely they form a complication of fibromata. If the entire uterine mucosa is hyperplastic, we have a fungous endometritis; but if the glands alone proliferate at the expense of the other substrata of the mucosa, if the physiological forms of the glands disappear, and there develops a diffuse tumor composed of glandular masses only, we have an adenoma. The latter may be a malignant tumor. However, carefully observed fatal cases with post-mortem appearances have not been described. On the other hand, a benign circumscribed glandular hyperplasia often develops; a polypus with broad base, or, by further growth, with pedicle, often forms in the endometrium. Cases of multiple polypi of that nature are also on record. Their surface usually is smooth, but they may also be villous, cockscomb-like; the latter especially when the polypus has crowded between the grooves of the arbor vitæ. On microscopic examination of these polypi, it is true, we also find such in which the glandular new-formation occupies the background. There are also soft tumors developed from the interacinous tissue of the mucosa and dilated vessels.

Fig. 144 represents a number of such soft polypi in the uterus.

Fig. 145 illustrates a polypus of that kind described by Rokitansky; it extended deeply into the uterine parenchyma.

The *symptoms* consist pre-eminently in menorrhagia and metrorrhagia. The serous discharges described on p. 174, of course, likewise occur.

The *diagnosis* of these polypi is formed by digital exploration. *In the case of all inconquerable hemorrhages of the uterus, the cavity of that organ must be palpated.* After dilatation of the uterus, the finger is introduced and feels the soft tumor escaping from the finger, and yet adherent. It is easier to arrive at a diagnosis when the adenoma projects into the vagina or, at least, is visible in the os. Then, of course, a carcinoma must also be borne in mind.

The *treatment* is immediately conjoined with the diagnosis. The tumor is removed in any manner. If it can be grasped in the snare of a wire écraseur (Fig. 146 and Fig. 143, p. 248), that instrument should be employed.

For the division of so soft a polypus this instrument suffices. The scissors (p. 178, Fig. 93, and p. 240, Figs. 140 and 141) may also be employed. Many such polypi bleed enormously on being touched. In these cases, as well as when the form of the polypus cannot be clearly made out, a large sharp spoon is to be employed (Fig. 37, p. 70). With vigorous pressure against the internal wall of the uterus, the polypus as well as the adjoining mucous membrane are scraped off. The finger having ascertained that the uterine cavity is quite empty, liquor ferri is to be injected or the vagina to be carefully tamponed; for not infrequently very serious hemorrhages occur after removal of these polypi.

FIG. 146.—Wire écraseur.

As such tumors are easily bruised and caused to slough, rapid methods of dilatation are to be preferred to tents for the operation of adenomata.

The *prognosis* is good. The removal of these polypi belongs among the most gratifying labors of the gynecologist.

E. The Free Hæmatoma of the Uterus.

The free hæmatomata, the *fibrinous polypi* of the uterus, are often mistaken for adenomata. In very rare cases, deposits of fibrin form around a remnant of abortion, membranes, or placenta, similar to the incrustation with uric salts of a foreign body in the bladder. Then we find in the centre the remnant of abortion, and around it concentric layers of fibrin. The irritation and dilatation of the uterus lead to continually renewed hemorrhages, the latter to enlargement of the "polypus." In this case, too, exploration and removal of the foreign body are indicated. The diagnosis can only be formed by examination of the tumor.

F. Tuberculosis.

Leaving aside the old literature from the time where the definition of tuberculosis was altogether different from what it is nowadays, very few cases of tuberculosis of the uterus are described.

Tuberculosis of the uterine mucosa is said to occur primarily. Secondary tuberculosis is observed with tuberculous peritonitis and with general tuberculosis. During the latter, there may be tubercle formation in all glands and mucous membranes, hence also in the uterine mucosa.

In tuberculosis of the uterine mucosa, the latter disintegrates, and secretes a whitish thin fluid which is not exactly sanious. The tuberculous mass may remain in the uterus without being expelled.

Characteristic *symptoms* do not exist or are masked by those of the the general tuberculosis.

Important for the *diagnosis* is the fact that tuberculosis, as opposed to the condition in carcinoma, attacks the tubal mucosa. In consequence thereof, a cheesy mass accumulates in the tubes and, dilating them, leads to palpable tumors. Thus it might perhaps be possible to diagnosticate tuberculosis of the uterine mucosa from the amenorrhœa, the discharge, and the palpable, dilated tubes.

G. Echinococcus.

There are but few authenticated cases of echinococcus of the uterus on record. The vesicles of the echinococcus were expelled spontaneously, as well as accidentally discovered at the post-mortem. Owing to the tendency to extend into the pelvic connective tissue, the *prognosis* is not favorable. No positive rules can be laid down for the operative method of removal with the few cases known. The object must be to remove the vesicles as completely as possible, and to cause the empty cyst to contract, with antiseptic precautions.

CHAPTER XIV.

DISEASES OF THE PELVIC CONNECTIVE TISSUE, THE PELVIC PERITONEUM, THE UTERINE LIGAMENTS, AND THE TUBES.

A. Parametritis.

Anatomy and Etiology.

BY parametritis we understand an inflammation of the connective tissue joining the peritoneum to the floor of the pelvis or to the uterus, rectum, and bladder. To be sure, the term "parametrium" is applied chiefly to the locality *close* to the lateral margin of the uterus between the layers of the broad ligament, limited below by the pelvic diaphragm and the lateral vaginal vault. However, as only in the rarest cases the inflammation or its product, the tumor, remains confined to this locality, it is necessary to find a common name for the inflammatory products in the pelvic connective tissue which are alike genetically. The most correct term would be *inflammation of the pelvic connective tissue*. But as the term parametritis is in general use and well expresses the primary form of the affection, we shall retain it. Other names are: pelvic cellulitis, extraperitoneal exudation or abscess, phlegmon of the broad ligament. As opposed to the florid peritonitides, parametritic abscesses were formerly called cold peritonitic abscesses. Only the most recent times have taught that abscesses occurring in the pelvis are to be separated into parametritic (connective tissue) and perimetritic (peritoneal). Although complications are exceedingly frequent, it is very instructive, especially for the beginner, to consider parametritis and perimetritis separately.

It would be sacrificing truth to a principle were we to assert that every parametritis is a traumatic affection based on infection. In the great majority of cases, to be sure, a lesion or the infection of the wounded surface is the etiological factor. This is not to be wondered at during labor. The cervix or the region of the internal os is greatly attenuated during the passage of the child. Should a laceration ensue here—and this is often the case—the parametrium will be opened, or at least the wound will extend into the immediate neighborhood of the parametrium. Similar to lesions during labor is the effect of injuries during minor operations on and in the cervix, pre-eminently plastic operations or the treatment with laminaria or sponge tents. Without antiseptic prophylaxis, severe parametritis is repeatedly observed as the sequel of the above-named manipulations.

There is hardly any gynecologist who has not formerly experienced accidents of this nature. Especially the treatment with stem pessaries was often followed by disease of the pelvic connective tissue. Should an ill-fitting, excessively large vaginal pessary not be removed, parametritis can likewise arise, usually complicated with perimetritis. But, undoubtedly, inflammations of the parametrium or of the pelvic connective tissue occur also outside of labor and without any lesion. In nulliparæ, both young girls and older women, we observe, though rarely, suppurating pelvic exudations of very obscure etiology. Cases of that nature can by no means be disputed. A number of them can be brought in connection with perityphlitis. Several times I have observed perityphlitis which finally changed into parametritis, and perforated into the rectum. In other cases, the parametritis seems to be able to develop from a hæmatoma of the pelvic connective tissue. Or both conditions complicate in a manner which cannot be more closely studied. At least, we find parametritic effusions of blood which finally suppurate, and parametritic abscesses which discharge blood-clots. What may have been the primary condition can at most be suspected in isolated cases from the age or the quality of the coagula of blood.

According to Bandl, a cervical inflammation easily propagates to the folds of Douglas. This is certainly probable. The nucleus of the folds of Douglas is uterine tissue. When the mucosa inflames, the underlying uterine parenchyma participates, and the inflammation extends as far as the folds of Douglas. Here, again, in the looser connective tissue, it finds a favorable spot for the formation of an exudation.

It is absolutely necessary for the understanding of the *anatomy of parametritis* to precede it by some more general explanations. If we imagine the peritoneum detached from the floor of the pelvis, the pelvic connective tissue is exposed, that is to say, the space is open, to the inflammation of which this chapter is devoted. Here we find, in the first place, great quantitative differences. The peritoneum is so closely applied to the body of the uterus that there is no separating layer of connective tissue between uterus and peritoneum. In the same way, there is little connective tissue in the sagittal (ideal) section between the peritoneum and the posterior vaginal wall, between the bladder and its peritoneal investment, as well as between the peritoneum and the rectum. A thicker mass of connective tissue fastens the uterus to the bladder (comp. Figs. 116 and 117, pp. 206, 207). From the bladder, however, the peritoneum, remarkably loosely attached, extends to the abdominal wall. An inflammatory connective-tissue tumor can be found in a *perfectly sagittal* direction only where connective tissue is accumulated preformatively—in front between the abdominal wall and the parietal peritoneum *anterior to the bladder*, and immediately in front of the uterus, between it and the bladder.

The greatest accumulation of connective tissue is found laterally at the uterus—the parametrium proper. The layers of the broad ligament are separated during pregnancy by the dilatation of the vessels. The connective-tissue or muscular nucleus of the broad ligament hypertrophies. After labor, the veins collapse, the peritoneal layers shrink and are but loosely apposed, the ligament can easily expand through inflammation and infiltration. If we imagine, then, a lateral sagittal section striking exactly the margin of the uterus and exposing the parametrium, we have in front, in the inguinal region, a portion of rather loose connective tissue between peritoneum and abdominal wall; then, alongside the uterus, the parametrium; and, posteriorly, the connective tissue in the folds of Douglas. Therefore, inflammatory tumors may arise *laterally* beside the uterus, also in front and behind. The lateral tumors will propagate to the above-mentioned connective tissue between bladder and uterus and surround in a semilunar form the side and the anterior surface of the cervix. Posteriorly the fold of Douglas expands so as to become more central and be palpable as a thick retro-uterine tumor. And, finally, the median, uterine part of the folds of Douglas may become infiltrated, the place where a ridge-like elevation of the uterus is almost always visible, so that a tumor is also attached to the posterior half of the uterus. This one, however, must be at a considerable height above the floor of the fossa of Douglas. Thus the uterus is eventually completely surrounded by masses of exudation. This exudation *originated laterally* and travelled around the uterus. The dislocation of the uterus is usually moderate, just because the infiltration hindered the expansibility of the ligaments and the mobility of uterus. On the other hand, if the inflammation and the formation of the tumor are confined to one side, the uterus may be displaced toward the opposite side.

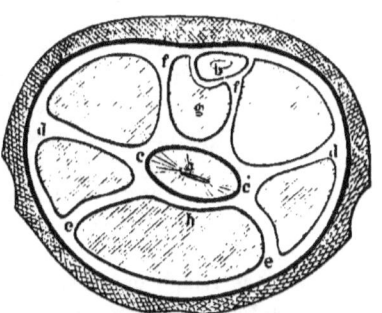

FIG. 147.—Schematic diagram of a horizontal pelvic section. *a*, uterus; *b*, rectum; *c c*, parametrium or lateral margin of the uterus; *c d, c d*, broad ligaments; *c e*, round ligament; *c f, c f*, the folds of Douglas; *g*, fossa of Douglas; *h*, vesico-uterine excavation, behind it the connective tissue between the bladder and uterus; *e, d, f, d, e*, superitoneal connective tissue.

As regards the dissemination of this infiltration or the course which may be taken by the product of this inflammation, the pus, a schematic diagram, Fig. 147, will most readily elucidate it. Assuming Fig. 147 to be a horizontal section of the pelvis, then *a* would represent the uterus;

at *c, c*, the parametrium, the primary inflammation usually arises; it may, if of inferior origin, travel along in the broad ligament, to *d, d*. Here, laterally, the tumor forms at the base of the broad ligaments as far as the iliac fossa. Or else the infiltration propagates upward—a rather rare case—and extends backward in the ligaments of Douglas to *f, f*. Then a tumor forms which eventually surrounds the rectum *b* and compresses it. Should the infiltration, the tumor, or the pus develop likewise at the margin of the uterus, but higher up in the neighborhood of the angle of the uterus, the infiltration migrates toward *e* in the round ligaments. This happens only in puerperal cases. But it is possible, too, that the infiltration, having extended from *c* to *d*, travels or descends subperitoneally at the pelvic brim from *d* to *e* toward the inguinal region. Every part left white in the figure, representing the connective tissue, may swell, *i. e.*, form a parametritic tumor, and finally the infiltration may be present *everywhere,* so that the uterus *a* is surrounded by inflammatory hard tumors.

Inferiorly these infiltrations are limited by the pelvic diaphragm, that is to say, by the levator ani and the tendinous layer above it, so that a parametritic tumor hardly ever penetrates directly into the vagina.

In the majority of cases, tumors form at the places designated, but they are again absorbed. However, should pus form, it will seek a passage out. It was precisely the circumstance that the final perforation occurs at places far distant from the uterus, and at a time when the *primary* affection, which perhaps had never been diagnosticated, had long disappeared, which for many years made the affection difficult to understand. It was thought that the pus had formed where it appeared. But this is by no means the case. Just as after infection of the finger-tip an abscess finally forms in the axillary gland, and is opened at a time when the finger and the lymph channels leading upward have long ceased to be affected, so the broad ligament is often merely *the road* along which the inflammation travels. Or, as palpebral abscesses remain after facial erysipelas, and peripheral abscesses occur after erysipelas of the extremities, when the disease has long run its course, so we find an extraperitoneal abscess, for instance, opening in the inguinal region, although the parametrium right and left is quite free and the uterus easily movable.

The points of outward perforation must be connected with the anatomical relations. Often the pus perforates into the uterus or the cervix. As, in deep abscesses at the bone, old cicatrized fistulous tracks often again melt down, and even the superficial cicatrix grows red and thin until it finally disappears and pus once more pours forth, so old lateral cicatrices in the uterus again melt down and the pus escapes outward through the vagina. The opening can be neither seen nor felt, at most it may be sounded. But the latter manipulation is to be decidedly cautioned against as useless and dangerous.

Particularly the puerperal parametritic exudations extending in the round ligament or from the iliac fossa to the inguinal region form larger, board-like tumors reaching to the umbilicus. Should purulent softening occur here, the pus generally perforates *above*, not below Poupart's ligament. But very rare cases occur in which the pus below Poupart's ligament chooses the same course as the ordinary psoas abscesss, the congestive abscesss in spondylarthrocace. Then the pus spreads along the inner surface of the thigh and there forms larger, tensely fluctuating tumors. Both forms I have seen only in puerperal parametritis.

The exudations seldom perforate directly into the vagina. But pus may also descend by the side of the vagina and reach the labium majus, so that from the point of perforation we may pass a sound upward along the vagina to the depth of twelve to fifteen centimetres.

Should the infiltration from the outset extend to the connective tissue between the bladder and uterus, the abscess may also perforate into the bladder. This is by no means a rare occurrence in non-puerperal parametritis. Abscesses may also perforate into the rectum, but of course only when the surroundings of the rectum were chiefly concerned in the formation of the tumor. Very rarely the perforation takes place beneath the glutei. The pus here chooses the way through the sciatic foramen. The abscess has also perforated through the obturator foramen, and equally rarely into the back near the quadratus lumborum.

The ureters are comparatively rarely affected. This complication is indicated more by the symptoms than by the objective conditions.

The fact that a parametritic exudation forms an abscess and opens externally is no proof at all that the exudation has suppurated *in toto*. On the contrary, single portions seem to melt down more frequently. Only in acute suppuration of the connective tissue in the puerperium, the exudation is observed to perforate, thus terminating the pathological process. But in general it is the rule that parametritic exudations soften merely in part. Not rarely but very little pus gets outside, and, in spite of the perforation, the form of the entire tumor is preserved. These cases are often very hard to understand, because no distinct diminution of the tumor and improvement of the symptoms are perceptible despite the obvious evacuation of pus. After evacuation of the pus, the point of perforation even closes again, to reopen after some time with or without external provocation. Thus the disease may last for years. In other cases the exudation has no tendency to suppuration, slowly extends all around the uterus which becomes firmly fixed, immovably held by the exudation. Such non-perforating exudations soften slowly, relapses again enlarge them, and the tumor persists for years, here and there changing its shape.

Symptoms and Course.

The symptoms of parametritis, of course, when we observe the *beginning* of the affection, are different from those which are to be described as caused by the exudation of long standing, the *final result* of the disease.

In the puerperium or after some therapeutic manipulation, the fever naturally indicated the beginning of some disease, previous to the occurrence of exudation. Very high temperature and initial rigor point to an inflammation. During early exploration, the infiltration, the tumor, is felt laterally at the uterus in the parametrium. It is by no means necessary to find a "tumor" at once. There may be at first merely an indistinct *resistance*. As the disease advances, more or less circumscribed tumors form. They can only be where they are anatomically possible.

The symptoms of a long-standing parametritic *exudation* are often surprisingly slight. Large hard tumors are found to surround the uterus without the patients believing themselves seriously sick. Often it is the perforation which first causes slight fever and pain so that the physician is called. More sensitive patients, however, complain of a sensation of pressure in the pelvis, difficult defecation, often quite considerable menorrhagia and vesical troubles. Especially the pressure upon the nerves of the pelvic floor and the sciatics causes neuralgia and paraplegia, so that not infrequently we meet patients who have been plagued for months with electricity for "sciatica," while they are suffering from a parametritic exudation.

A ureter, too, is often compressed. During every close observation, we frequently hear complaints which have to be referred to obstruction to the flow of urine from the pelvis of the kidney. To me it does not appear unlikely that an exudation may perforate into a dilated ureter and by this circuitous route find its way into the bladder. Once I have observed and verified by autopsy a complete atrophy of one kidney, the other functionating vicariously.

As there is no fever connected with long-standing exudations, the patients feel tolerably well, but they never present the appearance of a perfectly healthy, robust woman.

The *pains* during parametritis on the whole are inconsiderable. They increase especially if the patients do not take care of themselves; but then the pains may become as violent as in peritonitis, because the parametritis extends to the peritoneum. Even during strong combined pressure the exudations are often not sensitive. With *perforation* into the bladder and rectum the symptoms are very changeable; we see patients to whom every micturition causes the greatest agony, while others pass pus and blood with the urine without symptoms. Defecation, too, may be atrociously painful; in other cases again, a large quantity of pus is evacuated after the fæces without pain. Perhaps this is connected with the point of perforation: if it affect a region close to a sphincter, the nerves are irritated as in a fissure of the anus and the symptom of tenesmus appears. At least in a few cases I have been able to demonstrate this connection clearly. If the pus perforates through the external skin, it of course reddens and becomes sensitive to a circumscribed extent.

The *course* of parametritis is illustrated by the above-described anatomical conditions. There are quite *acute, puerperal parametritides* which are definitely at an end by abscess formation in from ten to twenty days. Such women again conceive without difficulty and are in no danger from later labors and puerperia.

But more frequently primary parametritis commences in the puerperium, after an abortion, or after a minor gynecological operation, quite acutely at first. The tumor and the fever render the diagnosis clear. Such a patient is under the impression that she is still a weak convalescent, being for weeks and even months in tolerable, but by no means perfect health. Suddenly, after a particularly profuse menstruation, some exertion, or a trauma, an abscess forms, the tumor is discovered, and even in case the primary affection has not been observed, the manner of its occurrence will be clear.

Unfortunately, in these more chronic cases the disease is not at an end with the formation of an abscess. Again and again the tumor increases and diminishes. I have observed a parametritic tumor for eight years.

In such cases the course is pronouncedly chronic, particularly in poor women who have neither the time nor the means and opportunity for rational treatment. I have known many a poor patient to be ailing for years and to be living in the hope of getting rid of her suffering through a small bottle of medicine. Eventually the general health suffers, death being brought about by fever or some consumptive affection. Quite rarely an inexplicable, sudden exacerbation is observed which has its cause in sloughing of the exudation. This is succeeded by sepsis and death.

Diagnosis and Prognosis.

In order to form an exact diagnosis of parametritis, it is necessary that an infiltration or a tumor can be demonstrated during the examination. Usually this is easy. On the other hand, it may be doubtful whether the tumor felt is really a parametritic exudation or whether the tumor should be interpreted otherwise. Quite early, parametritic infiltrations have a tensely elastic feel, about as if a stretched cloth is indented with the fingers. In the beginning we can hardly speak of a tumor, but rather of an increased resistance in the lateral vaginal vault. The lateral position of the tumor points to its parametritic origin. If the resistance is *first* in the fossa of Douglas, it can hardly be of parametritic origin (comp. below). In front the resistance lies usually rather higher than at the side, so that we must carefully feel high up, the bladder being empty, in order to discover it. It is to be urgently advised to introduce two fingers, as in that way alone positive information can be obtained concerning the relation of the tumor, or the resistance, to the uterus. Not rarely the tumor by the side of the uterus has almost the form of the latter, so that at the first moment there may be some doubt as to which is uterus and which tumor. But never does the tumor so continuously blend with the uterus that it should be quite impossible to feel a demarcation, a furrow between them. The tumors beside the uterus, which can be readily felt from the vaginal vault, may be got between the hands during combined examination, but often cannot be felt from without during abdominal palpation. Should the broad ligament be expanded, the tumor moves upward. Laterally to the uterus there is usually an oval, not quite regular, but never knotty body. In such cases the tumor may be the size of the head of a child or even that of a man, but always plainly lateral in its position. As has been stated under the head of symptoms, a swelling circumvalling the uterus must also have arisen in the pelvic connective tissue, that is to say, be interpreted as parametritic.

A parametritic tumor extending upward in the inguinal region is not thick above, often has quite a sharp margin, so that the abdominal coverings can be turned in behind it to some extent, as beneath the liver. Withal the tumor is very hard, board-like, and not sensitive to pressure.

In unilateral tumor, the uterus is felt completely crowded to one side. In parametritis arising primarily in front and extending between the abdominal wall and the bladder to the umbilicus, the uterus is often lifted more and more from day to day, so that on the fifth or sixth day of the parametritis it may be just reached with the tip of the finger from the vagina.

Of great importance for the diagnosis is the almost painless origin. The dispute as to whether a tumor the product of inflammation lies intraperitoneally or extraperitoneally is not so difficult to decide as is commonly assumed. If the tumor arose entirely without pain, if the tumor itself is not very sensitive to pressure, it must have developed in the pelvic connective tissue. Important, too, is the *lateral* position, and, besides, the fact that the tumor is found where according to anatomical conditions a tumor is to be expected in parametritis (comp. p. 267).

A differential diagnosis between parametritis and hæmatocele, that is to say, between extraperitoneal and subperitoneal hæmatoma, has a more academic value. We may diagnosticate the presence, the anatomical position, and the nature of the tumor, that is to say, whether it contains blood or pus, from the anamnesis and the group of symptoms. But the exploration *alone* does not furnish sufficient data. Owing to the slight influence of the diagnosis on the treatment, it should certainly be interdicted to try to render the diagnosis positive by the exploratory incision or puncture.

If the cases are not seen until after prolonged existence in debilitated patients, we must also bear in mind malignant neoplasms in the retro-peritoneal space. From time to time the specialist has cases with parametritic exudations sent to him with this diagnosis. From a single examination it is often barely possible to feel positive at once, but soon the diminution of the exudation or the reverse, the exacerbation despite every treatment, instructs the physician as to the condition.

A fibroma or an ovarian tumor could not easily be mistaken for an exudation both on account of its *form* and its *location*. Difficulties in the way of diagnosis arise when the parametritic exudation has existed for some length of time without manifesting itself, symptoms being caused only by the perforation into internal organs. Thus quite a florid peritonitis may be set up after perforation into the peritoneal cavity. This is very rare. But perforations into the bladder and rectum are frequent. As has been explained under the head of symptoms, the subjective phenomena are very variable, but the objective ones will be always the same. Should pus and masses of blood—the pus is always sanguinolent in these cases—be suddenly passed with the urine, and the latter again become quite clear in brief intervals; that is to say, should clear and purulent urine alternately appear at short intervals, there can be no vesical catarrh. The parametritic tumor, here usually situated *in front* of the uterus near the neck of the bladder, settles the diagnosis. The point at which the abscess opens into the bladder can hardly be determined. In two cases I have dilated the urethra according to Simon, but the *opening* could be felt neither in the urethra nor in the bladder. I had intended eventually to introduce a drainage tube so as to secure healing at last, but the fistula could not be found. It is also but rarely possible to feel the opening from the rectum. In one case, however, I felt a projection two centimetres high, on the top of which the opening was situated and could even be rendered visible by a rectal speculum. Generally, however, though we may prove by the symptoms that perforation has occurred, the fistula cannot be demonstrated.

Not always does fever indicate that perforation is occurring; it may also take place without any distinct symptoms. How difficult the diagnosis is at times was proved to me by a case in which I had diagnosticated a parametritic, suppurating exudation, and at the autopsy (death from pyæmia) was found a suppurating and sloughing, incarcerated, large myoma of the posterior cervical wall, penetrated by numerous fistulous tracks.

But if the diagnosis really cannot be formed, we must bear in mind that the treatment is plainly indicated even without diagnosis. Although the diagnosis may shift to and fro during the daily careful observation, we frequently get at the diagnosis by the results of the treatment.

The *prognosis* in general is good. Of course, there exist many dangers which have been touched upon above. But if the external circumstances permit of rational treatment, both recent and old parametritis are affections which are not incurable. The course, to be sure, is chronic, that is to say, we cannot remove a disease which has existed *for years* within *days*, but still the physician's exertions are usually crowned with positive results.

Treatment.

First it should be remembered that the *prophylaxis* consists in antisepsis when directing labors and abortions, as well as during the least gynecological manipulation. Then, the treatment is different when we observe the *genesis* of an exudation in order to prevent its spread and enlargement, and when we wish to remove an *old exudation*.

Should an exudation form, we shall try to prevent its enlargement by antiphlogosis. An ice-bladder on the abdomen and graduated injections are in place; success is also obtained by constant cold-water compresses according to Priessnitz. They have the advantage that they force restless patients to lie still. For this reason alone it is desirable to order them. Abstraction of blood from the vaginal portion (comp. p. 59) likewise belongs to the antiphlogistic apparatus. If there is high fever, it is treated according to general rules. From the first day on, sufficient evacuation of the bowels is to be carefully provided for. In parametritis, which often lasts very long, we must not forget the general health through the local affection. Frequently every local treatment is useless and without result, while it is possible, by the greatest care in nutrition, to secure an improvement of the *local* affection, a greater energy in the absorptive processes, *with* that of the general condition. For the rules for the employment of eccoprotics see p. 158. But even aside from the general condition, the mechanical irritation of a voluminous evacuation is injurious. Not rarely the defecation is excessively painful, especially in cases of perforation into the rectum.

Often, in a parametritic exudation which forms irresistibly, the sole art of the treatment consists in regulating the digestion, and causing the patient to take care of herself, by virtue of the medical authority.

To diminish *old exudations*, that is, cause them to be absorbed, we have different measures. The most effective are: *saline sitz-baths, hot injections,* and moist warm *packs* of the abdomen or of the pelvic region. Warm full baths are to be used with the greatest caution. Such a treatment should never be permitted without the most scrupulous medical direction. A full bath of one-half hour is nothing indifferent by any means. Many a patient thereby becomes nervous to the highest degree, excited, and after an ill-directed bath-cure is much more wretched than before, without the exudation having diminished materially. Less weakening to the body and more effective are *saline sitz-baths.* During their use old exudations often disappear in a few weeks. The physician must punctiliously prescribe every detail. With the sole direction " to take sitz-baths " hardly anything has been done. The patient had best bathe at night. To each sitz-bath of two pailfuls are added five hundred grammes of *previously dissolved* sea-salt or residuary salt (of salt factories, Mutterlaugen-salz).

The bath should not be less than 90° F. nor over 100° F., and we may vary between these extremes. In recent exudations we take cooler; in old, warmer baths. The patient is to sit in the bath dressed for the night. A mantle or a shawl covers the patient *and* the tub. The latter is placed close to the bed. The room during the bath should have a temperature of 68° F. In the patient's bed lies a woollen shawl, above that a linen cloth, upon both is a warming-pan. The patient remains in the bath for ten to twenty minutes. When she gets up—in which weak women must be assisted—she is hastily dried. Then the patient at once gets into the warmed bed. The linen cloth absorbs the rest of the moisture. The patient dries herself fully beneath the coverlet. Then the linen cloth, and after some time the woollen shawl, is removed.

At first two sitz-baths are to be used per week; later, in case the patient feels better, they are gradually increased to one daily.

The *vaginal irrigations*, enumerated as the second effective measure, are best taken by the patient in the saline sitz-bath. Previously a pail or pot of brine is prepared, at a temperature of 104–106° F. The assistant at the bath fills an irrigator, the patient introduces the tube into the vagina. Now brine is continued to be poured into the irrigator until *at least* five to eight quarts have been used. This method is very easily executed and has the advantage of keeping the bath-water constantly warm by the influx of hot water. The irrigations are certainly preferable to the employment of bathing-specula. Of course, matters may be so arranged as to render the assistance of a second person unnecessary.

The temperature of the irrigating fluid may also be increased. If the tube of the irrigator is very long, the passing water is cooled three or four degrees.

During this treatment, too, the patient must be under daily control, at least in the beginning. The hot-water injections are no panaceas by any means. Often this treatment must be again discontinued, after having been employed three our four times, on account of incipient fever or of pain. In many cases certainly we observe wonderful results. Exudations or remnants of exudation which have remained unchanged for years in spite of every treatment, disappear completely under hot-water injections within three or four weeks. Should the patient bear the injection well, the irrigation may be used twice daily. But after the procedure, even when not taken in the bath, the patient must always lie down, *well-covered*, for one or two hours.

The third measure are Priessnitz compresses. I urgently advise not to apply them to the abdomen only, but to wrap the wet cloth around the whole pelvis. These applications may remain *permanently*; they must form the substitute for every local treatment when the patient cannot move or is confined to bed by high fever. Should the brine cause eczema, water should be employed. Bath-mud is likewise used with good effect for these applications.

In old, stubborn exudations, a course of treatment may be combined from the three methods described. To be sure, at times the treatment of an old exudation extends over one year to eighteen months. Even then, though the symptoms are removed, a so-called remnant of exudation, perhaps to be interpreted as a cicatrix, or at least a permanent dislocation of the uterus referable to this cicatrix, remains behind.

Besides the measures named, a large number of other therapeutic proposals has been made. Thus, pre-eminenly, the employment of *preparations of iodine*. Tincture of iodine has been painted both on the abdomen and on the vaginal vault, according as it was hoped to act upon the exudation the more readily from one or the other point. Also the long-continued application of tampons with potassium iodide solution or iodized glycerin has been proposed. In the same way iodoform in the form of an ointment (2 : 10) is employed for rubbing into the abdomen and for the tamponade. I have never seen any positive effect attributable to these measures *alone*. Potassium iodide is also to be applied from the rectum, in the form of suppositories containing 0.5 gm. of potassium iodide in 2 gms. of cacao butter. Unfortunately the treatment of a parametritic exudation is often so protracted that there is ample time to employ all these measures. Only we must beware of ascribing accidental improvements to an occasionally applied remedy.

Quite recently *massage* of the abdomen has been proposed. I have no experience with it, but I should caution against believing methods to be good because they are modern. At any rate, massage should only be employed in very old remnants of exudation, as otherwise the manipulations are of course dangerous. These measures will probably have some psychical influence, as hysterical persons often believe the assertions of miraculous healers rather than the prosy veracity of a physician.

B. PERIMETRITIS.

Anatomy and Etiology.

Perimetritis is an inflammation of the serous membrane covering the uterus. But there is hardly a case in which the neighboring portions of the peritoneum are not implicated in the pathological process. Therefore, if we have been accustomed to designate the inflammations of the *entire* subperitoneal pelvic connective tissue by the name of *parametritis*, so we also call peritoneal inflammations which affect the peritoneum of the pelvic floor: *perimetritis*. More correct, of course, is the name pelvic peritonitis.

First I must make a few prefatory anatomical remarks.

In Fig. 148, taken from Luschka, the anatomical relations are represented. The preparation was made from a subject frozen in the recumbent position. Accordingly the uterus, following the law of gravity, has sunk *backward* and has applied itself to the parietal peritoneum. The fossa of Douglas is reduced to a narrow fissure (Fig. 148, 23). On the other hand, the strongly contracted bladder has detached itself from the uterus. On the right side, the peritoneum is removed, on the left it covers the pelvic organs. On the right, therefore, we see the subperitoneal space and in it, after elimination of the connective tissue and fat, the venous plexuses of the spermatic (18). Of the right broad ligament,

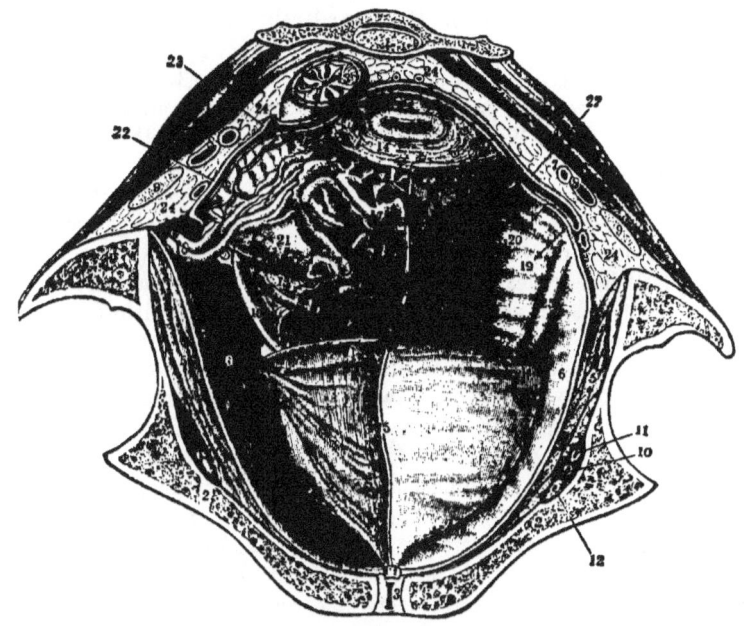

148.

Fig. 148.—Horizontal oblique section through the longitudinal axis of the pyriformes muscles and through the third sacral vertebra, after Luschka.

1, Third sacral vertebra. 2, 2, Horizontal ramus of the os pubis. 3, Symphysis. 4, 4, Pyriformis muscle. 5, 5, Internal obturator muscle. 6, Levator ani. 7, Superior gluteal artery. 8, Superior gluteal vein. 9, Sciatic nerve. 10, Obturator artery. 11, Obturator vein. 12, Obturator nerve. 13, Rectum. 14, Uterus; its left anterior side covered by peritoneum, its right side denuded of peritoneum. 15, Bladder. 16, Ureter. 17, Plexus of veins spreading between the peritoneum and the superficial muscular layer of the uterus. 18, Plexus of veins formerly inclosed between the layers of the broad ligament of the uterus and leading to the formation of the internal spermatic vein. 19, Anterior layer of the left broad ligament. 20, Posterior layer of the left broad ligament. 21, Posterior layer of the right broad ligament. 22, Posterior peritoneal wall of the fossa of Douglas. 23, Fossa of Douglas reduced to a narrow fissure. 24, Adipose layer of cellular tissue continuous with the subperitoneal cellular layer of the subperitoneal pelvic cavity laid open on the right side. 25, The left ureter lying underneath the peritoneum.

the anterior layer is absent, so that we see the anterior or inner surface of the posterior layer (21). In front of the uterus, likewise, are strong plexuses of veins (17). Therefore, should an inflammation arise in the localities described, on the right, or a product of inflammation be situated there, we should have *parametritis—inflammation of the pelvic connective tissue*. Should one of the veins represented rupture, a *hæmatoma of the pelvic cellular tissue* would arise.

On the left *side* we see a fossa containing the numbers 19 and 20; this is the left half of the vesico-uterine excavation, limited posteriorly by the broad ligament and the anterior surface of the uterus. Should an inflammatory product be situated here or in the fissure 23, between 20 and 22, the fossa of Douglas, and should it displace the uterus, 14, as far as the symphysis, 3, we should have an *intraperitoneal* or *perimetritic inflammation*. Blood effused in the same locality would form a *hæmatocele*.

The *etiology* of perimetritis is very manifold. In the first place, an enlargement, a venous hyperæmia of the uterus may implicate the peritoneal investiture. We must picture to ourselves that the excessively heavy uterus in its physiological dislocation injures the peritoneum, both at the perimetrium and at the places against which the uterus presses. The congestion, however, is the consequence of circulatory disturbances occurring with especial frequency during flexions and versions. Particularly in retroflexion, there will easily be a concurrence of hyperæmia of the uterus, stasis in the fundus, and interference with the physiological movements of the uterus. When the peritoneum is injured, inflammatory agglutinations occur. The latter are superficial at first, for there is no free space through which a pseudo-membrane may shoot from one point to another. For instance, should the uterus be adherent, not too firmly, posteriorly to the fossa of Douglas, and should it nevertheless artificially or spontaneously reassume its normal position, or should it enlarge during pregnancy, the superficial adhesion is stretched, it becomes a cord. The more the cord or the band is stretched, the less it is nourished. Thereby the pseudo-membrane is attenuated and finally extends as a cobweb-like, thin membrane from the uterus to the place where that organ has been adherent. The ovaries, too, if superficially irregular, enlarged, heavy, and descending, may reach spots where they may produce mechanical irritation or friction, for instance, between rectum and uterus. Then adhesions occur between the ovaries and the environments. Should the uterus subsequently assume its normal position, it will but imperfectly drag along the ovary agglutinated to other parts. Or else, an atrophying pseudo-membrane, extending along or over the ovary, may prevent the organ from reaching its physiological position. The same process takes place with the tubes. In this way there arise complicated dislocating influences which axially twist the ovarian ligament and the tubes, bend them toward one another, displace, flex, distort, and extend them.

Also chronic constipation and inflammation of the intestinal wall in typhlitis cause peritoneal hyperæmia with consequent adhesions. *Assuredly perimetritis may arise without any infection, simply through the circu-*

latory disturbance in dysmenorrhœa and chronic metritis. For perimetritis, in the more restricted meaning of the term, is a part of metritis. Just as in peritonitis the immediately underlying tissues—intestinal muscularis, bladder, uterus, etc.—do not remain free from inflammation and serous infiltration, so will the perimetrium not continue quite intact when the uterus in *chronically* inflamed.

In a similar way, perimetritis arises in consequence of acute circulatory disturbances. Thus when taking a severe cold during the menstrual period, there may be caused a quite acute inflammation of the uterus, its investiture, and the pelvic peritoneum. Although it may not be modern to believe in such things, this connection is but too frequently taught by experience.

Another etiological factor is furnished by *new-formations* in the peritoneal cavity. In them likewise the hyperæmia and the mechanical irritation are of importance. Thus perimetritic adhesions occur on hard dermoid cysts of the ovary and on firm subperitoneal uterine myomata, especially when, situated in the fossa of Douglas, they hinder the uterus in its physiological movements. In carcinomata proliferating into the body of the uterus, adhesions are always found, often so extensive that the fossa of Douglas is entirely absent. Even long-retained pessaries which press strongly against the posterior vaginal vault lead to agglutination of the fossa of Douglas. Moreover, in prolapse of the uterus, without and with inversion, numerous adhesions are often present in the deep funnel of the pelvic floor.

The parietal adhesions of large ovarian tumors have likewise been referred to mechanical influences, partly of external origin.

During ovariotomies, incipient perimetritis may often be observed, in the form of very loose adhesions on which may be demonstrated loss of endothelium, swelling and hyperæmia of the peritoneum, and slight deposits.

Blood coagula also excite the peritoneum to adhesions, exactly as do new-formations. Although blood-corpuscles and serum are absorbed, coagula of fibrin remain behind; they adhere and thus form the cause of perimetritis.

Moreover, perimetritis may occur after all intra-uterine manipulations, as after awkward sounding with firm, rough, or unclean instruments; after too large stem-pessaries the point of which imbeds itself into the fundus; after curetting of the uterus and enucleation of myomata from the depth of the uterine parenchyma; after treatment with sponge or laminaria tents. Why, indeed, in one case a parametritis ensues, in another the peritoneum becomes affected, or in still another both diseases occur simultaneously, will not always be capable of demonstration.

Furthermore, perimetritis is due, in the great majority of cases, to *gonnorhœal infection.* It may be that a gonorrhœal infection in woman runs an isolated course as vulvitis, vaginitis, and also endometritis (comp.

p. 93), but frequently the entire genital tract participates in the inflammation, from the vagina to the abdominal orifice of the tube, and including the pelvic peritoneum. Colica scortorum, formerly described as a neuralgia, is nearly always a perimetritis due to gonorrhœal infection. The connection may be threefold. The lymph-space of the peritoneal cavity inflames, as do the inguinal glands during the existence of an ulcer at the vulva or the penis. There is, then, an inflammation progressing in the lymph-vessels. But it is possible, too, that in virulent catarrh of the tubal mucosa the inflammation extends through the thin wall of the tube to the peritoneal investment, where it will spread and lead to adhesions. In favor of the view is the fact that, indeed, distortion of the tubes is found almost constantly in perimetritis. Besides, it is also likely that tubal catarrh produces virulent pus which finds its way into the peritoneal cavity, either because there is insufficient room in the tube or because a trauma mechanically expresses the tubal contents out of the abdominal orifice. In favor of this last-named connection we have those cases in which really the disease suddenly begins or exacerbates after a trauma, for instance, coition. But the dangerous character of tubal pus is demonstrated. Fatal peritonitis has frequently been seen after laparotomy in which a pyosalpinx ruptured and effused its contents into the peritoneal cavity.

Again, perimetritis is a complication of parametritis. More correctly expressed, a common injurious cause—infection—produces an inflammation both of the connective tissue (parametrium) and of the peritoneal covering (perimetrium or pelvic peritoneum). In the beginning and course, one or the other region may be implicated to a variable degree, so that the diagnosis hesitates or both diseases must be diagnosticated. Moreover, the final result of the pathological process appertains to both diseases, or at least to one more than to the other; purulent tumors occurring in which the extensive adhesions *above* the tumor prove the peritoneal, and the characteristic point of perforation or immediate seat on the nerves and vessels of the posterior pelvic wall, the subperitoneal, parametritic position.

The dislocations suffered by the tubes and ovaries by the adhesions cannot be even approximatively described in their multiplicity. Often there are only a few membranes which distort and fasten an ovary, often it is a veritable capsule through which the smooth, enlarged ovary gleams like the back of an extra-uterine fœtus through a ruptured tube. Or the ovary is so imbedded in the bands and membranes of the perimetritic products that it can be found only with the greatest difficulty. Thus the left ovary may lie in a curvature of the sigmoid flexure, the right may be adherent to the vermiform process in a mass of exudation.

Hitherto we have spoken chiefly of the *adhesive form of pelvic peritonitis*. It affects the majority of cases with the described etiology. In the puerperium, however, perimetritis and pelvic peritonitis often form the initial stage of *universal peritonitis.* The latter progresses with the formation of extensive exudations, and covers all the abdominal organs with a purulent deposit. In gynecological cases, the form of perimetritis advancing to universal exudative peritonitis is very rare; but we observe a quite acute, pernicious, septic peritonitis in consequence of infection or perforation. Formerly I have seen several cases of universal peritonitis due to the employment of sponge-tents, which ended fatally in five or six days. In one case the connection with ichorous endometritis and endosalpingitis could be demonstrated. Here all the intestines appear livid, the intestinal veins are filled to repletion, the abdominal cavity contains a slight amount of bloody, serous, thin, brownish fluid. Such cases have occurred now and then during nearly all gynecological operations previous to the use of antisepsis.

If, after what has been stated, a *universal*, purulent peritonitis is rare, there may be *circumscribed suppurations,* incapsulated patches of pus, abdominal caverns, the consequence of an exudative purulent pelvic peritonitis. To be sure, these cases, too, have remained behind mostly from the puerperium or after operations. Often they are connected with the above-mentioned extrusion of pus from the tube. Also intraperitoneal hemorrhages at times lead to suppuration. This pus generally lies at the deepest point of the peritoneal space, behind in the fossa of Douglas, unless that space has disappeared by adhesive peritonitis.

Inasmuch as such a confined abscess has a pyogenous membrane for its wall, the pus gradually augments, thus rendering the purulent cyst larger, tenser, more extensive. As the pressure increases in the cyst, the form of the abscess will more and more approach that of a sphere or ovoid. Thus the internal cyst-pressure deeply depresses the fossa of Douglas, forces the uterus close to the symphysis, and lifts the upper limiting membrane high up. Finally the purulent cyst becomes so large that it overtops the uterus above and laterally, and surrounds that organ like a parametritic abscess. In this way tumors arise which reach to the umbilicus and take up the greater part of the abdominal cavity.

The septum separating the abscess from the surroundings usually does not exist as a continuous membrane, but is formed of the surface of the intestines turned toward the cyst which are adherent to each other, and the interspaces of which are filled by thick membranes. It seems almost wonderful that the universally adherent small intestines are not absolutely disturbed in their peristaltic functions. For cases of "absence of the peritoneum" have been described which certainly are nothing but total agglutinations of all the small intestines in consequence of universal adhesive peritonitis.

However, the abscesses need not necessarily always lie below; an abdominal cavern may also lie in quite atypical situations, for instance, at the umbilicus or in the hypogastrium.

The illustrations show these relations. In Fig. 149 is represented an intraperitoneal exudation separated above from the peritoneal cavity by mutually adherent coils of small intestine. The exudation is seen to adjoin the rectum closely, so that a perforation may here easily occur, and that the finger may palpate a considerable extent of the exudation from the rectum and the vagina. Should the quantity of pus increase more and more, while the incapsulation persists, the form of the abscess represented in Fig. 150 will arise. Here the internal pressure has led to a rounding off in every direction. Posteriorly the abscess, surrounding the rectum, extends to the bone. The rectum passes through the abscess like a tube. In front, the abscess slightly overtops the uterus; a tense consistence can be felt everywhere.

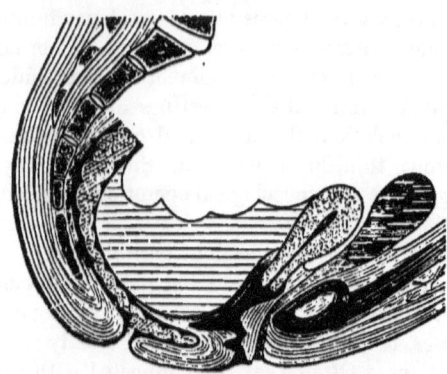

FIG. 149.—Large intraperitoneal exudation, limited above by adherent coils of small intestine.

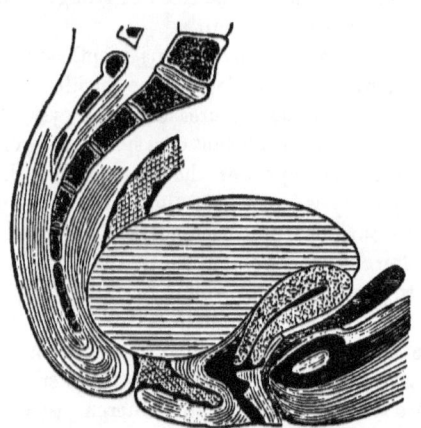

FIG. 150.—Intraperitoneal pelvic abscess, filling the whole of the lesser pelvis.

Several times the observation has been made that incapsulated peritoneal exudations *slough;* hence that on their artificial or spontaneous opening gases and offensive pus appear. This is likewise the case with psoas abscess, so that at the inner surface of the thigh abscesses are seen which evince the presence of air even before they are opened. Also abscesses forming on the inner wall of the pelvis in coxitis slough easily. *This untoward result is to be feared in all abscesses in the vicinity of the intestine.* Undoubtedly the septic germs come from the intestine. Although it may perhaps be impossible that bacteria and septic gases penetrate through the healthy intestinal wall, they can certainly pass through a diseased intestinal wall. As in universal peritonitis the upper layers of the uterus are found infiltrated, so also does the intestinal muscular is participate in an inflammation of the peritoneal covering. Through the thus infiltrated, inflamed intestinal wall bacteria or septic germs or bodies may penetrate into the abscess cavity, and thus the abscess sloughs. As perimetritic or peritonitic collections of pus are situated in closest proximity to the rectum (comp. Fig. 149), the danger of sloughing will be greatest in them. On the other hand, the spontaneous sloughing of an exudation will give rise to the conjecture of an intraperitoneal origin.

By the suppuration or sloughing of an exudation an *intestinal fistula* is also formed, both toward the external surface of the abdomen and toward the vagina, and perimetritic iliac fistulæ into the rectum and bladder have even been observed. The inflamed, paralytic portion of the intestine depends into the abscess cavity. Some fæces remain behind in the dependent portion. The after-coming fecal masses force that portion more and more outward, *i. e.*, into the abscess, solution of continuity occurs, and the old fecal fragments drop into the abscess cavity. Thereby the contents become ichorous, fever ensues, the wall of the abscess becomes inflamed, and perforation outward or into an adjoining organ follows; gas, fœtid pus, and a few old fecal fragments are evacuated. After the evacuation of the pus, the abscess contracts more and more, so that finally merely a fistulous track remains. Even the latter may close temporarily. Not infrequently, however, some more fæces exude in the depth, and the process is renewed without our being enabled to risk much interference, owing to the difficulty and danger which the locality presents.

The perforation of peritoneal abscesses takes place also directly into the rectum, the bladder, the vagina, or toward the external skin. In the latter case, the point of perforation may be situated a hand-breadth above the symphysis to the right or left, or even at the umbilicus. It will be evident, of course, that if the perforation takes place at a point characteristic of parametritis, the differential diagnosis can often not be made between parametritic or perimetritic, *i. e.*, between extra- or intraperitoneal origin.

Besides the *adhesive* and *purulent* forms of perimetritis, cases occur which may be called *serous* perimetritis. To be sure, the designation applies only in reference to the final product. An augmentation of the peritoneal fluid is possible for two reasons: either the transudation is increased, as in cirrhosis of the liver, heart-disease, kidney affections; or the absorptive capacity is diminished. The latter seems to be the cause in chronic serous peritonitis, without our being able, however, to state this connection as correct for *all* cases. When, owing to adhesive peritonitis, a portion of peritoneum becomes incapable of absorption, the fluid not taken up will gradually increase here. Hence serum may accumulate between pseudo-ligaments. Should, with increase of the fluid, the membranes either close by pressure, or adhere in consequence of exacerbations of the inflammation, sealed spaces will be formed in which a stagnating, physiological peritoneal secretion is inclosed. In this way serous cysts arise at and near the uterus. Such exudations are nearly always in the fossa of Douglas. Their long, unchanged persistence indicates embarrassed resorptive powers. It is very doubtful whether a purely purulent exudation may become serous; but inversely, a serous intraperitoneal cyst may become purulent or ichorous for the various reasons owing to which a purulent peritonitis usually arises. Such serous cysts have already been diagnosticated and operated on as ovarian tumors. They often reach a considerable size.

Symptoms and Course.

Perimetritis or pelvic peritonitis changing into *universal peritonitis* may run as malignant a course extra-puerperally as the most rapid puerperal septicæmic peritonitis. The symptoms are those characteristic of peritonitis—high fever; intense, continuous, spontaneous pains; sensitiveness to the slightest pressure; meteorism and constipation in consequence of intestinal paralysis; vesical paralysis, likewise due to infiltration of the musculature; rapid, weak pulse; enormously accelerated respiration; finally sopor and death.

In acute *pelvic peritonitis*, there is also fever and sensitiveness to pressure of the abdomen from the umbilicus down. When this inflammation becomes chronic, the fever ceases, but the sensitiveness to pressure remains for a long time or returns again with the least exertion. But there are not a few cases in which an acute onset cannot be demonstrated from the history, while the picture presented is still that of a pronounced perimetritis, which therefore exists in a chronic form. Its symptoms differ in degree. They consist in pain with the slightest exertion. When the patient walks on uneven pavement, slips, ascends stairs, is forced to walk fast, dances, rides in railway or other carriages, or even makes a hurried movement, pains occur. When quickly changing the position in bed from one side to another, a sudden intense pain succeeds. Even the slightest displacement of the uterus and the ovaries with every motion of the body causes symptoms. Rapid sitting down on a hard chair is as painful as rapid getting up. Coition is altogether impossible owing to the pain, especially in chronic inflammation of the fossa of Douglas (colica scortorum).

During the examination, frequently no tumor can be discovered anywhere, but every intended movement of the uterus is excessively painful. However, if there is a tumor, it will generally be situated in the fossa of Douglas or extend from there. The tumor will displace the uterus forward, and not laterally. Later, indeed, after the appearance of a very large tumor, the uterus may lie quite atypically, *i. e.*, laterally, and the sensibility to pressure, too, may gradually decrease.

To the symptoms caused by every pelvic tumor—bearing-down,

urinary and rectal difficulties—are superadded those of a retention of pus: hectic fever, great frequency of the pulse, anorexia, lassitude, loss of flesh, etc.

In serous cysts, nearly all symptoms may be absent, so that we are often surprised to demonstrate large tumors which exist and slowly disappear without causing any symptoms.

The adhesions displacing the tubes and ovaries cannot be felt directly; but an obscure sense of resistance, and a soft tumor escaping from and yielding to the finger, but always returning, proves their existence.

But it is strange that perimetritis is aggravated with especial facility during *menstruation*. There are cases in which all the results of careful treatment in the intermenstrual period are rendered nugatory by a sudden exacerbation during menstruation. But this is the case rather in the subacute than in the pronouncedly chronic forms of perimetritis.

Sterility, or at least impaired conceptive capacity is almost always present in perimetritis. Although the uterine or tubal catarrh may be the cause, the reason will probably more frequently have to be sought in the dislocations of the tubes and ovaries. Extra-uterine or tubal pregnancy is often caused solely by perimetritic displacement of the tubes.

In *septic cases* the course is equally acute as in the puerperium.

In *acute pelvic peritonitis* with residuary purulent exudations, the disease may last a very long time, yet the course in general is more rapid than in parametritis. Perforation into the abdominal cavity, sloughing, may speedily cause death; perforation toward the outside may bring about rapid cure.

In *chronic perimetritis* the condition of relative cure occurs by the formation of pseudo-membranes, that is to say, the inflammation has definitely disappeared, but the inflammatory products, the adhesions, remain to the end of life. Although they become gradually thinner and are loosened artificially or spontaneously, they do not completely disappear. The slowest course of all occurs in cases due to *gonorrhœal infection*. In them slight movements often cause renewed aggravations. In all probability the main affection here is a specific tubal catarrh which now and then evacuates its dangerous products into the peritoneal cavity. It is noteworthy that there are cases of this nature in which, despite agonizing suffering for years, neither distinct adhesions nor tumors appear.

As to the *terminus*, little need be added after what has been stated.

In sepsis and penetration of the pus into the peritoneal cavity death occurs. The same may take place from sloughing of an exudation, but such cases also often terminate favorably by perforation outward, after high fever for days augured the worst result. Small exudations are completely absorbed and disappear. But here as well as after adhesive peritonitis, sterility and dislocation of the uterus, tube, and ovary remain behind. Chronic gonorrhœal metritis really has no terminus, because it always exacerbates anew, in spite of every precaution and care in the treatment.

Diagnosis and Prognosis.

If the abdomen is very sensitive to pressure, if the uterus cannot be moved from the vagina without causing pain, if there are many subjective sensations of pain, the diagnosis is clear. An acute peritonitis at most might be mistaken for ileus. Renal and hepatic colic may at first sight lead us to suspect acute peritonitis. But the absence of rise of temperature and of great rapidity of pulse soon dispels every doubt.

Adhesive peritonitis is diagnosticated by attempted movements of the uterus in every direction. Thus the seat of the inflammation cannot rarely be accurately determined. For instance, if movement of the vaginal portion downward and forward is painful (comp. p. 184), there is inflammation of the folds of Douglas. And if in retroflexion the replaced uterus immediately returns to its position even in the knee-elbow position, the body of the uterus is adherent in the fossa of Douglas (comp. p. 199). If an ovary lies on the floor of the fossa of Douglas and remains there constantly even if the uterus is placed in anteversion, there is an adhesion at that spot between the ovary and the peritoneum. If the body of the anteverted uterus cannot be lifted, and if an obscure resistance is felt at one cornu (comp. p. 190), the uterus is adherent with this cornu, that is to say, adhesions inclose the tube and fix the uterus. With simultaneous sterility, this circumstance also indicates old adhesions.

It is more difficult to recognize with certainty the products of perimetritis, the *perimetritic exudations*, as such. In the first place, the *anamnesis* here will always demonstrate an onset with considerable pain. If a large inflammatory tumor at or around the uterus arises *without any symptoms*, it cannot be of perimetritic origin. When a peritonitis leads to exudation and the formation of tumors, to an incapsulated purulent patch, this inflammation has undoubtedly set in with considerable pain.

To be sure, in an old incapsulated exudation, the neighboring peritoneum is often so greatly altered that the exudation can be palpated by combined examination without causing pain. A parametritic exudation even, when in the process of suppuration or softening, at that time not rarely is more painful than a perimetritic exudation, because the adjoining peritoneum is not implicated in the inflammation till then.

The position of the tumors felt, of course, differs in small exudations and in the beginning. The parametritic exudation lies at first in the parametrium, *i. e.*, at the side or in front of the uterus; the perimetritic exudation, however, first fills the fossa of Douglas, and hence in the beginning is felt behind the uterus. Other distinguishing features are stated to be, the high location of the intraperitoneal, the low position of the extraperitoneal exudation. But these relations change with considerable increase in size, when also the peritonitic exudation descends deeply at the sides. We must always bear in mind that complications are easily possible, and hence that intra- and extra-peritoneal exudations may exist simultaneously. Should an exudation perforate at the upper half of the abdomen, it is peritonitic; these exudations are frequently filled with ichor, perforate during high fever, with much pain, and vivid reddening of the skin. Parametritic abscesses, however, perforate more inferiorly, more gradually, do not soften completely, and therefore disappear but partially after absorption.

The differential diagnosis may be very difficult when there is in the neighborhood of the uterus, as the final result, a cyst filled with serum; here it is often impossible, after a single examination, to make a positive diagnosis. If the tumor adjoins the abdominal coverings, it is permissible to make an exploratory puncture with the Pravaz syringe, under antiseptic precautions. Serous contents indicate a perimetritic tumor.

The *prognosis* generally is unfavorable. Life is endangered by sloughing, perforation into the peritoneal cavity, and by acute peritonitis generally. Health continues impaired during gonorrhœal perimetritis. Should the body long retain an accumulation of pus, suppurative fever and cachexia occur, even tuberculosis is possible. If only adhesions are present, the uterus remains dislocated, the tubes continue fixed, and the ovaries displaced. Connected therewith are many troubles, especially hysteria. Sterility and extra-uterine pregnancy should likewise be named.

Treatment.

As in parametritis, the *prophylaxis* of perimetritis consists in careful antisepsis during labors, abortions, and operative manipulations on the uterus. Furthermore, every virulent vaginal catarrh must be carefully treated from the beginning, so as to cure it, if possible, before it extends to the uterine mucosa.

When an *acute perimetritis* or pelvic peritonitis has set in, we endeavor first to combat it by antiphlogosis. We proceed here as in puerperal perimetritis, with ice, opium, abstraction of blood from the inguinal region and the vaginal portion. The old treatment, according to which, previous to the exhibition of opium, the bowels were emptied by a large dose of calomel (0.3–0.75 gm. = gr. v.–xij.), is to be urgently recommended in acute cases.

In *chronic perimetritis*, compresses to the abdomen are employed with good effect (comp. p. 274).

There have also been recommended, painting the abdomen, the vaginal portion, and the vaginal vault with iodine, or the introduction of tampons and suppositories with potassium iodide which is well absorbed by the vagina and the rectum.

Often the constipation forms the most troublesome complication and the main point of attack for the physician (comp. the rules given on p. 158).

In perimetritis, saline baths are much employed. And, indeed, a protracted "cure" in one of the many saline baths of Germany is often the only effective measure. These baths are: Hall, Königsdorf, Kreuznach and Münster-am-Stein, Dürrenburg, Kösen, Sulza, Wittekind, Halle a. d. S., Colberg, Oeynhausen. In perimetritis, full baths are to be preferred to sitz-baths, as the difficult getting up and the compression of the abdomen in the latter are often very troublesome. In bad cases, the curative effect may be increased by causing the patient to take protracted baths of one and a half to two hours. Subsequently to them, it is advisable to let her keep the bed for one or two hours.

Should there be at the same time a cervical catarrh or an erosion, the treatment should by no means be commenced with cauterizations during the existence of a florid or merely violent chronic perimetritis. The catarrh often disappears spontaneously. Not rarely I have freed patients of perimetritic pains by simply interdicting every vaginal exploration and manipulation for several weeks.

Opening a diagnosticated abscess is to be advised only when the natural process is hastened thereby or if a vital indication is furnished by sloughing. Without the necessity, without some special indication, I do not advise the opening. But if the skin is reddened, if there is a break in the tissue, if perchance we can feel the crepitation of gases present in the abscess, if everything points to the speedy perforation, the opening should be made. Thereby the duration of the pain and the high fever is not infrequently materially shortened. It is better to allow the abscess to perforate spontaneously into the vagina and bladder, for the antiseptic precautions during after-hemorrhages, etc., are here effected with great difficulty. Perimetritic abscesses often diminish rapidly after being opened.

The cavity must be carefully irrigated with a double-current catheter. Especial care is required in very cachectic persons and large abscesses, for in those cases, as during the opening of pleuritic exudations, death may occur quite suddenly, inexplicably. Strong pressure, the introduction of the finger or the sound, and excessive water-pressure during the irrigation are to be seriously discountenanced. But an antiseptic *puncture* with the Pravaz syringe never causes any danger.

In more recent times, these abscesses or serous cysts have been emptied with the *aspirator*. Extensive experience is still lacking regarding this method, and would seem to be difficult to accumulate, owing to the great individual diversity of the cases.

In chronic perimetritis it is particularly necessary not to forget the general condition through the local treatment.

C. HÆMATOCELE.

Anatomy and Etiology.

Since Nélaton had described hæmatocele as an *intraperitoneal* accumulation of blood behind the uterus, this disease has been made the subject of searching investigation by many gynecologists. The objective condition on touch necessary to the diagnosis was not rare; but the proof that the tumor felt was a coagulum of blood was furnished only in the most exceptional cases. For post-mortem results of hæmatocele are very seldom published.

We must first inquire, whence or how does the blood get into the abdominal cavity?

The most frequent source of the hemorrhage is the ruptured ovisac of a tubal or other extra-uterine pregnancy. In proof of this connection case reports and post-portem results are on record. Not only did menstrual disturbances precede nearly every case of hæmatocele, but previous pregnancy was also demonstrated. Should the hemorrhage be rather slow, the opening of the ovisac be not very large, or the quantity of blood be inconsiderable owing to very early interruption of the pregnancy, the hemorrhage arrests itself because the forming coagula plug the opening. Of the effused blood—which, following the law of gravity, lies at the deepest point, the fossa of Douglas—the fluid constituents are absorbed and a coagulum remains behind. The congealed mass *displaces the uterus forward, close to the symphysis.* Gradually the coagulum becomes progressively firmer; an uneven, slowly diminishing surface is distinctly felt; after absorption of the coagulum, the dislocated uterus finally resumes its former position.

Furthermore, an internal *varix* may also rupture. I have described a case of fatal internal hemorrhage from an eroded vein at the surface of a pregnant uterus.

Should the vein open toward the pelvic connective tissue, which is certainly possible, the hemorrhage would be extraperitoneal. But if the vein is close to, and its walls adherent to, the peritoneum, the hemorrhage would be intraperitoneal on rupture of the vein. For the production of the opening in the vein a trauma is of etiological importance, for exertions are now and then assigned as the cause. Such, for instance, is coition indulged in at the time of menstruation. Perhaps also a menstrual anomaly *per se* may lead to the formation of a hæmatocele. A hæmatosalpinx may discharge into the fossa of Douglas; at least we not rarely find hemorrhage outward and simultaneous formation of the tumor within. Moreover, the blood of a hæmatocele may have come from the ovary. The possibility of rupture of the ovary with effusion of blood into the abdominal cavity cannot be denied. In scorbutus, scarlatina, measles, purpura, etc., we find, as after difficult labors, bloody deposits on the peritoneum. But true tumors certainly form very rarely.

Quite arbitrarily a few authors assumed a peculiar form of peritonitis in which blood was "exhaled" from the peritoneum. Equally rarely would the formations of adhesions lead to hemorrhage, as Virchow endeavored to render probable from the analogy to pachymeningitis hæmorrhagica. Perhaps the connection is rather this, that women with adhesions are very liable to extra-uterine pregnancy, and hence that rupture of the tube is more likely to occur.

The second question to be discussed is, How does the retro-uterine tumor characteristic of hæmatocele originate? The main credit of having elucidated these relations is due to Schroeder.

In the case of a perfectly free effusion of blood, a retro-uterine tumor can be felt as little as in ascites. This is proved by cases of fatal internal hemorrhage, in which the blood can be demonstrated laterally by percussion as well as by the sense of fluctuation, but not by the feeling of a retro-uterine tumor. But if the hemorrhage takes place into an incapsulated space, that is to say, if the fossa of Douglas is incapsulated, the inflowing blood will of course distend the space and a tumor must form.

Withal it is not by any means necessary to assume an absolutely firm occlusion. Even if there is a passage through the intertwining and superimposed pseudo-ligaments, the blood will crowd these membranes together and thus effect the formation of a cyst. And should some blood escape, *the bulk of the mass, rapidly coagulating, will remain beneath the membrane.*

But perfectly free blood can also form a tumor, as I have described as early as 1873 and demonstrated by experiments. The blood in the peritoneal cavity does not remain fluid, but coagulates very quickly. Even in fatal hemorrhage, a few hours of life suffice to coagulate the blood. These coagula, of course, form a tumor at the deepest point of the peritoneal cavity—the fossa of Douglas. There they crowd the uterus forward and cause symptoms which are described farther on. Slowly or rapidly, according to the size of the tumor, it disappears, the greater part of the blood being absorbed, and only coagula of fibrin remain behind. But the blood-clots act on the peritoneum like every new-formation—*mechanically irritating,* so that a perimetritis ensues with resulting adhesions.

When the posterior side of the uterus is so agglutinated to the opposite peritoneum by superficial adhesions in the fossa of Douglas as to entirely obliterate that space, the blood must of necessity be located higher. Then it may also get in front of the uterus—forming a peri-uterine or rather ante-uterine hæmatocele. This must likewise be the case when the hemorrhage is so profuse as to flow forward over the uterus and the broad ligaments, after filling the fossa of Douglas.

Isolated ante-uterine hæmatoceles without agglutination of the fossa of Douglas are impossible. Cases formerly repeatedly described have been traced back by Schroeder, in a critical investigation, to an erroneous diagnosis, by proving them to be hæmatometra in a rudimentary atretic cornu.

Although thus far we have spoken only of *intraperitoneal* hemorrhage, and interpret hæmatocele in that way alone, we must not omit mention of the fact that *extraperitoneal* hemorrhages into the pelvic connective tissue are also possible and do occur. As the formation of hæmatomata, so-called thrombi, has been observed at the vulva, so also has it been in the parametrium. In recent times, several cases of parametritic hæmatomata have been described, and there is an inclination to the view that a large number, if not all, of *spontaneously* arising cases of parametritis are connected with such hemorrhages. Here, too, the distended broad ligament or the detached peritoneum may rupture subsequently, hæmatocele being the consequence of the hæmatoma. The etiology of these extraperitoneal hemorrhages is still obscure. Of course, this observation does not alter the picture of hæmatocele, to the nature of which the intraperitoneal hemorrhage always belongs.

Symptoms and Course.

The symptoms of hæmatocele are composed of those of acute anæmia and the rapid formation of an intraperitoneal tumor.

In larger hemorrhages of any kind, in "internal hemorrhage," syncope occurs; the latter is of *pathognomonic importance*. The nature of the pains is variable. In some cases they consist merely of a dull sense of pressure in the pelvis, in others again the patient at once lies down, owing to the excruciating pains. Invariably, however, touching the abdomen, and *especially the posterior vaginal vault*, is very painful. This sensibility to pressure is not as great as in peritonitis. Defecation is rendered difficult and painful. Micturition, too, is usually interfered with, because the uterus, pressed close to the symphysis, compresses the urethra or rather the inferior vesical segment.

Behind the anteposed uterus there is at first a tensely elastic tumor which by no means surrounds the uterus semilunarly as a parametritis, but is situated quite medially in the fossa of Douglas. The sensibility generally soon diminishes, the tumor seems to move higher up, becomes irregular, knobby. The hardening coagula can be slightly dislocated, sometimes producing a creaking sensation. The uterus moves more and more toward the centre. Finally, small tumors are behind the uterus; after a *single* examination, without exact observation of the previous course, it can hardly be determined whether they are the ovaries, remnants of exudation, or scybala. Eventually they, too, disappear and cure results, or else a perimetritis develops.

In the most favorable cases, a tumor cannot be felt externally. The tumor perceived in the vagina disappears in from three to four weeks.

Not rarely, too, cases of primary perforation into the rectum occur. Even after eight to fourteen days, masses of blood are passed per rectum without the appearance of suppuration, sloughing, or fever. This perforation may take place without any symptoms. We are surprised to find the tumor suddenly greatly diminished, and then only inquire after the stools which contain the characteristic small coagula.

The perforation may occur also into the bladder, vagina, or the peritoneal cavity, or into several places at the same time. Hæmatoceles treated by operative opening have run a particularly dangerous and malignant course. Rapid sloughing, sepsis and death, or long-continued debilitating suppuration and pyæmia ensued.

It is peculiar that often during a later menstruation a secondary hemorrhage occurs, similar to what we have learned to be the case in perimetritis. The duration of the disease will depend upon the size of the mass of blood. Tumors the size of a fist do not disappear in less than three or four weeks. Larger ones can be felt for months. Even if the case causes no alarm for weeks, the tumor may still suddenly slough for inexplicable reasons.

Diagnosis and Prognosis.

For the diagnosis the history is to be borne in mind. As above stated, menstrual disturbances precede. The onset in pronounced cases is so sudden that the patients are able to give the precise moment. Should there be syncope, or at least subjective symptoms of acute syncope —tinnitus aurium, scintillation before the eyes—this likewise indicates hemorrhage which, if it does not appear externally, must have taken place internally. If to this be superadded a tumor, the position and quality of which characterize it as an accumulation of fluid in the fossa of Douglas, the diagnosis is clear. Even the fact that a large tumor can appear in a few hours only through hemorrhage, assures the diagnosis. But it is certain that, if *sudden onset, elastic tumor, symptoms of anæmia, and absence of fever* belong to the diagnosis, the latter can be made but *very rarely.* Undoubtedly, quite pronounced cases of hæmatocele occur, in which there can be no doubt of the diagnosis even without palpable proof that there is blood in the tumor. But if the anamnestic data remain obscure, and the case does not come under observation quite early, the diagnosis must only be made with some reserve.

The diagnosis will hesitate most between perimetritis and hæmatocele, and it cannot be doubted that those authors who count hæmatocele among the most frequent diseases, have mistaken a large number of perimetritic exudations for hæmatocele. If the tumor has existed for some time and the history is obscure, it is quite impossible to form a positive diagnosis, for it cannot be *felt* whether a fluid contains more white or more red blood-corpuscles. The course, however, furnishes some guiding points. Continual absence of fever is in favor of hæmatocele, as is the transition from tense elasticity to a knobby, variously hard surface. A very great forward dislocation of the uterus presupposes preceding free mobility, that is, absence of adhesions, which could not be the case in perimetritis. The material dislocation, then, points rather to hæmacele. But the rapid diminution of the tumor especially indicates hæmatocele. Perimetritic exudations can be thought of only in disease of the peritoneum, but the *diseased* peritoneum has its absorptive power impaired. Hence an accumulation of pus can only be gradually inspissated, not *rapidly* absorbed. Therefore, rapid reduction in size and speedy disappearance of the tumor likewise are in favor of hæmatocele.

If a hæmatocele suppurates, it has changed to perimetritis and as such it runs its course with fever.

Furthermore, all tumors of the fossa of Douglas enter into the consideration, as fibroma, carcinoma, and ovarian tumors. In these, observation for a few days will render the diagnosis possible, unless the characteristic form and position of the tumor furnish the elucidation.

The differential diagnosis is more difficult in retroflexion of the gravid uterus complicated with perimetritis. In that case touch of the tumor from the vaginal vault is very painful, as is palpation; sounding must not be done on account of pregnancy and inflammation. The history is often obscure, because menstrual anomalies preceded in either case, and an incarceration of the uterus may occur rapidly, like a perimetritis. In such cases the forced examination may become very dangerous and even be followed by acute peritonitis and death. We should first ask ourselves the question whether the diagnosis is absolutely necessary *at once* for the treatment. Then chloroform should be given so as to arrive rapidly at a conclusion by combined exploration and palpation, unimpeded by the struggling of the patient. In narcosis we must proceed in elucidating the relations so as to obtain the diagnosis of in— of the uterus, or else the proof that no incarceration is pres—

—gnosis is clear from what has been stated. In interpreting usually as the consequence of a ruptured extra-uterine pregnave, by the formation of a tumor, the proof that the first hemorrhage, is past. The coagulum, however, like every taneous hæmatoma, will be absorbed slowly but surely.

On one hand, the position in the peritoneum will be very favor— absorption; on the other, special dangers will be caused by —ences—irritation of the peritoneum, agglutination, the form— —esions, sloughing, and perforation.

Treatment.

The operative treatment of hæmatocele has given very bad results, so bad indeed that even modern antisepsis has not caused the operation to be again attempted. In the mean time it has been empirically demonstrated that an expectative treatment gives a very good prognosis and that the tumors are absorbed without anything being done. *Hence the tumor of a hæmatocele is a "noli me tangere."* The cyst must be opened only when sloughing has occurred, that is, high fever, renewed fluctuation, projection into the vagina, rectum, or the external skin. The opening is made where the fluctuation is most distinct. Often this is an extraordinarily soft spot between hard surroundings.

In other cases we shall confine ourselves to ward off injurious effects or to prevent the enlargement of the blood-cysts. In quite recent cases, rest, morphine, and ice to the abdomen are at once to be ordered. In older cases, too, rest is necessary, for movements may easily be followed by dangerous after-hemorrhages, or at least the sensitiveness, the perimetritis, may considerably increase. Owing to the great pain caused by a voluminous defecation, *eccoprotics* are to be given from the beginning. This must not be neglected, as great displacement of the coagula during straining or the passage of a thick column of fæces may provoke renewed hemorrhage and largely increase the pains. The *diet*, too, must be selected with the view to the rest and the difficult defecation, in other words, all substances should be avoided which render the passages voluminous.

If perforation ensues, disinfection should be provided for. It is best to omit every examination or sounding or irrigation of the cyst from the rectum. For the fistulous opening will best prevent the entry of intestinal gases and sloughing. From the vagina, however, the cyst is to be carefully irrigated through a double-current catheter, with slight water pressure. In one case, I engaged the opening in a speculum and with a long pincette removed coagula of fibrin which were followed by pus. Owing to the sloughing, the cyst was drained merely into the vagina by means of a rubber tube six centimetres in length. The vagina was irrigated every two hours. Three days later, but little laudable pus escaped.

Any strong pressure upon the tumor is to be avoided. The treatment of the remnants of hæmatoceles coincides with that of perimetritis.

It is always advisable that the patient, even if she feels quite well, should remain abed during the next menstruation.

D. NEW-FORMATIONS IN THE PELVIC CONNECTIVE TISSUE.

Aside from inflammatory tumors, hæmatomata, and regionary metastases in uterine carcinoma, few tumors occur in the pelvic connective tissue.

An *echinococcus* may spread around the uterus like a parametritic exudation, and laterally behind the uterus it may extend upward in the connective tissue. The peritoneum likewise is dislocated far upward, so that an abdominal tumor arises.

Should spontaneous rupture into the rectum, the vagina, the uterus, or the bladder occur, the diagnosis is usually made from the escaping vesicles. An exploratory puncture or aspiration is also permissible. Chemical examination shows slight specific gravity, grape-sugar, scolices, and hooklets. In suppuration of the cyst, of course, only pus can be detected. Free incision, evacuation of the sac, and drainage under the strictest antisepsis has been followed by cure in several cases.

In the same way as myomata occur in the round ligament, their development is possible, too, in the broad ligament, owing to the connective tissue and muscular nucleus. Not rarely, myomata developed laterally in the uterus extend into the broad ligament like parametritic exudations. If these tumors are large, they crowd the uterus sideways and upward. Such tumors are at times very soft and spread like exudations, following the connective-tissue spaces.

Fibromata may also spring from the bone, as they do from the base of the skull. Several cases of dystocia have been described, in which large fibromata sprang from the periosteum or the pelvic synchondroses.

Enchondromata and *sarcomata*, *exostoses* and *cysts* of the pelvic bones likewise form large, partly very firm, partly softer tumors which can be felt from the vagina.

A rectal carcinoma may also grow forward and become a retro-peritoneal carcinoma or lead to secondary uterine carcinoma.

In all these cases exploration *per anum* is necessary. Thereby is formed the differential diagnosis whether a tumor springs from the sacrum or lies in front of the rectum in the fossa of Douglas.

It is particularly the reaction of a tumor on the general health which decides whether it is a malignant new-formation. Should a tumor grow uninterruptedly, become uneven, should ascites ensue, and the functions of neighboring organs be restricted, the diagnosis will not hesitate.

The treatment is symptomatic, as operative removal is usually impossible. But the lower half of the sacrum can be resected, so that a tumor springing isolated from it might be removed without danger.

Cysts of the broad ligaments, parovarian cysts, spring from the parovarium. Their diagnosis and treatment will be discussed in the following chapter.

E. Diseases of the Round Ligament.

When the round ligaments participate in pathological conditions of the uterus, they may be hyperæmic or œdematous. Considerable varicose expansions of the veins have been found in them. Whenever the round ligaments spring at different heights from the uterus, that side of the uterus upon which the ligament is attached farther down is less fully developed, hence the uterus is unequal, asymmetric, awry.

As the round ligaments always begin at the point of origin of the tubes, one of the round ligaments is the guiding line to the uterus in doubtful extra-uterine pregnancies; thus, if the round ligament leads to an ovisac *separated from the uterus*, that ovisac is the one rudimentary uterine cornu. But if the round ligament leads to the edge of the uterus at the median side of the ovisac, the uterus is perfectly normal, and the ovisac is formed of extra-uterine organs, the ampulla of the tube or the ovary.

As the round ligaments consist of uterine musculature, myomata must be liable to occur in them. Winckel has described such a case.

Under the name of *hydrocele of the round ligament* two pathological conditions are known: first, an abnormal continuation of the peritoneum into the inguinal canal and in front of the inguinal canal beneath the external skin. In this case a tumor filled with fluid, the size of a child's fist, arises in the inguinal region. As in hydrocele in male children, the canal occasionally closes, the fluid being thus separated from the abdominal cavity. Schroeder, however, states that he succeeded in pressing the serous fluid back into the abdominal cavity. That procedure and the application of a truss would constitute the treatment.

In the second form, a serous cyst develops in the midst of the round ligament and hence is situated above the inguinal canal.

F. Diseases of the Tubes.

Defects of Development.

Under the head of malformations of the uterus on p. 133 it was stated that, in rudimentary development of the female genitals, the tube may be completely absent or may form a solid cord. Among the congenital malformations belongs also the presence of a *second abdominal orifice of the tube* from which fimbriæ likewise project into the abdominal cavity. The tube is usually flexed at the abnormal ostium situated in the continuity of the tube.

Not rarely the musculature of the tube is absent at a circumscribed spot, there being a *hernia of the mucosa*, a small protrusion at the tube. Here, then, the tubal wall is composed merely of mucous membrane and peritoneum. According to Klob, these herniæ of the mucosa are important in the etiology of tubal pregnancy. Should the fecundated ovum lodge in such a pocket, it will remain there. The fact that a tubal pregnancy almost always ruptures early, points in these cases to a predisposing thin condition of the ovisac. Should tubal pregnancy be effected from any cause, the uniformly present muscularis would stretch to such an extent as to allow the pregnancy to terminate. This is very rarely the case. It certainly does appear remarkable that tubal pregnancy either leads *very early* to the rupture of the tube or, in very rare cases, goes to term.

Inflammation of the Tubes and its Consequences.

As has been repeatedly stated, the tubal mucosa participates in the gonorrhoic inflammation of the mucous membrane of the uterus. The pus may get into the abdominal cavity from the tube and incite violent perimetritis (comp. p. 279). There are cases even of interminable perimetritis, the cause of which is the pus again and again escaping from the tubes.

But dilatation of the tubes also occurs; of course, this affects only the abdominal portion. Under the head of perimetritis we have already mentioned the multiform *dislocations of the tube* due to adhesions, distortions, flexions, etc. They lead to constriction, even atresia of the tube. The latter occurs also by the fimbriæ adhering with their peritoneal surface.

If the tubal secretion cannot escape from the tube, it gradually distends, a pyosalpinx or hydrosalpinx forming.

In menstrual anomalies a tubal hemorrhage also occurs. In consequence of vicariating menstruation, a hæmatosalpinx forms in atresia and excessive distention of the uterus (comp. p. 144).

However, retention of secretion is possible also in not absolute occlusion, in mere constriction. The inspissated secretions may likewise be retained. The uterine ostium is so little dilatable that the internal pressure of the tubal cyst does not suffice for forcing the fluid into the uterine cavity. In very rare cases only has a periodical discharge of *dropsy of the tube* been observed, which form has been denominated *hydrops tubæ profluens*. That dilatations of the uterine part of the tube occur, though rarely, is proved by cases of accidental sounding of the tube (p. 31) or by the passage into the abdominal cavity of medicated fluids injected into the uterus.

The *form, size, and position of tubal cysts* are very variable.

Should the tube distend, it becomes too large for the ala vespertilionis, the peritoneal ligament of the tube. The peritoneum does not expand in conformity, but forces the tube to bend and flex still more than it normally does. Thereby the tube acquires the sausage form with indentations. At the abdominal end a sort of retraction occurs by the incurved fimbriæ expanding outwardly. However, both during laparotomies and in numerous preparations, I have seen smooth-walled, dilated tubes, running merely a slightly curved course. Even perfectly spherical, movable cysts may arise which can hardly be distinguished from small ovarian cysts during the examination.

The *size* is generally inconsiderable. Tubal cysts of from three to five centimetres diameter are frequent; large cysts, however, simulating ovarian tumors have been observed very rarely.

The *position* is almost always lateral. The cyst is usually movable. But a tubal cyst may also adhere to the uterus behind and in front, extending above the broad ligament, with or without fastening the ovary to the anterior surface of the uterus; or it may be situated upon the apex of the uterus.

The *diagnosis* of the various affections of the tube can be made only hypothetically *unless the tube is enlarged.* Thus in incurable, always relapsing perimetritis, in which gonorrhœal infection and endometritis have been diagnosticated, we justly suspect an endosalpingitis with escape of pus from the abdominal orifice.

All affections associated with *enlargement of the tubes* are to be recognized with our present combined methods. Often we feel by the side of the margin of the uterus the movable, club-shaped, or rosary-like tumor extending outward, so that no doubt obtains. Or a finger is simultaneously passed into both bladder and rectum, feeling for the angle of the uterus, while an assistant at the same time twists the uterus around its longitudinal axis by means of forceps applied to the vaginal portion and pulls it down.

During *castration* or in *laparotomies* generally, it is very important to pay attention to the tubes, lest perhaps a thin-walled pyosalpinx be ruptured and its dangerous contents be discharged into the peritoneal cavity.

Simple dislocation of the tube may be argued from aquired sterility after the termination of perimetritis. As we have stated above (p. 286), the final products of the inflammation may also here be occasionally felt.

The *prognosis* of tubal inflammation is unfavorable because treatment is impossible. Should the treatment of perimetritis not be followed by improvement, nothing can be done for the salpingitis.

The question whether it is justifiable to perform laparotomy with a view to extirpate the tube is still *sub judice.* The operation would be decidedly unadvisable during the existence of considerable perimetritis. But in the other cases the symptoms are not particularly urgent.

Should during a laparotomy the discovery be made that the cyst to be removed is a tubal dropsy, the latter should be *completely* extirpated if possible. For after making an opening in the tubal sac, the hole may close or become occluded by pseudo-ligaments, so that the dropsy recurs afresh.

New-Formations of the Tube.

Of new-formations, *tuberculosis* is to be first enumerated; it occurs here as a primary affection, and often fills the tube with large masses of cheesy tubercular products.

Carcinoma, however, has always occurred merely secondarily at the tube, starting from an ovarian carcinoma. *Fibromata* and *lipomata* and a papilloma of the tubal mucosa have also been described. The so-called *hydatid of Morgagni* is a small vesicle often hanging to a fimbria; it is of no practical importance. Now and then we find another vesicle on a pedicle perforating the broad ligament which springs from the parovarium.

CHAPTER XV.

DISEASES OF THE OVARIES.

A. VARIETIES OF DEVELOPMENT.

In abscence of the uterus, the ovaries are also lacking; in unicorn uterus, the ovary of that side is missing where there is no horn of the uterus. But an ovary with the tube may be snared off and have disappeared by torsion of the pedicle or by pseudo-membranes. In a rudimentary uterus, the ovaries are frequently rudimentary. They are of remarkably minute size and contain no Graafian follicles or mere traces of them.

We have stated on p. 225 that the tube and an ovary may easily get into a *hernial sac.* As in hydrocele of the round ligament, a diverticulum of the peritoneum forms, into which the ovary and the tube may drop or be drawn. Inasmuch as in the fœtus the ovaries and the tubes are situated very high, above the pelvic inlet, these herniæ arise in fœtal life. An inguinal hernia of the ovary is not rare. Much less frequently have crural herniæ, or descensus of the ovaries through the vascular opening of the obturator membrane and through the major sciatic notch, been observed.

Should the *ovary* lie in a hernial sac, it may, by swelling during menstruation, provoke pain. Should it degenerate into a tumor, the increasing growth will likewise cause distress. Careful examination must demonstrate, *first,* that the ovary is missing at its physiological seat; *second,* that the round body palpable in the inguinal hernia is the ovary. By sounding and moving the uterus, by bimanual palpation of the pelvis, the correct diagnosis will be arrived at. Should it cause distress, the ovary is to be removed operatively. Should it be merely inflamed or suppurating, it is to be treated antiphlogistically or the abscess opened externally.

Under the name of *accessory ovaries* have been described small appendages consisting of ovarian stroma, either attached to the ovary by a short pedicle or separated from it merely by a slight constriction. *Supernumerary ovaries* are very rare; in a few cases *two* ovaries were found on *one* side, or else the third ovary may be behind or in front of the uterus.

B. Inflammation of the Ovaries.

Acute inflammation, suppuration, and softening of the ovaries occur in the puerperium, usually as a concomitant phenomenon in septic or at least exudative peritonitis. Outside of the puerperium, acute oöphoritis likewise appears as a partial manifestation of an acute febrile general disease.

Slaviänsky found that, in acute infectious diseases and in peritonitis in the environment of the ovaries, the follicles inflame; in the former case, the primordial, and in the latter, the ripe follicles. Both the cells of the membrana granulosa and the ova disintegrate; the theca folliculi likewise participates in the inflammation. The practically important result may be the loss of all the follicles and hence sterility. Slaviänsky calls this the parenchymatous form of oöphoritis, and by interstitial oöphoritis he understands an inflammation or infiltration of the connective-tissue stroma, such as leads, for instance in the puerperium, to complete suppuration. Older authors understand by parenchymatous oöphoritis an inflammation of the entire ovary, and separate therefrom an inflammation of the follicles—the parenchymatous inflammation of Slaviänsky.

In the interstitial form, of course, cures with cicatricial contraction as well as suppuration are possible. Patho-anatomical results are also on record which indicate both processes. Thus quite atrophic ovaries have been found imbedded in peritonitic adhesions.

In the interstitial form, as in the puerperium, the ovary may suppurate. As the fact that such ovarian abscesses perforate into the bladder has been described as especially characteristic, I would point to the uncertainty of the diagnosis in general. It will hardly be possible to completely relinquish the suspicion that these cases were parametritic or perimetritic exudations. However, even a small ovarian cyst may suppurate under the influence of puerperal infection or peritonitis.

The *treatment* of acute oöphoritis is that of acute pelvic peritonitis. There is no method of treatment having a direct influence on the ovaries.

Chronic oöphoritis is often assumed in a condition which equally admits of the diagnosis of perimetritis. When we feel the somewhat enlarged ovaries, with concomitant irritability of the ovaries and their surroundings, we justly assume that this enlargement and irritability are referable to inflammation. We regard as the anatomical final result of chronic oöphoritis an atrophic condition—the granular atrophy in which, like the liver in hepatic cirrhosis, the ovary is contracted and indurated. Hypertrophy of the ovaries may also be induced; in this the hyperplasia of the intersitial part is particularly obvious. Very often the ovaries are dislocated in this condition, so frequently that *dislocation of the ovaries* has been described as a disease *per se*. Should the ovaries be displaced backward in retroflexion of the uterus, they usually lie laterally *above* the uterus, in the same manner that they otherwise lie laterally *behind* the uterus. Not till after the reposition of the uterus, therefore, are one or both ovaries often felt laterally behind the uterus in the fossa of Douglas. Through the thin vaginal wall even cicatrices and isolated irregularities on the surface of the ovary can be detected with surprising distinctness. According to our preceding explanations, the ovary, if situated between the rectum and uterus, must injure the peritoneum. Thus are caused losses of endothelium of the peritoneum, agglutinations, and permanent adhesions.

The ovary may, by its weight alone, get into the fossa of Douglas, even if the uterus be normally situated. That the weight is the factor of etiological importance is shown by the elevation of the ovary with gradual diminution.

The *symptoms* of chronic oöphoritis consists in pains during coition, defecation, and violent movements. The fact that menstrual anomalies are likewise frequently present particularly points to changes in the ovarian parenchyma. The *course*, in the presence of firm adhesions, is chronic. Not rarely, however, it is observed that, under appropriate treatment, the ovaries resume their normal position with the disappearance of all symptoms. Still, numerous hysterical phenomena, reflex neuroses, etc., may gradually ensue, so that the case becomes less favorable.

The *diagnosis* should be made only when the enlarged ovaries are felt. If they lie in the fossa of Douglas, they might be mistaken for hard fecal masses. The compressibility of the intestinal contents, as well as their disappearance after an enema, confirms the diagnosis.

Furthermore, the differential diagnosis from remnants of exudations or hæmatoceles is often impossible. In such cases it is necessary to search for the ovary at its normal site. It may be felt—by no means always—at the upper angle of the uterus, with two fingers, by counterpressure from without. The entire lateral margin of the uterus as far as the linea innominata is subjected to conjoined palpation. Thus the cylindrical, movable ovary is generally got between the fingers. The ovary escapes from between the finger-tips, when healthy is not sensitive to pressure, and nothing else can easily be mistaken for it, on account of its form, consistence, and mobility. However, should it be impossible to feel the ovary with certainty, owing to the rigidity of the abdominal coverings, not even the result of the treatment will settle the diagnosis of a small tumor of the fossa of Douglas. For just as a heavy, prolapsed ovary elevates itself, so does an exudation disappear under appropriate treatment.

Other tumors do not enter into the question, simply on account of their size.

The *treatment* is that of perimetritis (comp. p. 287). Easy defecation should especially be provided for. Abstraction of blood proves likewise advantageous. Should no treatment be successful, and should the reflex neuroses or hysterical affections referable to the ovarian disease urgently demand a change of the condition, nothing will remain but *castration.* Owing to the, at best, not inconsiderable danger of that operation, it would have to be carefully determined in every case whether the conditions justify a hazardous procedure. Inconsiderate, hasty action is to be earnestly cautioned against. It may happen that the ovaries are so placed that their removal is impossible. Hence Hegar justly demanded that the ovaries should be felt previous to the operation.

If castration is to be done, it is performed from the abdomen. After the abdominal cavity is opened, the ovary is seized, ligated, and cut off with the scissors. The operation may be exceedingly simple, but also enormously difficult. The latter is the case in nulliparæ with firm abdominal coverings and where there are perimetritic adhesions.

C. Hæmatoma of the Ovary.

A hemorrhage may occur into the ovarian tissue, and the blood-cyst thus formed may rupture. In this way an intraperitoneal hemorrhage arises—a hæmatocele. But hæmatomata of the ovary also occur in which that organ is enlarged to a diameter of five to eight centimetres. Although these hemorrhages at times are connected with general diseases (Bright's disease, scurvy, typhoid fever), they are difficult of explanation in other cases. Schultze found a hæmatoma at the dissection of a newborn child.

In some cases a vicariating hemorrhage may be assumed in dysmenorrhœa. For instance, I performed castration for excessive dysmenorrhœa and metrorrhagia with fibroma of the uterus. In one ovary were found two cysts with bloody contents, the other ovary was changed to a hæmatoma; on both sides hæmatosalpinx and atresia of the tubes by pseudoligaments existed. The patient died. The uterus contained two myomata, was of stony hardness, and its cavity was so compressed that barely a thin sound could enter it.

There are no *symptoms* peculiar to hæmatoma. The *terminus* probably is usually atrophy, recovery, and acquired sterility. In a few such cases it appeared noteworthy to me that the menopause set in prematurely with the atrophy.

The *diagnosis,* during life, could be formed only if the rapid occurrence, and especially the gradual atrophy of the cyst, were observed. In the differential diagnosis, extra-uterine pregnancy particularly enters into the consideration. Especially the fact that, with the formation of a hæmatoma, atrophy, and cessation of the ovarian function, a dysmenorrhœa may disappear, makes the *prognosis* seem not unfavorable.

D. NEW-FORMATIONS OF THE OVARIES.

The ovary consists of a glandular parenchyma—the follicles—and a purely connective-tissue stroma surrounding them, in which blood and lymph vessels and nerves extend. The larger surface of the organ is covered with *germinal epithelium*, genetically the same as the epithelium forming the ova. Perhaps the inter-connection with the parovarium is effected in such a manner that the germinal epithelium furnishes *merely the ovum* for the Graafian follicle, while the granulosa cells are derived from the parovarium. At all events, the latter may be followed into the hilus of the ovary. The hilus, or the lower, smaller half of the ovary, is implanted into the peritoneum (comp. Fig. 5, p. 7). The various tissues produce the new-formations characteristic of them, to which are superadded mixed tumors and abnormalities.

We distinguish therefore:

1. Simple retention cysts. Follicular dropsy.
2. Epithelial tumors, derived from the glandular or germinal epithelium (granulosa cells):

    Cystic cylindro-cellular adenoma, to wit: glandular and papillary tumors. Carcinomata.

3. Tumors derived from the connective tissue: fibroma, sarcoma, angioma, etc.
4. Dermoid tumors.

1. *Dropsy of the Graafian Follicle; Hydrops Follicularis.*

Physiologically every Graafian follicle is a small cyst. Menstrual congestion causes this cyst to swell and to rupture. Rindfleisch correctly points out that simple diffusion does not explain this process. He assumes that the cells of the membrana granulosa physiologically form some expansible colloid material which, swelling under the influence of menstrual congestion, causes the follicles to rupture. Pathological absence of this material or its superabundance produces tumors. Thus, in the absence of this material, the granulosa cells secrete fluid, but no expansible material; consequently the follicle, though gradually enlarging, does not rupture. The final result is a retention cyst with serous contents—*dropsy of the Graafian follicle.*

Scanzoni has attempted a similar explanation by assuming imperfect congestion to be the cause of the non-rupture of the follicle, and referring this imperfect congestion to general hydræmia. Other authors plainly speak of a catarrh of the Graafian follicle. Perhaps also an irritation from without, for instance, neighboring peritonitis, effects the cyst-formation, or the development of membranes prevents the rupture of an otherwise healthy follicle. Rokitansky has demonstrated that even after the rupture of the follicle during ovulation it may close again and then develop into a cyst.

Assuming the cells of the membrana granulosa to be derived from the parovarium, dropsy of the Graafian follicle would be genetically the same as a parovarian cyst. In favor of this view are the similar contents.

Small cysts have been found even in the fœtal ovary, and in later life they are extraordinarily frequent. Sometimes they form an agglomeration of several small and large cysts. In the small ones we still find the ova—a fact directly in proof of their origin from Graafian follicles. Very rarely the cysts reach a considerable size, up to that of a man's head. They have very thin walls and are often ruptured accidentally, for instance, during laparotomies. The ovarian tissue between them is normal, menstruation and pregnancy occurring. When the pathological process takes place in many Graafian follicles, a larger, multilocular abdominal tumor may arise. These cysts of course are benign, and it is not improbable that spontaneous cure occurs very frequently by rupture. The latter is to be referred to accidental trauma or the increasing internal pressure.

2. *Cystomata: Adenoma, Papilloma, and the Carcinomata.*

During the development of the ovary the ova form from the germinal epithelium covering the surface and extending into the stroma of the ovary. From this germinal epithelium, really a mucous epithelium, are also derived the cells of the membrana granulosa, which indeed, according to a different view, are formed from the medullary cords extending from the parovarium into the hilus. Should an "ovum-tube" or the epithelium derived from the medullary cords fail to reach the physiological development, it changes pathologically into a tumor, perhaps by the stimulus of an irritation. The cylindrical epithelium and its matrix develop into an atypical glandular formation covered with cylindrical epithelium. Accordingly, all ovarian cystomata would have to be referred to a faulty fœtal development of the ovary.

Other authors, basing on observations, assume that incomplete glandular tubes form also in later age, and degenerate into tumors. Lastly, however, it is by no means proved that a completed Graafian follicle cannot likewise change into an adenoma.

The essential feature of these tumors, therefore, is the atypical proliferation of the cylindrical epithelium; the soil on which the epithelium arises will obviously grow too, so that larger, quite irregular adenomata and papillomata develop. A separation of the two species is impossible. Should—a very rare occurrence—the growth start from the superficial germinal epithelium, *papillomata sessile upon the ovary* will arise. Should the growth commence in the ovarian stroma, *cystomata* appear; and should the source from which the development of the tumor springs be situated quite deeply, in the hilus, the papillomatous masses proliferate downward into the pelvic connective tissue as well as upward into the ovary. The first form, the papilloma upon the surface of the ovary, is very rare. Nor does it spring from the internal wall of a small ruptured or perforated cyst, but as clearly takes its origin from the ovary as does a cauliflower excrescence from the vaginal portion. As the superficial germinal epithelium is the same as that in the ovary, this origin presents nothing wonderful. It is strange that the superficial papillomata are decidedly malignant. They incite the peritoneum to exudation, and themselves secrete fluid. They infect the peritoneum by detached epithelia forming secondary papillomata on the peritoneum, and frequently degenerate into carcinoma. The second form, the adenoma or papilloma arising within the ovary, usually forms a cystic tumor and is the most frequent new-formation of the ovary. These cystomata are very multiform. In the first place, there are unquestionably large, perfectly unilocular cysts with colloid contents. The epithelium may cover the wall as a sin-

Fig. 151.—Schematic diagram of the relations of growth of an ovarian cystoma.

a, Superficial cylindrical epithelium; *b*, connective-tissue cyst-wall; *c*, superficial germinal epithelium; *l*, a protrusion of the cyst-wall in form of a gland, produced by the adenomatous proliferation of the epithelium; *d*, vessels; *e*, *h*, two cysts formed by adhesion of papillæ, colloid degeneration of the epithelia, incipient atrophy by pressure of the septum producing a single cyst; *g*, glandular projection outward, forming a very thin-walled small secondary cyst adjoining the main cyst; *i*, epithelial indentations simulating uterine glands. At *h*, commencing colloid degeneration of the epithelium with consequent cyst-formation. At *k*, an extensive papillomatous proliferation has perforated the cyst-wall, so that one portion of the proliferating papilla, *n*, lies still within the cyst, while the other portion, *o*, has proliferated outward; *m*, two papillæ growing one toward the other, agglutination, incipient cyst-formation.

gle-layered cylindrical epithelium, or express the energy of its growth by universal indentations resembling uterine glands (Fig. 151, *i*). But should the growth of the wall and of the contents or the epithelium not proceed *pari passu*, the surface of the cyst, as it were, does not suffice for the epithelium formation, the epithelium proliferates inward, and outward too; but more rarely in the latter direction, for the thick cyst-wall must here be first perforated. More frequently the epithelium, dragging along its matrix, forms lower and higher tree-like anastomoses (Fig. 151, *m e h, n o*); they may proliferate through the entire cyst papillomatously, or be merely attached to the internal surface in the form of a few low protuberances. In the development of these excrescences the single portions so appose themselves as to form closed spaces behind them (Fig. 151, *m*). Or else, the excrescence covers and occludes some of the above-described glandular formations; or, with great internal pressure in the cyst, the epithelia grow more toward the outside into the cyst-wall, where they often plainly represent acinous glands (Fig. 151, *i*). In all cases, of course, the epithelium investing a closed space secretes or degenerates exactly like that covering the surface of the papillæ. Should the secretion or the product of degeneration—the colloid—be profuse, the inclosed space must become spherical, hence, as it were, forming a new cyst within the old or in the wall of the mother cyst. Should such a retention cyst arise more in the depth of the cyst-wall, the latter may be thinned toward the outside so that a quite thin-walled translucent cyst externally adjoins (Fig. 151, *g*) the otherwise thick-walled cyst *a b c*. In general colloid degeneration, moreover, the single glandular epithelia disappear, so that at the spot where they originally extended into the cyst-wall nothing remains finally but a small lump of colloid—a minimal colloid cyst. According to the predominance of the formation of papillæ extending inward, or the production of glands, these new-formations, according to Waldeyer, have been called papillary or glandular cystoma. If the papillary excrescences arise with the first development and interlace extensively, the entire tumor becomes more firm, solid, microcystic, multilocular. Should but few larger, secondary cysts arise, during their agglutination within a larger cyst the septa may atrophy (Fig. 151, *e h*); then one secondary cyst opens into the other, the contents mingle. Fragments of the septa, as well as vessels coursing within them, persist. All the conditions combine. In some cases, we find a large unilocular cyst, and, at a single spot only, a few papillomatous proliferations. In other cases again, there is a confused mixture of larger and smaller, ruptured and closed secondary cysts, with various metamorphoses of their contents.

In order to elucidate these conditions I have introduced Fig. 151, on p. 306, which is intended to be schematic merely.

The size of the entire cystoma may far exceed that of the uterus near term. As to the shape of the cysts, the smallest and the largest are usually round or ovoid and regular. Those ranging in size between that of the uterus in the fourth month and that of the largest cysts, exhibit more irregular forms. We can often distinguish, particularly, two larger cysts which, however, are intimately connected by an extensive surface. Bilateral disease is very frequent.

While the tumors thus far described arise within the ovary and hence are situated intraperitoneally, an extraperitoneal location likewise occurs. This is especially the case in tumors of a papillomatous character. These tumors perhaps spring from the parovarian segment of the ovary, *i. e.*, near the hilus, in that part of the ovary which lies within its peritoneal insertion. They represent the third form of the adenomata or papillomata. The discovery of vibratile epithelium in these tumors made their parovarian origin appear probable to Olshausen. However, vibratile epithelium is now and then found in all ovarian tumors. Pathognomonic of these tumors is an especially malignant character, due to the manner of their growth. Usually they arise bilaterally, proliferate downward, and lie so close to the uterus and extend so deeply into the pelvic connective tissue that their removal is often quite impossible (comp. below).

Cystomata contain very different fluids. The color is usually grayish, becomes yellowish by the admixture of pus, and by the admixture of blood brown or dirty greenish. The consistence is equally variable. We find quite glass-like, firm, tough gelatin which must be torn out by the hand, but more frequently the contents are fluid and escape through a trocar. In the multilocular form, thin and thick fluid may be present in the same tumor. Quite watery fluid is found in parovarian cysts and in dropsy of the Graafian follicle. The tough colloid masses consist of the cell-contents of the epithelial cells of the inner surface which have changed into colloid.

Carcinomata likewise belong among the tumors springing from the epithelial formations of the ovary. They may arise as carcinomatous degeneration in the papillomata described on p. 305. They then infect the peritoneum and lead to ascites. The ascitic fluid is derived partly from the secretion of the tumor, partly it is peritonitic serous fluid which remains unabsorbed on account of disease of peritoneum.

In some of these carcinomata were found peculiar inorganic bodies soluble in acids; therefore they were called psammous carcinomata. In the same way carcinomata form in the adenomatous cysts discussed sub 2. How near these adenomata or cystic papillomata are to carcinoma is proved, outside of this degeneration, by a case of Klebs' in which a carcinoma grew in the remnant of a cyst left after ovariotomy. Carcinoma was observed also in a dermoid cyst (comp. below). Besides, ovarian carcinoma proper occurs in two forms; first, as a parenchymatous diffuse cancer which enlarges the ovary considerably, but leaves it its form. In these cases were found the various species of cancer—scirrhous, medullary, and colloid, or a combination of the latter two. And, second, as a distinctly glandular carcinoma which, as it were, eats up the ovary and represents a slightly uneven tumor of rapid growth.

It is peculiar that ovarian carcinoma frequently occurs bilaterally and at a very early age.

3. *Tumors of the Ovarian Connective Tissue, Fibroma, Sarcoma, etc.*

As the ovarian stroma contains no muscular fibres, myomata could not possibly be observed. Should they be reported nevertheless, the mistake is caused by a subperitoneal uterine myoma having migrated into the region of the ovary. Even if the ovary is completely absent and its place clearly occupied by a myoma, it will be impossible to assume an ovarian myoma with certainty, for the snared ovary may be absorbed and the uterine myoma may have separated from the uterus.

Also regarding *fibromata* of the ovary there is little unanimity among authors. By some, "fibromata" are described which are no new-formations in the ordinary sense of the term, at any rate no distinct tumors, but merely hypertrophies of the stroma; other "fibromata" are better designated by the name of sarcoma or fibro-sarcoma. Without, however, entering into critical explanations, we shall reproduce the views to be found among the majority of observers.

Under the name of fibromata have been described hyperplasiæ of the connective tissue which, in some cases, are to be considered cicatricial residues of puerperal processes. The ovary becomes enlarged by this new-formation of connective tissue. In especially characteristic cases, all the follicles have fallen a prey to the cicatricial formation, owing to which fact, of course, sterility exists. The inflammatory origin is indicated by the frequent complication with residues of a pelvic peritonitis. Some authors also describe small fibromata which they trace back to a corpus luteum. The demonstration of the follicular membrane as the limit of the tumor supported this interpretation. In other cases, though an incapsulated fibroma cannot be demonstrated, still a certain spot is found in the ovary which represents a small connective-tissue tumor. Here, too, the remaining portion shows inflammatory alteration, both by deposits upon the ovary and by chronic inflammatory processes within. Exceptionally, a few very large fibromata of from four and a half to thirty kilograms (ten to sixty-six pounds) have been described. In the larger fibromata greatly dilated vessels are found. Also cysts—softening cysts with fatty diffluent contents and smooth-walled cysts with serous contents (probably dilated follicles)—have been observed; in the latter, low cylindrical epithelium was likewise found. These larger tumors may calcify, both at the centre and on the surface. Of course, in such a case especially careful examination must be made, so as to disarm the suspicion that we are perhaps dealing with an ill-nourished, calcified, subperitoneal myoma separated from the uterus.

The *sarcomata* of the ovary are usually spindle-celled sarcomata and to some extent are closely related to the fibroma or the simple hyperplasia of the ovarian stroma. In proportion as these tumors grow, the Graafian follicles disappear, but the latter may persist both in the primary form and change into cysts. In that event tumors arise which anatomically should be designated cysto-sarcoma. Processes of softening, especially fatty degeneration, lead to breaking down and cyst-formation. Should a vessel be implicated in this process, it may become thrombosed or be opened. In the latter case hemorrhage occurs into the tumor. The cyst may also rupture, so that the blood escapes into the peritoneal cavity.

Round-celled sarcoma is found more rarely than spindle-celled, but cases are on record of pure round-celled sarcoma as well as of that form mixed with spindle-celled sarcoma. Mixed types approaching carcinoma have also been observed, like a tumor designated by Spiegelberg as "hemorrhagic carcinomatous myxosarcoma."

Should the Graafian follicles participate in the enlargement of the tumor not merely passively, but actively, by proliferating idiopathically, such a tumor is to be denominated adeno-sarcoma. The latter form seems to lead to the formation of metastases more easily than the firmer forms of sarcoma. Metastases have been described in the pleura, the peritoneum, and the rectum.

The occurrence of these tumors has been observed at every age. Even congenital ovarian sarcomata have been described. Sarcoma does not by any means appear to occur more frequently bilaterally than unilaterally. Ordinarily these tumors are not larger than a man's head, often smaller. Pronouncedly cystic formations or combinations with adenomatous forms reach an enormous size.

Besides the species of tumor described, we find mention of cases of tuberculosis, one case of enchondroma and angioma of the ovary.

4. *Dermoids of the Ovary.*

The dermoid is a tumor containing constituents of the cutaneous organ, inclosed within internal organs. Formerly, *dermoids of the ovary* were explained as the remains of an ovarian pregnancy. Then, also, a kind of parthogenesis was believed in. The hypothesis was set up, that there was a formative energy peculiar to the ovary which led to the partial or incomplete development of an ovum. Especially the fact that nerve substance was likewise demonstrated caused the simple explanation of inclusion of small portions of the cutaneous organ to appear insufficient. But since His has demonstrated that the primitive trace of the germinal gland forms from the axis cord, in the genesis of which the upper germinal plate also participates, it is not to be wondered at that constituents of all germinal plates may be found in the ovary. Therefore, every dermoid cyst is congenital, and statistics computing their frequency at various ages are obviously nonsensical. On the other hand, it cannot be denied that dermoid cysts may enlarge only with the beginning of menstruation so as to cause symptoms and lead to their detection.

Ovarian dermoids are usually smooth-walled cysts with a bunch of hair which is no longer attached to the skin, and filled with an atheromatous mass, *i. e.*, a lump of fat. The latter, perhaps, has congealed only in the cold preparation taken from the cadaver. In many cases the hair is inserted into the skin. Plates of bone are also found; they not rarely bear teeth. Their frequently enormous quantity (three hundred) proves that teething proceeds continuously. A preparation has been described in which was found a deciduous tooth half detached, and a second one growing beneath it. Gray nerve substance has also been demonstrated in dermoid cysts, as well as smooth, but never striated muscular fibre.

The wall of the dermoid may calcify, or it may thin and soften so that the contents are effused into the abdominal cavity. Suppuration and hemorrhage also occur within the cyst proper. In still other cases, further neoplastic processes take place in the cyst-wall. Small cysts in the wall have been described as retention cysts of the sudoriparous and sebaceous glands, and carcinomata springing from these glands. Thus *mixed tumors* are produced which are difficult to interpret. This is also the case when in one ovary a cystic cylindro-cellular adenoma and a dermoid cyst are simultaneously present.

The Relation of Ovarian Tumors to the Neighborhood; Pedicle, Growth, Adhesions, Metamorphoses in the Tumor, Rupture of Cysts.

As the ovary projects free into the abdominal cavity, a tumor developing in the ovary must likewise do so. Unfortunately we possess no investigations which determine the type of the peritoneal limit on ovarian tumors. Should a tumor arise from a Graafian follicle, or at least from an incipient Graafian follicle, the whole tumor will be covered with germinal epithelium. But should the ovary be inserted more deeply into the broad ligament than is usual (Freund); should the tumor form from the parovarian segment at the ovary, the medullary strands; or should, perhaps, one of Pflüger's tubes extend too deeply, the developing tumor will be intraligamentous in situation.

In ovarian tumors arising in the intraperitoneal part, the tumor must be connected with the uterus or the broad ligament in the same way as the normal ovary. The connection between the tumor and the pelvic organs, *the pedicle,* hence will be composed of the ovarian ligament, the broad ligament, and the tube.

Traction of the heavy tumor, sinking first posteriorly, then anteriorly, stretches and lengthens the ovarian ligament. The tube is drawn toward and above the tumor and lies close to it, being attached by the greatly elongated but narrowed ala vespertilionis. At times, too, the ala vespertilionis is altogether lost, so that the tube, thickened, injected, and materially elongated, lies immediately upon the tumor.

Therefore, when the tumor is lifted out, the "pedicle" is formed of the broad ligament in which are seen the two thicker cords, the tube and the round ligament; between them the broad ligament is so thin that a thread of the ligature can be forced through it with a blunt instrument. Should the tumor not grow downward at all, but altogether intraperitoneally, the pedicle at times becomes five to ten centimetres long, can be easily constricted with a thread and in that condition is barely as thick as a lead-pencil.

In other cases, however, the tumor also grows extraperitoneally, namely, at first immediately approaches the uterus between the layers of the broad ligament. On lifting out the tapped and emptied cyst, the "pedicle" is formed by the end of the cyst. In these cases the uterus is displaced laterally and elevated; it may even, like the tube, be pulled toward the tumor and be drawn out, *i. e.*, elongated. A ligature placed ever so deeply or close to the uterus around the pedicle includes the lowest segment of the cyst-wall. Then we have "a thick pedicle," and a portion of the cyst remains below the ligature, but it usually atrophies and disappears, cure being perfect.

Large vessels are not always found in the pedicle; often we are even surprised to see but few nutrient vessels in the thin pedicle—a proof that the enlargement arose from colloid degeneration of the epithelia, from passive expansion. On the other hand, again, band-like veins, one centimetre broad, run toward and into the cyst-wall.

Should the tumor, for reasons above given, grow still more *extraperitoneally*, downward into the broad ligament, the latter becomes unfolded. The tumor extends into the pelvic connective tissue according to the laws governing perimetritic exudations. Thus the tumor, displacing the bladder, reaches the abdominal wall in front, like a puerperal parametritic infiltration (comp. p. 267). Should laparatomy be performed in this case, no peritoneum will be found in front, because the tumor, growing upward, inserted itself between it and the abdominal wall. At the posterior side of the pelvis the tumor also gets beneath the peritoneum. In rare cases the rectum is displaced by the tumor. The sigmoid flexure, or the cæcum with the vermiform appendix, lies upon the tumor which, therefore, has grown behind these organs. The tumor may even press the mesentery apart, having thus a knuckle of small intestine above it, and the unfolded mesentery surrounding the extraperitoneal tumor.

Special difficulties are caused by tumors developing *partly intraperitoneally, partly extraperitoneally.* Here the dilated ring of the ovarian peritoneal appendix divides the tumor into two, generally unequal halves. Of course, the subperitoneal part of the tumor greatly elevates the broad ligament—the constricting ring of which causes a sort of stricture in the tumor.

Ovarian tumors extending free into the peritoneal cavity must, following the law of gravity, assume different positions recalling the dislocations of the pregnant uterus in retroflexion. In the first place, every ovarian tumor, provided the ovary in question was not previously adherent in front, will lie at the side of the fossa of Douglas. Should the tumor grow larger, it will not have room enough in the pelvis and will fall over in front, usually without symptoms, like a spontaneously replaced retroflexed uterus. In the erect posture, therefore, the tumor will seek its support at the anterior abdominal wall. Thereby the pedicle is stretched and elongated. The upper surface of the tumor comes to be in front. The tumor turns a quarter of a circle; the uterus, formerly in front of the tumor, is now crowded backward. Therefore the connection between uterus and tumor must turn backward at the uterus, forward at the tumor. A spiral is formed. Should the tumor grow irregularly, for instance, should the posterior upper quadrant develop most strongly, the tumor in its forward flexion will have to turn its upper surface still farther forward. When the patient lies down, the tumor would have to become retroverted. But if the pedicle is long drawn out and the tumor greatly flexed forward, the intestines lying upon the tumor and the intra-abdominal pressure prevent its resumption of the former backward flexion and hence it gravitates laterally and backward beneath the pedicle, producing torsion of the latter. When the patient rises again, the tumor drops forward as before, and subsequently repeats its backward turn beneath the pedicle. In this way *several* torsions may be added to the *first*, thus completely twisting the pedicle together. Of course, the process will be possible only with small and firm cysts, because those with flaccid walls may lie in any position, and in them the fluid rather than the tumor changes its place. Moreover, the pedicle must be long and thin, otherwise it will not twist. Therefore, torsion of the pedicle is observed with special frequency in firm tumors, *e. g.*, dermoid cysts, and those with thin and long pedicles.

Torsion of the pedicle has different consequences: first, it may cut off the supply of nourishment, so that the cyst atrophies. Such atrophied cysts have been found at autopsies. In other cases, the sudden occurrence of torsion may cause immediate death of the tumor, and this again, sloughing, softening, rupture of the cyst, and fatal peritonitis. Or else, the torsion is more gradual; then the thin-walled veins are compressed, but not the arteries. Phenomena of congestion appear in the tumor—it may enlarge suddenly; soften œdematously, or a hemorrhage may take place into its mass.

Should no torsion ensue, and the tumor continue to grow into the abdominal cavity, it will lie close to the abdominal wall like the pregnant uterus. Flaccid unilocular cysts lie in the peritoneum like ascites. Tumors consisting of two large cysts may lie more across the abdomen, like the pregnant uterus in transverse position. In very large cysts, the abdomen is universally expanded in front, the ribs and the xiphoid process are bent outward.

Large, and especially hard, firm tumors are nearly always *adherent* at various points. Should the superficial epithelium of the ovary be lost, adhesion to the peritoneum will be easily effected by an inflammatory traumatic process. The tumor may adhere by a large superficial extent to the parietal peritoneum in front. Adhesions are also often observed to the spleen, liver, the intestines, and the omentum. Should two cysts grow, whether one from each ovary or both from one ovary, they may fasten the omentum between them.

Superficial adhesions at times are so loose that they may be separated like the placenta from the uterus, but very firm *superficial adhesions* also occur, and still firmer *cord-like adhesions* which can only be severed with knife or scissors.

Of pathological alterations in the tumor, we have already become familiar with congestion and hemorrhage during torsion of the pedicle. Other causes may likewise incite *hemorrhage* into the tumor, like *destruction of the septa* (comp. p. 306), during which a vessel coursing through the wall may be ruptured. Or if it should remain intact for the present, it will be torn during the subsequent expansion of the cyst which lengthens the distance between its starting-point and its terminus. The anatomical proofs of such vascular remnants have been repeatedly demonstrated with sanguineous cyst-contents. Finally, hemorrhages into the cyst occur also after tapping. It will be impossible to decide whether the hemorrhage came from a wounded vessel or ensued in consequence of the *horror vacui*, after the removal of the glandular contents, when the cyst dilated through the elasticity of its wall.

Moreover, a cyst may *suppurate* or *slough*. Nowadays, we shall not be surprised if this occurs after puncture. Even a Pravaz syringe may carry enough infectious material into the tumor to cause decomposition of the cyst-contents. Another cause, torsion, has been mentioned above. Also intestinal adhesion, inflammation, perforation, and entry of intestinal contents into the cyst lead to sloughing. In some cases, sloughing has occurred quite spontaneously. Not rarely whitish flakes are found in the cyst-contents, which under the microscope prove to be pus.

Easier of explanation than this spontaneous pus-formation are a number of more *regressive processes* occurring in the cyst-wall. Thus, with or without torsion of the pedicle, we find, in the cyst-wall as well as on its inner surface, *concrements of lime* which, when the hand is passed along, are not rarely demonstrated as fine granules. *Fatty metamorphosis* of the cyst-wall also occurs; this leads to spontaneous rupture. The epithelium of the cyst, likewise, not rarely undergoes fatty degeneration; in the fluid obtained by tapping, we find both single fatty cells and larger portions of epithelium, still attached to its base, undergoing fatty degeneration. Still more frequently smaller and larger portions of the epithelium show colloid degeneration, enlargement, distention. On careful microscopical examination of the epithelium of an entire agglomeration of cysts, we find beautiful cylindrical, cuboidal, and flattened (by centrifugal pressure) epithelium; also, in *the same* cyst, epithelium in total fatty and colloid degeneration.

Softening of the cyst-wall may here and there be connected with fatty degeneration, but if the entire cyst is so soft that every attempt to loosen adhesions tears the wall, it is usually in a state of inflammation or infiltration. This is not meant to imply that there are not also unchanged cysts which in some places are so thin that it is impossible to lift them out without rupturing some of the single cysts.

Not only a preceding softening or inflammation or perforation of the cyst-wall by penetrating papillomatous proliferation leads to escape of the cyst-contents into the abdominal cavity, but also an actual *rupture of the cyst.* This usually is the consequence of a trauma. Of course, the disposition to rupture will be caused by the above-described conditions. On the other hand, a very thick-walled firm cyst may rupture in consequence of a trauma, as I have witnessed in one case. The patient had fallen down-stairs.

The cyst may also rupture into the rectum. This is occasionally the case in smaller cysts adherent in the fossa of Douglas. The small intestine may also become agglutinated and the perforation take place into it. The cyst-contents were even evacuated into the stomach, were vomited, and the tumor disappeared. Very rarely the tumor opens into the bladder, vagina, or upon the external skin. If the latter, the most frequent point is through the umbilical ring. If ascites complicates, the umbilicus is not rarely protruded by it in the shape of a bladder.

Very difficult to explain are those cases in which the fimbriæ of the tube become attached to the tumor in such a way as to cover the point of rupture. Then the cyst-contents escape outward through the tube and uterus. This happens both spontaneously and after physical exertion. The rupture has even occurred with a certain regularity every four weeks at the time of menstruation. As these cysts are formed of an ovarian cyst and the dilated tube, they are called tubo-ovarian cysts.

Symptoms and Course.

Small ovarian tumors up to the size of the fist cause symptoms only during peri-oöphoritis (comp. p. 302). Should the tumor grow, much will depend upon whether it is soft with lax walls or firm and hard. The former bend around all the organs, and may reach the size of a man's head without giving rise to any symptoms. Firmer tumors, however, dermoids, sarcomata, fibro-sarcomata, irritate the peritoneum by pressure. Should the tumor become adherent to the peritoneum, this agglutination will lead to ever renewed irritations, further dislocations, and impaired movements. Should an ovarian tumor lie in the depth of the pelvis, it must force upon the uterus a position which resembles that of a pathological anteversion or anteflexion; hence similar symptoms will be present. In small, still retro-uterine tumors, we also frequently find vesical tenesmus.

The tumor likewise presses backward, causing sciatica by pressure on the nerves, pelvic pains, and difficult defecation. Should the tumor extend into the abdominal cavity, the patient at first feels just as well as a gravida with a uterus of the same size. The above-described symptoms may subsequently disappear as in spontaneous reposition of the pregnant retroflexed uterus. Only a sense of pressure in the pelvis will be present, as in nearly all affections of the genital organs.

Should the size of the tumor exceed that of a pregnant uterus, the physiological functions of the organs in the abdominal cavity are disturbed. The stomach is unable to take in a proper quantity of nutriment. Digestion is retarded. Constipation exists both mechanically and consensually. The impaired appetite and assimilation, with the consumption of large quantities of nutritive material for the construction of the rapidly increasing neoplasm, cause a disproportion between receipts and expenditures—the patient becomes marasmic.

Respiratory disturbances also set in. The base of the diaphragm becomes broader, the vault is pressed upward so that the lung is directly compressed. Thereby its internal surface is diminished, the number of respirations must increase, in short, dyspnœa occurs. The dyspnœa becomes greatest in the complication with pregnancy; here the cyst is to be repeatedly tapped to avert the impending pulmonary œdema. The type of respiration becomes purely costal. The lower ribs are bent outward. This is best seen after removal of the tumor, during the operation.

The altered intra-abdominal pressure also leads to symptoms in the *circulatory organs.* The tumor may press directly on the iliac vein, thus retarding the return of blood from the lower extremities, with consequent œdema. The impeded influx of the blood from the lower extremities into the abdominal cavity, in which the pressure is increased, likewise leads to symptoms of stasis in the legs, exactly as during pregnancy. In all cases of considerable œdema of the lower extremities it will be necessary to make a careful differential diagnosis between congestive œdema and a complication with renal and cardiac diseases.

With nearly every larger abdominal tumor there are symptoms of an *irritation of the peritoneum;* thus isolated regions of the abdomen are found very sensitive or friction sounds like those of pleuritis sicca are demonstrated.

As to the symptoms characteristic of the several forms of ovarian tumors, the following is to be remarked. In *dropsy of the Graafian follicles* there is often an absence of all symptoms, as well as in *parovarian cysts,* even if large. Especially they may exist for years without any impairment of the general health. In *dermoid cysts*, however, symptoms arise through metamorphosis of the neighboring peritoneum. With firmer tumors, as, for instance, *sarcoma* and *carcinoma*, as well as with *superficial papilloma* (p. 305), the formation of ascites comes into the foreground. In all malignant and rapidly growing tumors the general health soon suffers. The mental depression imparts to the debilitated patient a sad facial expression which has been thought to be specific and is described as *facies ovarica*.

In the course of ovarian tumors a number of incidents occur which cause quite definite symptoms. We have above spoken of *torsion* (p. 313). In the most favorable case this occurrence takes place so gradually that the tumor, thought it cannot continue growing, is still nourished sufficiently to prevent its necrosis. Should in these cases other vicariating sources of nutrition be lacking, *e. g.*, by adhesions, the tumor atrophies without symptoms. On the other hand, with nutrition by adhesions, the tumor may continue to grow, although it successively detaches itself from its old source of supply. Such "non-pediculated" ovarian tumors have been repeatedly found. With sudden torsion of the pedicle, pains, collapse, symptoms of incarceration, and phenomena resembling ileus appear. Should blood be effused into the cyst, symptoms of acute anæmia occur. The unnourished cyst may also become gangrenous.

The dangerous proximity of the intestine, however, leads to suppuration and sloughing even without torsion of the pedicle, so that with the collapse high fever suddenly ensues. *If there be merely continuous great rapidity of the pulse, we must suspect purulent cyst-contents.* If the cyst-wall soften and the hemorrhage or the pus require more room, the cyst will rupture, the contents being effused into the abdominal cavity and leading to acute septic peritonitis.

Should the formerly intact tumor *rupture*, in the most favorable case a greatly increased quantity of evacuated urine will indicate that the contents have been absorbed by the peritoneum and been removed by way of the circulatory organs. In many cases, the rupture causes absolutely no symptoms. But should hemorrhage occur simultaneously, symptoms of collapse will be evinced. Schroeder believes that the absorbed cyst-contents may act toxically. At all events the symptoms are often alarming. In the direct examination, the altered shape of the abdomen would also enter into consideration.

Inasmuch as in the majority of cases but one ovary is affected and the tumor, moreover, develops from a small spot, while intact ovarian parenchyma lies near, there are no typical *menstrual anomalies*. For the same reason conception is possible.

The *course* is slow. An ovarian tumor never grows as rapidly as the pregnant uterus. In order to reach the size of the latter, the tumor usually consumes one and a half to two years.

Dropsy of the Graafian follicle reaches no considerable size. The internal pressure leads to atrophy of the secreting membrane; cysts of but three to four centimetres in diameter often remain constant.

There is a lack of observations respecting *superficial papilloma*. It terminates fatally by the development of ascites or carcinomatous degeneration.

In *papillary cystoma*, life terminates after from three to five years under the above-described marasmus. Torsion, rupture, hemorrhage, and sloughing may cause death sooner.

In all of these tumors spontaneous cure is possible. Even cystomata are arrested in their growth, when the nutritive supply is cut off by torsion. At least Virchow and Rokitansky have found remnants of tumors in the form of agglomerations of cysts with colloid contents imbedded in pseudo-membranes. Atrophy by calcification also occurs.

In *carcinoma* of the ovary, the rapid growth of the tumor with simultaneous wasting of the body is as characteristic as in all other carcinomata. Besides, ascites is always considerable. After every tapping the abdomen quickly refills, and death occurs at the farthest two years after the discovery of the tumor.

The solid tumors have a variable course, according to their character. Hardly anything definite can be said regarding the course of *fibromata*. In the case of *fibro-sarcoma* or *sarcoma*, the debilitating influence of a neoplasm is not as prominent as in carcinoma. Only if considerable ascites require repeated tapping, do the patients soon become debilitated. Otherwise these tumors with ascites may persist for a comparatively long time.

Although the *dermoids* are tumors which, owing to their nature, have a tendency to remain constant, they do at times enlarge in their course. At the beginning of puberty the increased fluxion may lead to the development of additional cyst-contents. Thereby the less resistant portions of the cyst-wall are expanded to the tenuity of paper. In connection therewith, rupture may easily occur. The contents of dermoid cysts are particularly dangerous to the cavity of peritoneum.

The course of *combination tumors* cannot be predicted. Of course, with rapid growth, the course will resemble that of adenoma and carcinoma.

In *pregnancy*, the mutual influence is variable. The tumor may grow rapidly under the influence of gravid hyperæmia, but may also remain unchanged. The encroachment leads to dyspnœa demanding palliative aid by tapping. During labor a large tumor may rupture into the peritoneal cavity without causing symptoms. Despite enormous distention of the abdomen, labor has often progressed favorably. Small, especially hard and very movable tumors not rarely get under the presenting part of the child in the fossa of Douglas. Then the os is high up in front, and the parturient canal is occluded. If reposition be impossible, labor may be so impeded as to require the Cæsarean section.

Diagnosis.

Small ovarian tumors lie at the posterior terminus of one of the oblique pelvic diameters, but soon sink down so as to occupy the centre of the fossa of Douglas. *Combined examination* must determine the absence of the ovary from its physiological locality and the quality of the discovered tumor. As the vagina is very thin posteriorly, it is not rarely possible to feel the characteristic form—the ovary with protuberances and small cysts.

In its further growth the tumor fills the fossa of Douglas and dislocates the uterus forward. Then, during the diagnosis, *all tumors of the fossa of Douglas* must be considered. Inflammatory exudations and hæmatoceles have characteristic symptoms, position, and form. But in complications between an ovarian tumor and an exudation, the differential diagnosis may cause great difficulty if the physician has not observed the case from the beginning. Thus, after the perforation of a perimetritic exudation which I had treated for weeks, I found colloid masses in the pus evacuated per anum. Repeatedly have the contents of dermoid cysts been unexpectedly evacuated, after peritonitic affections, from the vagina and even from the bladder. The diagnosis could not be made previously. It is also often impossible to recognize as such an adherent small cyst surrounded by chronic exudations. *Uterine myomata* may spring from the posterior cervical wall and likewise lie in the fossa of Douglas. However, their consistence protects them from being mistaken for small cysts, though firm ovarian tumors at times are at least as hard as myomata. In these cases, the differential diagnosis is soon cleared up by ascites speedily occurring in malignant ovarian tumors.

Carcinomata grow rapidly and often become uneven and irregular. But in case of adherence of uterine myomata, ovarian sarcomata, or carcinomata, together with tenderness on pressure, subjective pain, and many of the frequently occurring symptoms, certainty is often obtained only after prolonged observation. This is especially true also of *extra-uterine pregnancy.* Should here the ovisac not rupture early, the comparatively rapid growth and the subjective and objective signs of pregnancy will furnish certainty in from three to six weeks. In *retroflexion of the gravid uterus* mistakes are likewise possible and have happened to very experienced gynecologists! Outside of the signs of pregnancy, combined examination after evacuation of the bladder will elucidate the case.

Besides ovarian cysts, other cystic formations, such as *dropsy of the tube,* enter into the consideration. If small, we feel the characteristic cylindrical tube springing from the lateral margin of the uterus. But here, too, the differential diagnosis may be impossible, not only in slight, but also in greater distention of the tube.

Cysts of the broad ligament displace the uterus laterally, though but slightly; otherwise they behave like unilocular ovarian cysts. In all the cases mentioned, except the possibility of retroflexion of the gravid uterus, *puncture with the Pravaz syringe* is permissible. In innumerable cases of doubtful tumors, I have aspirated the fluid with the Pravaz syringe, under the most careful antisepsis, and never witnessed any bad results. As to the fluid, see below.

Should the tumor be larger and be situated in the abdomen, the examination is made *lege artis:* By *inspection* the general form is determined. An ovarian tumor increases the abdominal circumference *below* the umbilicus. Often the outlines of several cysts are distinctly projected on the external skin. In *ascites* the abdomen is lower and broader. In *pregnancy* the umbilical region is prominent, cicatrices of pregnancy and pigmentation of the linea alba are distinct. In large ovarian tumors the skin is very tense, the hairs are separated, the veins become distinct owing to the loss of flesh, even phleboliths can be seen and felt in very old tumors.

By *palpation* we frequently feel the large adjoining cysts through the thin abdominal coverings; but in obese persons any closer distinction may be impossible. Then we try to excite fluctuation. In ascites the wave impulse strikes most readily, but also in loose-walled ovarian cysts, especially the very thin-walled, unilocular ovarian cysts and in echinococcus, the fluid fluctuates rapidly and easily. Should the cyst be multilocular, or the contents firm, no fluctuation is obtained. It may be absent also in thin-fluid contents if the cyst is very tense. Greatly distended bowels (tympanites abdominis) likewise render the fluctuation indistinct.

Should the small firm tumor be floating in profuse ascites, the tumor is moved between both hands—*ballottement.* Thereby we not infrequently ascertain which end of the tumor moves most readily, that is, whether the tumor is adherent above or below. Should the relations be too indistinct on account of the ascites, palpation is performed after tapping (comp. p. 323).

Then *percussion* will show the relation of the intestines to the tumor. In an ovarian tumor, the middle of the abdomen yields dulness; but at the sides, the site of the ascending and descending colon, the sound is resonant. In ascites, the inflated intestines float, so that the intestinal tone is demonstrated *above,* and a dull sound *at the sides.* But should ascites complicate the ovarian cyst, the region of the colon also sounds dull; and, if the ascites be great, the intestines are within it and are not demonstrable close beneath the abdominal wall. But should the depressing finger crowd the ascitic fluid away, *deep percussion* will elicit the intestinal tone. It is also very important to ascertain the relation of the fluctuation to the results obtained by percussion. If the wave impulse readily strikes at the point where the intestinal sound was obtained, it indicates ascites.

Then *palpation and percussion* are performed *in various positions of the body*. By elevating the pelvis, we demonstrate whether fluid gravitates into the vault of the diaphragm, whether a tumor in the fossa of Douglas is adherent there and immovable, whether a tumor is firmly fixed in the pelvis or can be displaced upward. The knee-elbow position is also required at times, to demonstrate whether a tumor can be removed from the fossa of Douglas.

In percussion, the *lateral position* is especially important. Very loose-walled, perhaps formerly ruptured and half-empty cysts lie close to the posterior abdominal wall, as does ascites. With a colored pencil we mark the dulness on the external skin and place the patient on the side. In ascites the upper side must be free or yield the intestinal sound. In ovarian tumor this is not the case. Conclusions may be drawn also from the rapidity with which a change in the result of percussion is effected by the altered position. Ascites changes its position rapidly.

Old adhesions may make the diagnosis difficult. Thus I have seen a case in which a loose cyst could be demonstrated by percussion and palpation after the patient had walked for some time. In the morning, however, the cyst was not there, but the ascitic fluid had escaped through a small opening from a sacculated dropsy of the peritoneum into the other peritoneal cavity. An exploratory incision subsequently proved the absence of a cyst.

Mensuration is of importance in doubtful tumors. The measurement is taken from the umbilicus to the symphysis, to the xiphoid process, from one spine of the ilium to the other, and of the greatest circumference of the abdomen, the results obtained at various times are compared, and thus the growth of the tumor observed.

Auscultation is unimportant. At most the placental souffle which is heard in pregnancy and large myomata, but very rarely in ovarian tumors, would be considered. Furthermore, in doubtful tumors, we might investigate whether intestinal sounds can be heard in a tumor apparently adjoining the abdominal coverings; this would indicate hydronephrosis. Friction sounds in adhesive peritonitis have been mentioned above. Auscultation is important for ascertaining or excluding pregnancy.

Then *vaginal exploration* is performed. In the case of smaller tumors, the uterus usually lies in front, pressed against the symphysis, and is so far elevated that the finger at times can barely reach the vaginal portion. The uterus lies similarly in extra-uterine pregnany, but generally quite medially. In ovarian tumors the uterus may be displaced laterally, toward the healthy side. An extra-median position of the uterus has been especially found in parovarian cysts. Later, the uterus lies more frequently behind the tumor (comp. p. 313). Often, however, the uterus cannot be felt at all. Should the sound likewise fail to demonstrate it, it is often impossible to make the diagnosis of the position of the uterus even by the rectal examination. Small secondary cysts attached externally may simulate the uterus.

We next attempt the *artificial dislocation* of the tumor by palpation with one hand, then with both, and by the combined methods of examination, if possible by the aid of an assistant. By displacing the tumor we succeed in demonstrating that it is movable, *i. e.*, not firmly adherent. This is especially of value in those abdominal tumors which can be simultaneously felt in the abdomen and the fossa of Douglas. Should we fail to move the uterus separately, or should it follow every motion of the tumor, and should the tumor not escape from the fossa of Douglas when the abdomen is lifted, a firm adhesion exists at that point which renders the operation very difficult, even impossible to complete. We often succeed in insinuating the flat hand between the lower surface of the tumor and the crest of the pubis while an assistant vigorously draws the uterus upward. In this way the fingers of the inserted hand are at times able to feel the pedicle or the connection of the tumor with the pelvic organs. Inversely, we let the assistant draw the uterus down with forceps (Fig. 15, p. 36), twist it to and fro, and thereby try to ascertain the relations of the tumor to the uterus during these movements. During the simultaneous upward pushing of the tumor and downward traction of the uterus, the structure of the pedicle must be stretched. Should these manipulations prove absolutely negative, we have to deal with adherent or parovarian cysts. In doubtful cases it is necessary, besides, to examine from the rectum and vagina, as well as from the rectum and abdomen. Simon's rectal exploration (comp. p. 29) has been again abandoned; it was hoped that it would furnish particular information regarding the pedicle.

By the above-described methods it was intended to demonstrate *adhesions*. To be sure, now and then cord-like adhesions will be palpable, especially after artificial diminution of the tumor (tapping). Superficial adhesions cannot be diagnosticated in larger tumors almost filling the abdominal cavity. The abdominal coverings are so movable one on the other that, should the peritoneum adhere to the tumor, this fact will have but little influence on the mobility. In that case, the peritoneum, the musculature or the fasciæ, and the skin are displaced relatively to each other. As the adhesions do not contra-indicate ovariotomy, the diagnosis is of no practical importance. Only the repeatedly-mentioned adhesions in the fossa of Douglas must be recognized.

Should doubt still exist regarding the nature of the tumor, and the operation therefore not yet be decided on, *tapping* from the abdomen is resorted to. We have stated several times that puncture with the Pravaz syringe is not dangerous. But even tapping with a larger trocar has few dangers. For there are patients who have been tapped a hundred or more times before ovariotomy was done. The best instrument for this purpose is Thompson's trocar (Fig. 152), which I employ as shown, provided with a T-tube. On withdrawing the stylet *b*, the tube of the trocar is closed air-tight above in the handle *a*, so that the fluid escapes through the lateral tube *h*. The T-tube *d* is filled with carbolic solution; at *e* is a pinch-cock, and another at the upper end of the tube *e*, not shown in the cut. The whole tube is filled with disinfecting fluid. The tube *f* is likewise filled with fluid, and below is immersed in a vessel containing carbolic solution. Now the point of puncture is chosen. It is best to tap after catheterization of the bladder, in the linea alba, midway between umbilicus and symphysis. Before inserting the instrument, careful percussion and palpation are repeated, to make sure that the tumor is closely adjoining. Should all fluctuation be absent in the linea alba, while it is plainly perceptible at some other point, the fluctuating spot is chosen for the puncture. Then, *after scrupulous cleansing* — the patient being in the dorsal position — the trocar is inserted and the stylet drawn up. Should the fluid escape below, an assistant presses the end of the tube together and a clean vessel is placed underneath. The end of the rubber

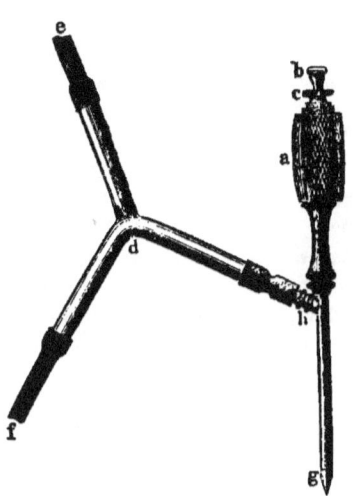

FIG. 152.—Thompson trocar, with T-tube. *a*, handle of the trocar; *b*, handle of the stylet; *c*, screw for taking the instrument apart during cleansing; *d*, T-tube of glass; *e*, upper, influx opening; *f*, lower discharge opening; long rubber tubes are attached to *e* and *f*; *g*, point of the stylet; *h*, lateral discharge opening, freed by withdrawing the stylet *b g*.

tube must be below the level of the fluid, especially toward the last, and when the escape is slow. During a deep inspiration—or with the motions in vomiting—there is often a negative pressure in the cyst, but in the method described no air is drawn in. As the examination after complete evacuation of the cyst may be very valuable for the diagnosis, all the fluid is to be withdrawn in tapping. To this end the patient is so placed, toward the last, that the trocar is in the lowest part of the abdomen. Moderate pressure is permissible, strong compression and kneading of the abdomen are to be interdicted. In ascites and solid tumors, too, the relations of position and form of the tumor, as well as its connection with neighboring organs, often become clear only after removal of the ascites.

Should nothing at all escape, small lever motions are made with the trocar, it is advanced as far as possible, and gradually drawn back. Should no drop of liquid be obtained, the tumor is either solid or its contents are viscid, not fluid. Should the fluid pass first in a full stream and then stop suddenly, the punctured, perhaps small secondary cyst is either emptied or a coagulum blocks the trocar. In these cases, *and particularly when the contents are ichorous*, water is made to flow into the cyst by the influx tube, irrigating it, like an inflamed bladder, in this manner, by alternate opening of the ends of the T-tube.

Among the *dangers of tapping*, sloughing in consequence of infection and the entrance of air, formerly stood at the head. Both may be avoided by the method described. Hemorrhage in consequence of the puncture is possible in a very vascular solid tumor, but causes no special dangers (comp. p. 314).

The fluid obtained by tapping is of great diagnostic importance.

The fluid from *dropsy of the Graafian follicle and that from parovarian cysts* is clear as water, contains sodium chloride and but little albumen, and deposits no sediment. It may be void of any microscopic elements. Very small quantities of vibratile epithelium and cylindrical cells have been found. But it may be possible, especially after preceding tapping, that chocolate-colored fluid with coagula of fibrin, *i. e.*, sanguineous fluid, escapes. Specific gravity 1,002–1,008; but when mixed with blood it may rise to 1,023.

In the *papillary* or *adenomatous cystomata*, colloid material, thick fluid, and also thin fluids are evacuated. The color, if sanguineous, is red, brown to black; if purulent, yellow; if containing masses of cholesterin, opalescent. Acetic acid precipitates mucin, which is the product of epithelia changed into colloid masses. Furthermore, but not always, albumen may be demonstrated in various modifications. Boiling and the addition of acids precipitate the albumen. Great importance was formerly attached to paralbumen. Microscopically we find the characteristic cylindrical cells, at times also vibratile cells; besides, cells disintegrated or in fatty degeneration, large globular transparent elements with nucleus displaced or altogether absent. Detritus, shrunken and recent red blood-corpuscles, flakes of pigment derived from them, pus-corpuscles, and plates of cholesterin. Specific gravity 1,015–1,030.

Free *ascites* yields a thin, pale-yellow to greenish-yellow fluid. Exposed to the air, a gelatinous coagulum of fibrin forms; much albumen is precipitated on boiling. Microscopically we find white blood-corpuscles or amœboid cells. But in cases in which the ascites is due to superficial papilloma or carcinoma, or even to carcinosis of the peritoneum, cylindrical cells are also present in the sediment. In intraperitoneal hemorrhage, the fluid becomes brownish, later greenish. The fact that the reactions which were thought to be characteristic of paralbumen occurred also in hydronephrosis and ascites, robbed paralbumen of its importance. Either other substances likewise react on the methods employed for the demonstration of paralbumen, or else paralbumen is present in many different fluids; the former is probably more correct. Specific gravity 1,006–1,015.

The contents of a *hydronephrosis* are very variable. In the thin, clear fluid the well-known products of regressive metamorphosis have been demonstrated—urea, creatin, tyrosin, and crystals of leucin. But these substances may also be completely absent, namely when the kidney has ceased functionating for years. Then more purulent contents, cholesterin, even a completely inspissated whitish mass is found. If the contents are thin-fluid, the specific gravity is 1,005–1,020.

For the fluid from *cysto-fibromata* (comp. p. 229), the lemon-yellow color and spontaneous coagulability are characteristic. But neither is constant. Microscopically we find only cholesterin and white blood-corpuscles, with the occasional admixture of red blood-corpuscles. The quantity of fluid is usually small. The specific gravity is about 1,020.

The *fluid of echinococcus*, usually as clear as glass, contains no albumen, but sodium chloride, succinic acid salts, grape-sugar, inosite, and leucin. Of great importance is the microscopic examination, in which can be demonstrated, though not always, scolices, hooklets, and membranes containing chitin bodies. The examination becomes more difficult in suppuration of the hydatid cyst, as the hooklets disappear during suppuration. Specific gravity 1,008–1,010.

In *dermoid cysts* the trocar is generally not employed, owing to the slight extent and hardness of the tumor. But the diagnostic puncture is of great value in inflammation of a dermoid cyst with high fever. In that case the puncture would have to be done with the Pravaz syringe. We find cells of epidermis and crystals of cholesterin; the fluid is thin, ichorous, dirty-brownish, and smells, especially in intestinal communication, cadaverous and fecal. Lighter-colored pus is found in suppurating exudations.

The *sac of an extra-uterine pregnancy*, before term, often contains much liquor amnii; the latter gradually disappears subsequently, or is inspissated to a thick mass. The diagnosis can probably always be made from the anamnesis and combined examination. The hope of demonstrating in the amniotic fluid, obtained by the Pravaz syringe, or even in larger quantities of it, microscopic characteristic constituents (lanugo, scales of epidermis) was not realized by myself or others.

The *changes in the tumors* are diagnosticated by the symptoms described on pp. 316 and 317. Sudden peritonitic symptoms, apparent or real swelling of the tumor, shock, and symptoms of acute anæmia point to *torsion* or hemorrhage into the cyst. *Rupture of cysts* may be suspected after a fall on the abdomen, or during labor, when collapse or peritonitic sensitiveness immediately succeed. Exceedingly often, for reason above explained, colloid escapes without symptoms into the abdominal cavity, where it can at times be demonstrated by the peculiar colloid crepitation.

Suppuration of the tumor at once manifests itself by fever. Considerable increase of temperature may also exist for a long time without other symptoms, and during laparatomy we find merely one or more cysts of a multilocular tumor purulent. When the suppuration is limited, the *high* fever gradually ceases. But great frequency of the pulse, and slight rise of the temperature at night, always continue for a long time. In sloughing or rapid necrosis of the entire tumor the symptoms are very alarming. High septic fever makes the fatal issue appear imminent.

The *diagnosis of the various species of ovarian tumors* has been repeatedly touched upon. *In follicular dropsy*—slow growth, persistence of small irregular tumors forming an agglomeration of cysts, watery fluid, no influence on the general condition. *In cystoma*—rapid, unlimited growth, very uneven, in the main round, but still irregular tumor of a consistence differing in portions, colloid contents. The general health always suffers when the tumor is of the size of the pregnant uterus. Unilocular, lax cysts with fluctuation are also frequent. Here the size will furnish the differentiation from follicular dropsy; the contents, from parovarian cysts. In favor of *carcinoma* we have ascites, the surface with small protuberances, the immobility of *larger* tumors, the cachexia. Diffuse carcinoma which enlarges the ovary uniformly like *sarcoma*, cannot be distinguished from the latter, but the ascites here, too, proves the malignancy. The examination of the ascitic fluid for cellular elements is of importance. *Parovarian cysts* are unilocular, hence they show fluctuation, the uterus is often dislocated laterally, the cyst cannot be separated from the floor of the pelvis, the contents are limpid, possibly include vibratile epithelium.

The diagnosis respecting the *bilateral development of the tumor, and the question whether the tumor springs from the right or the left ovary*, is of little practical importance. Only in small tumors, especially if solid and malignant, their bilateral presence may be often demonstrated. There the demonstration is of great value, as malignant tumors arise by preference in both ovaries simultaneously. Large tumors, following the law of gravity, must always lie in the middle. But at times it is possible to demonstrate intestinal resonance to a greater extent on one side. In that case, this is the normal side. The dislocations and combined methods described on p. 321 often elucidate the diagnosis regarding the origin of the cysts.

Repeated allusions have been made as to the *differential diagnosis*. During the examination, we must bear in mind, generally, every tumor in the peritoneum, in the abdominal coverings, in the pelvic organs, and every accumulation of fluid in the peritoneal cavity. The beginner will do well to take into consideration, *quite schematically*, every possibility, and, with the assistance of the anamnesis and all diagnostic methods, not to believe even the most improbable to be impossible. Errors have nearly always arisen when, in the diagnosis, every possibility was not taken into the consideration, and quite a number of mistakes would be avoided by merely bearing in mind the likelihood of their being made.

Therefore, proceeding quite systematically in the differential diagnosis, inflammations—*parametritis* or *peritonitis*—and accumulations of blood—*hæmatocele* and *hæmatometra*—can usually be easily excluded by the anamnesis, combined examination, course, and puncture with the Pravaz syringe.

In the same way, *fecal accumulations, retention of urine, gastrectasia, tympanites and pseudo-tumors,* as well as *climacteric obesity,* ought to present little difficulty. We must be aware that the greatly distended sigmoid flexure may reach to the right side, and that in malignant new-formations ascites may occur also at and in the intestine. The evacuation of the rectum by clysters, according to Hegar, and rectal exploration will secure elucidation. Retention of urine, in consequence of incomplete evacuation of the bladder, may lead to vesical dilatation continuing for weeks. Catheterism, required for every examination, removes the possibility of a mistake. It appears hardly credible that in an intended ovariotomy the dilated stomach has once been opened. When suspecting that spasmodic contraction of the recti muscles simulates a tumor, palpation must be done in narcosis. The differential diagnosis is by no means always easy in climacteric obesity. Here, owing to the thickness of the abdominal coverings, percussion yields almost no result, and the relaxed abdomen simulates indistinct fluctuation. However, with conscientious examination, a mistake should be impossible.

We now come to *paraperitoneal tumors.* Cysts and large accumulations of pus have been found in the sheath of the rectus. The urachus forms preperitoneal cysts, and an echinococcus may grow upward anteriorly, also preperitoneally. There were observed, moreover, fascial sarcomata at the aponeurosis of the obliquus externus, fibromata, and myxomatous fibromata at the posterior sheath of the rectus. As these tumors always develop toward the outside, and cause the skin of the abdomen to project, an error is hardly possible. Lipomata may get under the external skin, especially at the umbilicus, by hernial perforation of the fasciæ; but they may also be situated on the parietal peritoneum and the omentum. Their consistence, the almost imperceptible enlargement, and the undisturbed general condition render the diagnosis clear even after a brief period of observation.

The *peritoneum, per se,* produces pre-eminently tubercles and carcinoma; in both the ascites is considerable. Tumors of the peritoneum, carcinoma, fibroma, sarcoma, and lipoma, are particularly distinguished by their great mobility and their multiple development. It is essential to demonstrate whether *a circumscribed* tumor is really present, whether the tumor is connected with the uterus, whether both ovaries are present, and whether the tumor is malignant. Until an exploratory incision is made, a probable diagnosis only can often be formed. The high rectal exploration might here become necessary.

Proceeding now to the *tumors of the other abdominal organs*, those of the uterus first come under consideration. Subperitoneal *uterine myomata* with thin pedicles have repeatedly led to mistakes. In tumors the size of the pregnant uterus or larger, of course the position can decide nothing. Thus, *cysto-myomata* have been very frequently diagnosticated as ovarian tumors, and been operated upon. In the same way, *tubal cysts* lead to mistakes (comp. p. 297). In *hepatic tumors*, the connection with the liver can be demonstrated, tumor and uterus can be moved separately. Errors have been committed only in *hydatid cysts* filling the entire abdominal cavity and adherent below. The characteristic fluid and the hooklets are landmarks of diagnostic importance. It appears remarkable, too, that smaller sections of the large tumor fluctuate very easily, while in composite cystomata such free fluctuation in single cysts is not of common occurrence. The so-called hydatid thrill is no certain symptom. Great dilatation of the *gall-bladder* by carcinoma or numerous gall-stones, complications of hepatic carcinoma and ascites, have likewise been mistaken for ovarian tumors. Tumors of the *spleen* spring from the left side above, where they are clearly adherent. Of course, the possibility of *migratory spleen* must be borne in mind. Of greater importance are the not rare *renal tumors*. A renal tumor may enlarge beyond the linea alba and be so movable as to be taken for an omental tumor. In three cases I have convinced myself of the great mobility. Hæmaturia, *which is always present, at least for a time;* the peculiar, almost soft consistence; the irregular surface, and the connection with the region of the kidney, remove all doubt. To be sure, a floating kidney may also undergo carcinomatous degeneration. I have found such a one quite centrally located. The central position of the comparatively small tumor, the presence of both ovaries, and the absence of any connection with the uterus rendered the diagnosis very difficult. The autopsy showed a large renal carcinoma in a displaced kidney. Echinococcus of the kidney, too, has been several times observed. *Hydronephrosis* is of especial importance. Here the diagnosis is formed particularly from the long standing. The demonstration that the tumor is not connected with the genital organs can be furnished both by the high rectal exploration and by examination after puncture. Simon has also felt the calyces of the kidney.

In *carcinoma of the colon*, palpable tumors arise ; if they are complicated with acites, doubts might well be entertained. But the absence of any connection between the uterus and the tumor, the generally high position, the vomiting, constipation, and cachexia will soon dispel all doubts. Indeed, a colloid carcinoma may infiltrate and infect the connective tissue quite rapidly. In that case, indistinct tumors are felt in the fossa of Douglas and on the whole abdomen which may incite to an exploratory incision.

A *pregnant uterus near term* could hardly be mistaken for an ovarian tumor, but errors have been frequently committed in the presence of complications. The pregnant uterus has been taken for a secondary cyst and been incised or punctured. Whoever bears the possibility in mind and examines carefully in every direction could scarcely err. The same is true for *extra-uterine pregnancy*. Of decisive importance, after the death of the child, is here the crepitation of the bones of the skull, and the fact that the uterus is almost invariably situated medially in front of the ovisac. Retroflexion of the gravid uterus has been discussed above.

There can be no doubt that we have by no means come to the end of our knowledge regarding abdominal tumors. Questionable cases, or those in which the interpretation is probably erroneous, are still reported from time to time. But we may hope that, with the rapidly increasing quantity of material, all tumors at all likely to occur will be soon characterized. Rather obscure are some cases of *uterine cysts*, in which there was perhaps a serous peritoneal cyst adjoining the uterus (comp. p. 284).

Treatment.

The treatment consists in the removal of the tumor—*ovariotomy*. The operation should be desisted from only when the tumor is small, does not grow, *i. e.*, is benign, and when a large malignant, firm tumor is adherent throughout the lesser pelvis. In the latter case, the patients are generally already cachectic, and besides the tumor could not be removed. Advanced age *alone*, however, never contra-indicates ovariotomy.

Although in most cases we may be able to choose our time, immediate operation will be required if there be threatening symptoms, *e. g.*, high fever. Pregnancy is no contra-indication, but preferably we should operate in the first half rather than the second. Should labor be hourly expected, tapping would have to be done for urgent dyspnœa. This is also necessary in inoperable tumors with fluid contents, whenever respiratory difficulties arise from hyperdistention of the abdomen.

For *ovariotomy* we require a number of special instruments. These are, 1st, the trocar. The best instrument is that devised by Spencer Wells (Fig. 153). The trocar is thick because the fluid is often viscid. With the handle, fastened above by a bajonet joint, a blunt tube is protruded from the sharp tube after puncture, in order to prevent injury of the cyst-wall by the sharp point. The lateral spring clamps are intended to fasten the cyst-wall to the trocar, lest the cyst glide off and its contents escape into the abdominal cavity.

After the pattern of this instrument I have had constructed a simpler and far cheaper trocar (Fig. 154). The protecting internal tube is omitted, hence we must be careful not to cause injuries. The point *a* has the same form and size as that of the trocar Fig. 153; the cyst-fluid escapes from *c*. An attachment is affixed at the side against which the thumb is pressed during the insertion. This attachment I had made hollow and provided with a rubber tube. When the cyst-fluid is ichorous, compression is made at *c* and carbolic acid injected into the cyst through *b*. Thus the cyst is irrigated with a disinfectant, so that no sanious matter can get into the abdominal cavity during the removal or laceration of the cyst.

2. We require convenient artery forceps, which are now constructed in the shape of scissors. Fig. 155 represents Spencer Wells's forceps which I use in an enlarged, stronger form, provided with rather broader blades. Fig. 156 illustrates the more delicate artery forceps of Köberlé and Péan. At the point it is provided with hooks like a volsellum so as to grasp and hold firmly.

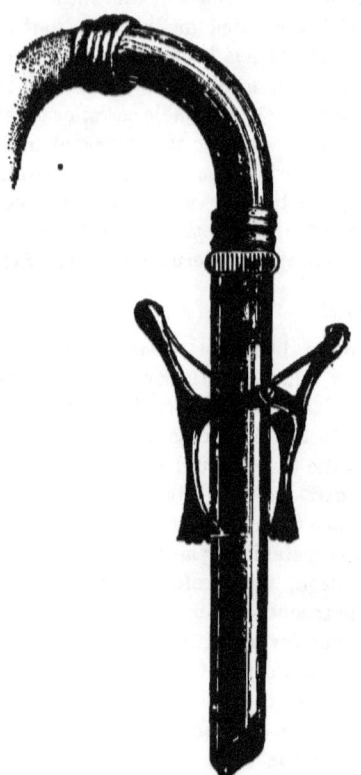

FIG. 153.—Trocar of Spencer Wells.

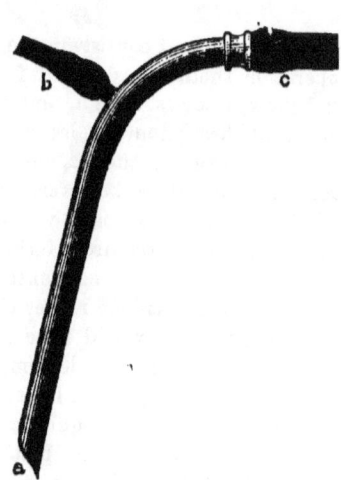

FIG 154.—Trocar for the evacuation of ovarian cysts.

3. We need large forceps for grasping and holding the cyst (Figs. 157 and 158).

Nélaton's instrument, although it holds more firmly, can only be employed for evacuated cysts. The teeth are so sharp that, in grasping more solid tumors, profusely bleeding wounds are caused. Spencer Wells's instrument thoroughly compresses the grasped portions. For instance, if a second puncture has to be made, Spencer Wells's forceps holds the first opening sufficiently together.

4. For the disposal of the pedicle, either Köberlé's wire écraseur (Fig. 143, p. 248) or Billroth's forceps (Fig. 159) are employed.

Billroth's forceps, as will be seen from the figure, may be fixed according to the thickness of the grasped tissues. The serrated blades of the forceps, curved on the flat, firmly compress the pedicle of an ovarian tumor.

Besides, there are required for an operation a number of aseptic sponges which are constantly preserved in five-per-cent carbolic acid solution, and most scrupulously cleansed of blood and tenacious colloid masses after every operation. A sponge dropping on the floor during the

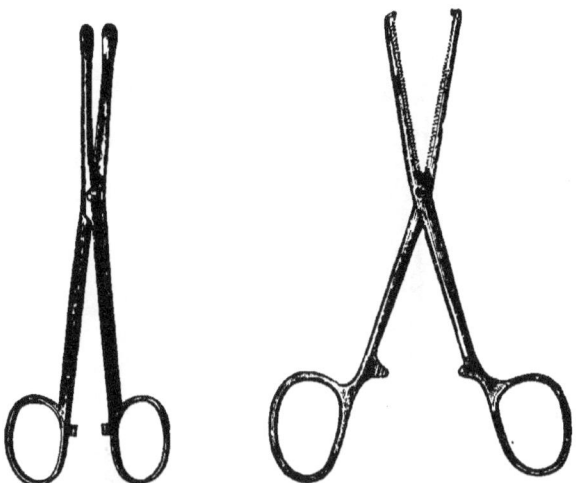

FIG. 155.—Artery forceps after Spencer Wells.   FIG. 156.—Artery forceps after Köberlé and Péan.

operation is no longer used. A number of sponges are separately kept for cleansing the peritoneal cavity after the operation is completed. The sponges must always be counted before and after the operation, lest a sponge may inadvertently be left in the abdominal cavity. A like precaution is necessary with the instruments.

Principles applicable to every ovariotomy are: most scrupulous asepsis, the least possible number of instruments and assistants. The most rapid and gentle operating possible. Never put your hand needlessly and aimlessly into the abdominal cavity; the pedicle should always be dropped.

After all antiseptic preparations (comp. 53), the *abdominal incision* is made. In order not to be hindered therein by subsequent hemorrhage, we penetrate between the recti. In nulliparæ, the median line should be looked for at the upper end of the incision; in multiparæ, it is more easily found. Above the peritoneum we first encounter a layer of adipose tissue which is dissected off to a slight extent with forceps and knife or Cooper's scissors. The peritoneum having been found, it is lifted with volsella and opened. Ascitic fluid now frequently escapes. With or without the hollow sound the incision is then enlarged sufficiently to allow of the introduction of the hand. The whitish ovarian cyst presents itself. In the upper angle of the wound the trocar is vigorously inserted into the cyst; a place having been chosen by inspection which is

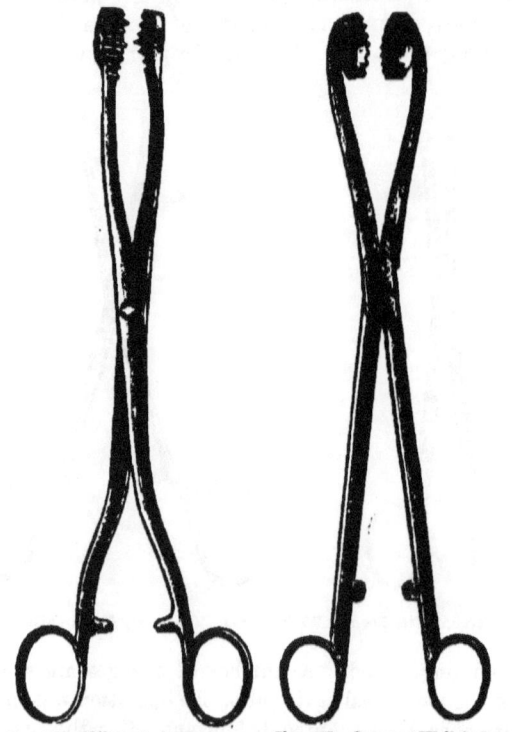

Fig. 157.—Nélaton's forceps   Fig. 158.—Spencer Wells's forceps
for the withdrawal of ovarian cysts.

free from large vessels, and which has shown distinct fluctuation on palpation. Even while the cyst is being emptied, it is grasped with forceps (Figs. 157 and 158) and drawn upward. The hand aids herein. Caution is necessary, lest perhaps the large tumor pop out, drop down, and drag at its point of attachment. While now an assistant holds the tumor in such a way as to prevent the entrance of fluid into the abdominal cavity, the operator feels for the pedicle and unfolds it, while the lifting traction on the cyst is remitted. If the pedicle be thin, a double, strong silk thread is passed through a non-vascular portion of its centre by means of a blunt needle, a pincette, or an ordinary needle. The thread is divided, and made to constrict on both sides the halves of the band-like pedicle. Then the tumor is cut off with the scissors or the knife, while an assistant keeps the ligated ends of the pedicle taut, so as to maintain the cut surface well under observation. With the scissors and pincette the stump is trimmed smooth and short—about one centimetre in length—and watched for a little while to see if it bleeds. The knots must be tied very firmly so that the stump can neither bleed nor secrete serum. Then the threads are cut short, the pedicle is replaced, and the omentum laid over the intestines.

Should the other ovary be likewise changed into a tumor, it is removed like the first. But only when the tumor is distinctly felt from without by palpation of the abdomen, it is necessary to enter the abdominal cavity, palpate it, and perform ovariotomy. Small enlargements, which could be diagnosticated only by direct palpation of the tumor, should be left unheeded. *Any superfluous penetration and manipulation in the abdominal cavity render the prognosis worse.*

The abdominal wound is closed with button sutures. It is immaterial whether the margin of the peritoneum is included in them or not. The wound must be perfectly coapted; should it gape here and there, superficial sutures are inserted. The suture is covered with borated lint, carbolated cotton, or other material. A typical Lister dressing merely represents a compressive dressing. Should an infection have taken place, an aseptic abdominal dressing is of no use, and if no infection has occurred, it will be impossible subsequently, through the well-closed, very rapidly agglutinating abdominal wound. Already after three or four days, every dressing, even if very tight at first, is loosened through the emaciation and emptiness of the bowels, so as to be valueless as an antiseptic dressing. However, compression of the abdomen is very important in all cases of preceding great distention; nor can it be immaterial whether the blood flows unhindered into the pelvic organs free from pressure, or whether a compression of the abdomen renders it artificially anemic. Pressure, here as everywhere, is the most effective

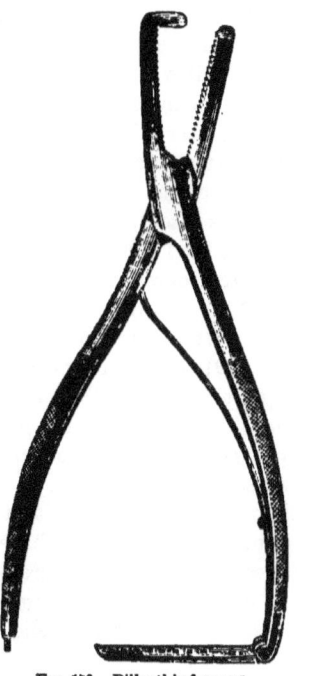

Fig. 159.—Billroth's forceps.

antiphlogistic. Cold is also particularly important in the form of ice bladders.

Schroeder opens the tumor with the knife, inserts a pair of Nélaton's forceps (Fig. 157) into the incision, and with them draws the opening outward. For the protection of the intestines a cloth is laid under the tumor and is removed after the sutures are inserted into the abdominal wound, before they are tied.

As to the *disposal of the pedicle*, the opinion of operators differs. For a time, the pedicle was largely fastened in the abdominal wound, that is, extraperitoneally. It was hoped thereby to remove every wound from the peritoneal cavity. Still a good deal can be said against this method. By the traction, on the one hand, the uterus is drawn upward; on the other, the abdominal wound is drawn in so deeply that the pedicle is often barely visible and cannot be manipulated or disinfected. Tetanus, flexions of the intestines, subsequent abdominal herniæ and gangrene of the pedicle, later infection, detachment and gliding back into the abdominal cavity have been observed. Therefore, of late, dropping of the pedicle is more in use. If the pedicle is very fat or œdematous, or if the pedicle proper is absent, so that the folded lower end of the cyst represents the pedicle, the ligatures must be especially firm. With this object, Olshausen often compressed the pedicle with Koeberlé's wire constrictor (p. 248, Fig. 143), and then laid a catgut ligature into the groove made by the wire. We succeed more rapidly with Billroth's forceps (comp. p. 333, Fig. 159); it compresses the pedicle firmly. Particularly in the case of large tumors, after the application of the forceps the tumor is cut off above the forceps and the pedicle ligated below it in as many portions as appears necessary. From the grasped portion of the pedicle the fluid is expressed by the forceps, so that, as long as the ligatures are drawn very tight, the pedicle can neither bleed nor secrete serum. Should the pedicle be quite mushroom-shaped and thick above the ligature, it will at times be possible, after careful trimming, to draw the peritoneum over the wound with a few marginal sutures. Keith seared the pedicle with the hot iron and had excellent results with this method.

Some difficulties during the operation should be known, so as to employ the proper remedies. In the first place, the tumor is often not found. It is so intimately connected with the peritoneum that the latter cannot be lifted off separately. Not rarely even an operator has detached the parietal peritoneum under the impression that he was separating adhesions. In such a case, we must go as high up as possible, cut around the umbilicus—of course on the left, on account of the round ligament of the liver on the right side—and thus seek to reach the fundus of the cyst. Should this fail, we must carefully penetrate into the depth and evacuate the cyst. Then, entering the cyst, we must try to detach it, or else, content with having opened it, endeavor to secure its atrophy under antiseptic dressing.

Very rarely are the *superficial adhesions* so extensive and firm; more frequently they can be loosened by stroking with the flat hand downward along the tumor. This loosening is commenced while the fluid is escaping, the resistance during traction with the forceps indicating the location of the adhesions. The trocar has the advantage that it is but little in the way during the loosening of the adhesions. Excessively firm adhesions, especially the cord-like ones and those to the omentum, should be sought to be rendered accessible to inspection; should vessels run along the cord or should it be very thick, it is doubly ligated before its division. Catgut may be used for this purpose, because it has little to hold here, and because too many silk ligatures irritate the peritoneum. Intestinal adhesions must be separated with great care. Should the small intestine appear adherent to the tumor, we must try to ascertain positively whether we have to deal with adhesion or whether the ovarian tumor perhaps is situated between the layers of the mesentery. In the latter case the operation is impossible, because the small intestines becomes gangrenous if detached for a great distance from the mesentery. Should the large intestine lie in front of the tumor, there may be a hydronephrosis or an echinococcus of the pelvic connective tissue. Then, in the former case, at least a partial extirpation will be attempted; in both cases, drainage effected and *after-treatment strictly according to Lister carried out.*

The *adhesions in the lesser pelvis* have been explained on p. 322. Should they be firm, it would be exceedingly dangerous to loosen them. In such cases, drainage of the fossa of Douglas has been repeatedly practised with success. A further opening of the fossa of Douglas by an incision would be more correct (Bardenheuer); for the intestines place themselves at the openings of the drainage tube and rapidly incapsulate it. The flexions of the tube and the difficulty of asepsis of the vagina render drainage scarcely commendable. It would be more correct to place a number of glass tubes, to act as levers, through the abdominal wound to the floor of the fossa of Douglas and thus to establish asepsis by irrigation.

Schroeder obtained the best results by cutting off the upper part of the cyst and sewing the rest into the abdominal wound. The sac, thus separated from the peritoneum, rapidly diminished. This procedure would also commend itself for parovarian cysts. But the latter are generally so conspicuously thin-walled that a pedicle can be formed directly from the cyst.

Should tenacious cyst-contents fail to flow off, enough is scooped or dragged out to permit the passage of the tumor through the abdominal wound. In the case of numerous small cysts this would consume too much time, and the destruction of the large number of septa causes too great a loss of blood, besides rendering the operation untidy. Multiple ligatures, likewise, serve no purpose, as the threads or wires cut the tumor. Therefore, the abdominal incision should be made as large as possible, and the tumor sought to be lifted out by hand. Rapidity is necessary, as the tumor often breaks and bleeds; but care is equally important, lest adhesions be torn or portions of the tumor left in the abdomen. The latter cannot be avoided with friable cysts.

The mode of operation in the case of *solid ovarian tumors* does not essentially differ from that described.

Should the cyst-contents be ichorous, the wall is usually very friable; then we must endeavor, after puncture, to disinfect the contents as much as possible by irrigation and rinsing of the cyst. This procedure consumes barely three or four minutes. After that the friable cyst is carefully removed.

Should an *intraligamentous* tumor have migrated beneath the pelvic connective tissue; should a malignant tumor have infected the peritoneum; should we find, therefore, a multiple carcinoma of the peritoneum; should there be a micronodular neoplasm adherent all through the pelvis; should the case be one of intestinal carcinoma, or should perhaps no tumor at all be present, the abdominal wound is again closed. Such attempts at operation are also made intentionally, as "exploratory incisions."

After every uncleanly operation, *i. e.*, when cyst-contents or blood have escaped into the abdomen, the "toilet of the peritoneal space" closes the ovariotomy. With the sponge, all adhering coagula are removed from the omentum, intestine, or parietal peritoneum. Then a sponge is firmly compressed with four fingers of one hand, and introduced into the fossa of Douglas. By relaxing the grip, the sponge absorbs the fluid. Often, when nothing can be seen above, the fossa of Douglas contains a large quantity of fluid. Should much of the cyst-contents have escaped into the abdomen, should universal adhesions have been severed which continue to bleed somewhat, the abdominal cavity is washed out. Five or six litres of warm salicylic solution (100 : 0.3) are allowed to flow into the abdomen, and are most simply discharged in the latero-abdominal position. Any kneading and pressing of the abdomen might again provoke increased hemorrhage. The fluid must finally escape clear.

When the abdomen has been very greatly distended, the quilled suture is to be preferred to the button suture, but suppuration frequently occurs in the long suture tracks. Should the intestines interfere with the suture, by inflated loops inserting themselves between the wound, a carbolized cloth or sponge is to be laid upon the intestines. The sutures are then inserted above the cloth or sponge, which is carefully removed subsequently, and the sutures tied.

After the operation, we remain expectative. Only liquid nourishment should be given until the fifth day. Should septic peritonitis ensue, the patient is lost. The attempt has been made to save the patient by opening the wound and irrigating the abdominal cavity, but ineffectually. However, should the sudden occurrence of acute anæmia indicate hemorrhage from the pedicle, an additional ligation of it would be required.

The *prognosis* is good when forty-eight hours have passed aseptically, *i. e.*, without fever or increase of the pulse. The more rapidly and the higher the temperature rises, the more unfavorable is the prognosis. In general, ten per cent of those operated on die. In fortunate cases, the patient remains abed only ten to fourteen days.

CHAPTER XVI.

HYSTERIA.

Eliology.

In the great majority of cases, hysteria is a reflex neurosis due to irritation of the genital organs (Romberg). Neither the occurrence of hysteria in children nor in old women speaks against this view. In children, it would have to be placed parallel to precocious menstruation, in old women we would be justified in assuming that through the prolonged abnormal nervous activity the functions of the peripheral nerves as well as those of the central organs have become pathological. The fact that hysteria has been found with perfectly normal genital organs has no weight nowadays. For in extirpated ovaries absolutely nothing was found that could be interpreted as pathological, although the gravest "hysterical" symptoms had completely disappeared after castration. Hysteria must be designated as a "neurosis," because the majority and the most evident of the symptoms are coupled with nervous functions, even with certain nerve-tracks.

The irritation leading to the reflex neurosis, to hysteria, is usually justly relegated to the ovary. For hysteria has been found with defect of the uterus. To be sure, in hysteria a number of symptoms are observed which are identical with certain subjective symptoms of pregnancy. In the same way, we have stated before that indications, at least, of hysteria occur almost invariably with dislocations of the uterus and with chronic perimetritis. Thereby, however, no proof is furnished that the irritated uterus is likewise to blame for the hysteria. For during pregnancy changes occur in the ovaries which are also found in the circulatory disturbances and dislocations of the uterus.

A vast number of gynecological patients do not suffer from hysteria. It is necessary, therefore, to assume either generally predisposing or special etiological factors. Among the former we count heredity. To be sure, the observations here are inexact, because, owing to the disposition to imitation of children in general, the child copies the hysterical symptoms of the mother partly intentionally, partly spontaneously. Besides, hysterical mothers do not educate their children correctly, and thereby, as it were, plant the germs of the future disease. Furthermore, the *female nature* is of etiological importance. Psyche and soma are more dependent on each other in woman than in man. Certain bodily changes peculiar to woman, depending on the procreative faculty in the most comprehensive sense, are followed by equally certain mental impressions. The pathological condition of the mind reacts on the body of the woman to a much greater extent than on that of the man. In the same way, a pathological bodily condition extends itself to the entire mental life and the disposition much more directly than in the case of male individuals.

Among the special etiological factors belongs pre-eminently physical weakness. So-called acute hysteria is observed more frequently in connection with prolonged sickness, in women who have become greatly weakened by long-continued lactation, or in such as have suffered general malnutrition in consequence of multiple labors, abortions, or hemorrhages. Altogether, an explosion of hysteria is possible with every chronic gynecological affection. Prolonged irritations of peripheral nervous regions lead to reflex neuroses. Thus we often find *enlargement and dislocation of the ovaries, perimetritic, parametritic, or idiopathic displacements of the uterus* in hysterical women and virgins. The disappearance of the symptoms after the cure of the disease frequently proves the connection directly. I have seen a case of complete cure of the gravest hysterical symptoms after menstruation had prematurely and definitely disappeared with the atrophy of an ovary the size of the fist. Freund has described a "parametritis atrophicans" as important. I have seen hysteria develop gradually with special frequency in chronic gonorrhoic perimetritis. All authors report, and every physician has observed, that *amenorrhœa, menstrual irregularities,* hence also disturbances of ovulation, are exceedingly frequent in hysteria. *Dysmenorrhœa* likewise plays an important part, be its cause what it may.

Depressing mental influences intensify, as it were, the predisposition, the nervous taint. Also erroneous education, excessive coition, masturbation, sterility, impotence of the husband, and an unhappy marriage generally, are enumerated as of etiological importance.

It is surprising that hysteria is infectious to some degree. A number of *endemics* have been described. In these the sight of certain hysterical symptoms, for instance, fits of laughter, crying, screaming, or convulsions, caused other individuals to become affected with the like hysterical symptoms.

Symptoms.

The symptoms, in the first place, concern certain nerve-functions; the perverted nervous activity again exerts its depraving influence on the whole individual. We observe, in hysteria, *hyperæsthesiæ and neuralgiæ.* For instance, on one-half of the body every tactile sensation is felt more intensely than on the other. The slightest touch of the skin causes pain. Thus such a patient may be unable to comb her hair, and always walks about with dishevelled hair because the slightest draught incites violent headache (hairache). There is a sensation as if the hair were continually bristling, and as if each particular hair were painful. This symptom occurs with special frequency in very anæmic, and also in nursing women.

In *mastodynia* the mammary gland is both spontaneously painful and especially so at every touch, even the slightest friction of the clothing. The breast has been amputated in order to terminate the unbearable condition.

Moreover, the *joints* are at times *hyperæsthetic*, particularly the knee and hip joints. With this, even a swelling of the integument is said to have been observed, so that the condition may be mistaken for an inflammatory articular affection. In a case of hysterical articular hyperæsthesia, the pains were so intense that the patient demanded amputation, and the amputation was even performed.

Parenchymatous pains designate a condition very frequent in hysteria which might most appropriately be termed *hysterical muscular rheumatism*. Here all motions of the entire body, of one or two or more extremities, are *painful*, with a co-existing rigidity of the muscles.

Hyperæsthesiæ of the organs of special sense are purely subjective sensations and difficult to demonstrate as such. Hallucinations and perversions of smell or hearing occur. An abnormally acute sense of taste, recognizing the components of every dish; an incredibly sharp ear; or an idiosyncrasy against certain colors, odors, or sounds, are repeatedly mentioned.

As to the character of the *hysterical neuralgiæ*, the pain is often of quite a temporary nature and largely subject to mental influences. With great excitement, either joyful or mournful, the most violent pains occur suddenly or disappear abruptly.

Universally known is the hysterical headache which but rarely affects the whole head. The name, *clavus hystericus*, characteristically designates the very intense pain confined to a very small space. It is situated in the occiput, in the region of the sagittal suture, unilaterally in the forehead, sometimes also in the orbits so that the eyeballs appear painful. When it has existed for years, the forehead is usually corrugated toward the root of the nose, the eyes seem to lie very deeply, are surrounded by brown rings, the orbital fat is absorbed. *Hysterical tooth- and face-ache* likewise occur.

Intercostal neuralgia is very frequent. The nucha and shoulders (omalgia, brachialgia) are also often the seat of nerve-pains. As in *spinal irritation*, certain vertebræ are very sensitive to pressure. This form of the pains is so frequently observed that some investigators have identified hysteria with spinal irritation. If the entire spine is painful it is called *rachialgia*. More frequently the sensitiveness is confined to the lowest part of the spinal column. Under the name of *coccygodynia* a peculiar neuralgia has been described which is not rarely to be referred to luxation of the coccyx by a fall or through labor. Still, cases of coccygodynia occur which develop, in the presence of other hysterical symptoms, in a manner which forces us to interpret them as a local hysterical phenomenon. In complete coccygodynia there is both spontaneous pain and sensibility to pressure. The latter manifests itself especially by such patients being unable to sit for any length of time and suffering great pain when sitting down and arising.

Sciatica and lumbar neuralgia are also observed. Under the names of *ovarie, hysteralgia,* and *hyperæsthesia of the peritoneum* there were formerly described some hyperæsthesiæ and neuralgiæ which need no longer be assumed nowadays. Chronic inflammatory conditions of the pelvic peritoneum, or at least chronic nutritive disturbances in consequence of pathological circulatory relations and displacement of the pelvic organs, will be the source of these pains. On the other hand, the introitus of the vagina is often exceedingly painful. Despite a roomy vulva and vagina the patients scream during the examination. Still there is no spasm of the sphincter, hence this hyperæsthesia must not be identified with vaginismus.

As opposed to the hyperæsthesiæ and neuralgiæ, *anæsthesiæ* and *analgesiæ* occur in hysteria. A patient, for instance, has so little sensibility in the hand that she does not know whether she holds an object or not. Even a sense of muscular consciousness has been assumed because many hysterics execute movements only when they control the movement with the eyes. When blindfolded, however, there is the consciousness of the completion of an intended movement without its being actually completed. This anæsthesia also occurs at the conjunctiva and the pharynx, so that deception must be excluded. Diminished electrical excitability was demonstrated in the muscles and nerves. Under *analgesia* we understand the strange condition in which, though the touch is perceived, pain is not felt. That in chloroform narcosis every incision can be distinctly felt and heard without perceiving the pain, I have experienced on my own person. Something similar takes place in hysteria. In this category belong the notorious cases of hysterics who endured every pain, and not only permitted, but even provoked operations without complaining in the least of pain. One hysteric introduced needles beneath her skin and gradually had three hundred and eighty-nine of them removed by operation. Both the anæsthesiæ and the analgesiæ are generally confined to certain nerve-tracks.

In the *organs of special sense* functional disturbances likewise occur; thus *amaurosis* has been repeatedly observed in hysteria, more rarely hysterical deafness. The latter was connected with intense tinnitus and clavus.

An abnormal *sensation of heat* occurs in the menopause in the shape of flying heat, in hysteria in a more intense degree. The abnormal *sensation of cold* is probably more the consequence of bodily rest and indoor life. Especially hysterics who cannot be induced to go out become more and more sensitive, and finally cover themselves in the most ludicrous manner with all sorts of warm stuffs. Spontaneous complaints are also made of *formication* over the entire skin or in the extremities. In nearly every hysteric the skin is very indolent. The patients have a surprisingly dry, but slightly flexible, lax skin.

The *stomach* and the *intestines* are often distended by gas, and in connection therewith a peculiar hysterical colic exists. The patients also have a sensation of fulness, so that they believe they have a swelling or tumor and assert that it is impossible for them to close their dresses or tie their skirts firmly.

Intractable, spasmodic, long-continued *vomiting* is likewise observed in hysteria. Withal the stomach can be quite healthy and tolerate even very heavy nutriment. Indeed, such patients often eat nothing for days, and then suddenly voraciously swallow large quantities. In the same way many hysterics drink so much that a hysterical polydipsia has been spoken of. The appetite manifests itself at unusual hours. Hardly a single hysteric eats regular meals like other people. Connected with this we find, of course, digestive disturbances, especially constipation. In other cases, although vomiting does not occur, a permanent *hiccough*, *singultus*, renders the condition at least equally oppressive. Particularly during exitement and even when merely intending to speak, every conversation is rendered impossible by the singultus. Equally disagreeable is a continual *yawning*, and the subjective sensation of a rising ball, *globus hystericus*.

At times the *urinary secretion* is altered. After hysterical spasms often an enormous quantity of urine is evacuated, while on the other hand positive retention occurs, the condition having been designated as hysterical ischuria. Here, too, there were disturbances of innervation.

In connection with the *heart* different symptoms appear, such as very intense *palpitation*. Many patients also described the condition as a *spasm of the heart*—the heart, as it were, ceases to beat, or strikes the thorax quite irregularly and with varying intensity. Here we usually have, probably, a disturbance of innervation of the vagus, for in all pronounced cases there is an ample evacuation of urine at the end of the attack. The *sphincter of the bladder* is also at times spasmodically contracted. Thus some hysterics had to be catheterized for years.

In relation to the *respiratory organs*, a spasmodic hysterical cough has been described in many ways. It ensues after an intolerable tickling in the larynx which incites the cough. The latter is toneless, hoarse, and annoys the patient excessively. In some cases the cough resembled pertussis—ever recurring paroxyms led to vomiting. Actual dyspnœa—hysterical asthma—and notably rapid respiration—hysterical apnœa—

have also been repeatedly observed. The hysterical spasm of the glottis has even been the cause of death.

In the motor sphere, *paralyses, contractures* and *convulsive conditions* occur. In the paralyses, the muscular contractility is normal, but the sensibility is reduced; the often sudden disappearance of paralyses after strong mental impressions is wonderful. There are on record the histories of many patients who have lain in bed, paralyzed, for years and suddenly recovered. The prognosis in these cases is not unfavorable. The paralysis affects arms and legs. To some extent, even contractures occur, so that the picture of hemiplegia and paraplegia is presented. Ptosis, strabismus, and facial paralysis have likewise been observed. Many patients even assert that they are completely paralyzed, have themselves fed, and lie motionless in bed, without any mental occupation, for months. Several cases are on record of *somnolence* even to apparent death and of somnambulistic conditions.

The slightest degree of *convulsions* is the peculiar hysterical restlessness very often appearing in the form of restlessness of the legs—*anxietas tibiarum*. With equal frequency we encounter the complaint of a peculiar sense of weakness and trembling in the forearms. The latter manifests itself especially by the patients being unable to perform either manual labor or to write. Whether *vaginismus* (comp. pp. 100 and 341) can be interpreted as a local hysteria is very doubtful in spite of the ever repeated assertion. Altogether, sexual relations, dyspareunia, nymphomania, or individual aversion to and anæsthesia during coition play but a subordinate part.

Fits of crying and laughing, often in rapid alternation, also fits of screaming, vociferations, belong to the picture of pronounced hysteria.

Actual convulsions, with or without consciousness, are a grave symptom of the most serious *hysteria*. These attacks, called *hystero-epilepsy,* have altogether the character of epilepsy. It is well known that in several patients it was possible to provoke these attacks by pressure upon the ovarian region.

I have observed a hysteric patient who always falls insensible from this pressure as if narcotized, remains in a state of lethargy for two or three hours after, and does not regain her speech until a much later time. Whenever a word is to be spoken, all the facial muscles begin to tremble without a word being produced. In some cases the extremities remain in a forced position after the hystero-epileptic attack, so that the patients continue for hours in distorted attitudes.

In long-standing hysteria, the abnormal nervous activity nearly always leads to mental disturbances—the reflex neurosis changes into the reflex *psychosis*. The peculiarly rapid change of disposition, the permanent mental depression of hysterics are well-known. Characteristic is the craving to be admired and to be considered an interesting case. The greatest sacrifices are made to this passion. Thus hysterics have inflicted wounds on themselves and irritated them until the condition became dangerous to life or even an amputation was necessitated. Frogs and worms, fæces and urine have been swallowed in order to appear interesting by the vomiting of these objects. With this wonderful strength of will in endurance, often a great relaxation and lack of energy in other things goes hand in hand.

A very hysteric patient nearly always presents the appearance of a confirmed invalid, even if emaciation be often absent. The consciousness of being sick, the above-mentioned etiological conditions of mental depression produce a melancholy and helplessness expressing themselves in every motion and the entire demeanor. In slighter grades of hysteria, on the contrary, the patients are often remarkably sprightly, even interesting. Indeed, nearly all modern novel heroines are described with clearly marked hysterical traits.

The *course of hysteria* is chronic, its duration unlimited. Complete cure is possible. Hysterical girls are at times thoroughly cured by marriage, maternity, and maternal cares. In the climacteric period hysteria ceases; *usually, however, very gradually*. But it may also continue to the most advanced age. Although, in some cases, paralyses, convulsions, and the graver symptoms generally, cease gradually, still nervousness, "irritable weakness," as well as slight physical and mental capacity, remain. With great excitement and renewed depressing influences, grave symptoms may again recur, even after they had ceased for years.

Diagnosis and Prognosis.

It is essential here to know all the symptoms sufficiently so as not to be led, perhaps, to seek grave organic lesions at the root of the hysterical symptoms, rather than to form a diagnosis. Thus careful observation for weeks is often required so as to exclude spinal affections in the case of rachialgia, etc. In the same way careful electrical examination is necessary in contractures and paralyses. Arthritic pains, hysterical ischuria, and polydipsia have also been mistaken for diseases with similar symptoms.

It has been already stated that many a hysterical patient, in order to appear interesting, invents the most incredible things and continually speaks untruths. We must, therefore, be careful and not believe everything. On the other hand, the very fact that such falsehoods are brought forward is pathological. It is especially important to separate the symptoms of hysteria from those of the morphine habit. Inasmuch as both often concur, and as all morphine eaters lie and are possessed of little sense of honor, such a case requires the most careful study.

The *prognosis* is bad in the grave forms. In the slighter hysterical phenomena, the prognosis depends on the possibility of detecting the etiological connection and of removing the causative agencies. In general, those cases in which a single symptom is pronounced, even if to a high degree, are prognostically more favorable than cases of slight multiple hysteria.

Treatment.

According to our representation, in especially grave cases, it would be justifiable to remove that organ to the irritated condition of which the reflex neurosis of hysteria is due; just as a reflex epilepsy has been cured by nerve-stretching or the excision of a cicatrix. Therefore, the removal of the ovaries, *castration*, would be valid as rational treatment. And indeed, in recent times, this course has been by pursued Battey, Hegar, and others. However, castration is an operation endangering life, while generally hysteria does not jeopardize it. Hence only when the condition caused by the reflex neurosis is such as to directly threaten life, or if the health is undermined and life thus indirectly endangered, would the heroic treatment by castration be justified. Therefore, castration would be indicated in cases of hystero-epilepsy, hysterical spasm of the glottis, cough, or intolerable dysmenorrhœa; moreover, perhaps, in cases in which the outbreak of a psychosis could be assumed as certain should the insufferable condition continue.

It is clear that, in proportion as it is possible to rob laparotomy of its dangers, such operations will be more permissible. Those who interpret hysteria, not as a reflex neurosis, but as a multiple nervous affection, will have to differ regarding the justification of castration.

The difficulty of castration is not to be underestimated. In parous women, that is, those with lax abdominal walls and movable pelvic organs, castration can be performed as in play. With firm, fat, tense, virginal abdominal coverings, however, with slightly movable uterus, the operation may be exceedingly difficult. Acute or subacute perimetritis contra-indicates castration. It is also to be cautioned against when so many pseudo-membranes are present that in the preceding combined examination the ovaries cannot be demonstrated distinctly, and isolated. In these cases the removal of the ovaries would be a matter of chance, and laparotomy, therefore, be unjustifiable. *However, the question of the permissibility of castration in hysteria requires the empirical proof of good results rather than a logical justification by scientific reasoning.*

In *coccygodynia*, the coccyx has been circumcised subcutaneously or completely extirpated. Should the coccyx be luxated or displaced, should it be quite isolated and not sensitive to pressure, the operation is justified. But should, as is often the case, the sensibility extend also to superior portions, the pains will return sooner or later after the extirpation. The technique of the operation is simple—a longitudinal incision is made, the small bone is grasped with forceps and torn out, firm adhesions being separated with knife and scissors. A clean, merely cutting "excision" is difficult, owing to the deep situation of the bone. The rectum, the only part endangered, is protected by the introduction of a finger and the control exercised by the latter.

There are no specific remedies for hysteria. Every ascertained affection of the genital organs will be sought to be removed. Furthermore, all the functions of the body must be regulated, *the body must be brought to normal conditions.* Should this be impossible under the domestic relations, the patient must be brought to an institution devoted to nervous diseases. Assimilation, digestion, sleep, out-door exercise, reading, work of all kinds—in short, nothing should appear unimportant to the attending physician. The patient is to be treated in a painfully attentive manner, with conscientious care, and tireless perseverance, so that the mode of life for every hour is prescribed. Should it be really possible to control the patient accurately both in psychic and somatic relations, results will be secured.

Narcotics should never be employed in hysterical patients without necessity. The largest percentage of morphine eaters is furnished by hysterics, and here again the blame attaches to many physicians who are ever ready to exorcise the evil hysterical spirit with the hypodermic syringe.

In the line of *baths*, the cold-water cures are justly in favor. The diminished activity of the skin points to the utility of baths.

INDEX.

Abdominal cavity, ascitic effusion into the, 320.
    cavity, effusion into the. from ovarian tumors, 315.
    cavity, purulent effusion into the, 271, 281, 285.
    cavity, tumors of the, 326.
    tumors, differential diagnosis of, 326.
Abortion due to retroflexion, 197.
Abscess, cold peritonitic, 265.
    extraperitoneal, 265.
    of the bladder, 108.
    of the cervix, 169.
    of the vulva, 74.
    ovarian, 301.
    parametritic, 271.
    pelvic, intraperitoneal, 282.
    perimetritic, 281.
    uterine, 153.
Abstraction of blood, 59.
Acid, nitric, as a caustic, 64.
    pyroligneous, application of, in endometritis, 166.
Acne of the vulva, 77.
Adenoid of the vaginal portion, 168.
Adenoma, cervical, 168.
    malignant, 251.
    of the ovary, 305.
    of the uterus, 262.
Adhesions due to cauterization, 65.
    of ovarian tumors, 311, 335.
    of the tubes, 297.
    of the uterus in perimetritis, 286.
Æolipile, the, 65.
After-treatment in gynecological surgery, 46.
    of ovariotomy, 336.
Ala vespertilionis, 6.
Amaurosis, hysterical, 341.
Amenorrhœa, 19.
    after hæmatoma of the ovary, 303.
    in hysteria, 303.
Amman's stem pessary, 186.
Ampulla of the oviduct, 6.
Amputation of the cervix for erosions, 151.
    of the cervix for hypertrophy of the vaginal portion, 151, 215.

Amputation of the cervix for inversion of the uterus, 224.
    of the cervix for prolapsus, 215.
    of the cervix in chronic metritis, 66, 151.
    of the vaginal portion for cystic degeneration, 176.
    of the vaginal portion with the galvano-caustic knife, 66.
    Schröder's supravaginal, 259.
Anæsthesia in hysteria, 341.
Analgesia in hysteria, 341.
Anamnesis, 23.
Anteflexion, castration for, 188.
    of the uterus, 179.
    of the uterus with prolapse, 210.
    pessary, 188.
    with col tapiroid, 180.
    with posterior fixation, 180.
Anteposition of the uterus in hæmatocele, 289.
    of the uterus in ovarian tumors, 311.
    of the uterus in perimetritis, 292.
Anteversio-flexion, the result of a myoma, 189.
Anteversion of the uterus, 188.
    the result of abortion, metritis, and perimetritis, 189.
"Anticipated climax," 248.
Antisepsis during laparotomies, 53.
    gynecological, 44.
    rules for prophylactic, 44.
Antiseptic treatment of gynecological wounds, 46.
Anus præternaturalis vestibularis, 71.
Apnœa, hysterical, 342.
Ascites, differential diagnosis of, 320.
Asthma, hysterical, 342.
Astringents for vaginal injections, 56.
Atresia ani et vulvæ completa, 71.
    ani, prognosis of, 73.
    ani vaginalis 71.
    ani vestibularis, 71.
    hymenalis, 142.

Atresia of the tubes, 297.
  uterina, 142.
  vaginalis, 142.
Auscultation in abdominal tumors, 321.

Bacilli, 67.
Bartholinitis, gonorrhoic, 76
Bartholin's gland, anatomy of, 2.
  gland, inflammation of, 80, 81, 94.
  gland, new-formations of, 84.
Baths in hysteria, 346.
  in metritis, 158, 159
  in parametritis, 274.
  in perimetritis, 288.
  saline, in perimetritis, 288
Billroth's forceps, 333.
Bischoff's colporrhaphy, 219.
Bladder, anatomy of, 8.
  carcinoma of the, 115.
  catarrh of the, 107.
  diphtheria of the, 108.
  diseases of the, 104.
  diverticulum of the, in prolapsus, 211.
  drainage of the, in fistula, 119, 126.
  ectopia of the, 104.
  fissure of the, 104.
  hemorrhage of the, after fistula operation, 126.
  inflammation of the, 106.
  inversion of the, 105.
  lesions of the, 115.
  new-formations of the, 113.
  perforation into the, in ovarian abscess, 301.
  perforation into the, in parametritis, 270.
  perforation of the, traumatic, 115.
  tuberculosis of the, 115.
  villous tumor of the, 113.
Blood, abstraction of, 59.
  abstraction of, in metritis, 153, 159.
  abstraction of, in perimetritis, 287.
  accumulation of, in atresia, 142.
  accumulation of, in hæmatocele, 289.
  accumulation of, in hæmatoma, 291.
  cysts of the ovaries, 303.
  vessels of the female generative organs, 9.
Bozeman's catheter, 48.
  speculum, 89.
Braun's syringe, 68.
  syringe, method of employing, 67.
  syringe, modifications of, 68.
Byrne's speculum, 66.
  tenaculum, 66.

Cachexia, cancerous, 254.
Carcinoma of the bladder, 115.
  of the cervix, 249.
  of the ovary, 305, 308.
  of the uterus, 249.
  of the vagina, 99.
  of the vulva, 81.
  papillomatous, 250.
  villous, 113.
Carcinosis of the peritoneum, 250.
Caruncle, urethral, 128.
Carunculæ myrtiformes, 2.
Castration for anteflexion, 188.
  for hysteria, 345.
  for myoma, 248.
Catarrh of the bladder, 107.
  of the cervix uteri, 169.
  of the internal surface of the uterus, 161.
  of the vulva, 74.
Catheter à double courant, 48.
  Bozeman's, 48.
  double current, 48.
  Schücking's, 51.
Catheters, Fritsch's, 48.
  uterine, 47, 48.
Cauliflower excrescence, 250.
Caustics, the application of, 61.
Cauterization in uterine carcinoma, 257.
  method of, 64.
  of the cervix for erosion, 175.
  of the internal surface of the uterus, 67.
Cautery, actual, the employment of, 65.
Cellulitis, pelvic, 265.
Cervix Schroeder's division of the, for the elucidation of the anatomy of prolapsus, 209.
  uteri, affections of the, 168.
  uteri, amputation of, in prolapsus, 215.
  uteri, amputation or excision of the, 66, 151, 215.
  uteri, catarrh of the, 169.
  uteri, cystic degeneration of the, 173.
  uteri, dilatation of, instrumental, 41, 43, 138.
  uteri, dilatation of, with laminaria tents, 38.
  uteri, dilatation of, with sponge tents, 39.
  uteri, erosion of the, 169, 177.
  uteri, excision of, in conical vaginal portion, 139.
  uteri, excision of, in hypertrophy, 139, 215.
  uteri, follicular erosion of the, 171.
  uteri, glandular hyperplasia of the, 173.
  uteri, hypertrophy of the, 139, 169, 180.
  uteri, incisions into the, 137.

INDEX. 349

Cervix uteri, mucous polypi of the, 172.
    uteri, myoma of the, 228.
    uteri, papillomatous erosion of the, 171.
    uteri, stenosis of the, 136.
Chadwick's gynæcological table, 34.
Clamp, the, in ovariotomy, 332.
Clavus hystericus, 340.
Climacteric, 19.
"Climax, anticipated," 248.
Clitoris, anatomy of, 2.
Coccygodynia, 340, 345.
Col tapiroide, 139.
Colic, hysterical, 342.
Colica scortorum, 284.
Colpitis, 92.
    acute universal, 94.
    emphysematosa, 95.
    gummosa, 95.
Colpohyperplasia cystica, 95.
Colpo-hysterotomy, 260.
Colpo-myomotomy, 239.
Colporrhaphy, Bischoff's, 219.
    Hegar's, 220.
    Simon's posterior, 219.
    Winckel's, 220.
Columnæ rugarum, 3.
Combined examination, 27.
    examination, position of the hands during, 27.
Conception, theories of, 21.
Condylomata, broad, of the vulva, 75, 79
    pointed, of the vulva, 81.
Conical excision of the vaginal portion, 139.
    vaginal portion, 139.
Consensual phenomena, 24.
Contents of ovarian cysts an aid in diagnosis, 324.
Convulsions, hysterical, 343.
Corpus luteum, fibroma springing from, 309.
Cruciform union after amputation of the vaginal portion, 216.
Crura clitoridis, 2.
Curette, dull, 70.
    sharp, 70.
Curetting of the uterus, 69, 167.
Cumulus proligerus s. ovigerus, 8.
Cusco's speculum, 33.
Cystitis, 106.
Cystocele, 204, 205.
Cystoma, ovarian, schematic diagram of the relations of growth of an, 306.
Cystomata of the ovary, 305.
Cysto-myoma, 229.
Cysts of Bartholin's glands, 84.
    of the broad ligaments, 295, 320.
    of the cervix uteri, 173.
    of the ovaries, 304.
    of the parovarium, 295.
    of the round ligaments, 296.
    of the vagina, 99.
    of the vulva, 84.
    tubal, 297.

Decidua, menstrual, 162.
Defectus uteri, 134.
Dermoids of the ovary, 310.
Descensus, isolated, of the vaginal wall, 204.
    of the vaginal wall with descensus of the uterus, 206.
    primary, of the uterus, with inversion of the vagina, 207.
Development, history of, of the female genitals, 130.
Diagnosis, general, 23.
Diagram of the fœtal genitals, 130.
    of the surface freshened in fistula operation, 123.
    of the suture in fistula operation, 124.
Dilatation of the bladder, 113.
    of the cervix, 38, 41.
    of the rectum, 28.
    of the uterus, forcible, 42.
    of the uterus, mechanical, 41.
    of the uterus, Schroeder's method, 42.
    of the uterus with tents, 38.
Dilator, Schultze's, 41.
Dilators, Fritsch's uterine, 41.
Diphtheria of the bladder, 108.
    of the vulva, 77.
Discharge in cauliflower growths, 252.
    in cervical catarrh, 168.
    in endometritis, 153.
    in mucous polypi, 262.
    in uterine carcinoma, 252.
    in uterine sarcoma, 261.
    in vaginitis, 96.
Dislocation of the ovaries, 301.
    of the tubes, 298.
    of the uterus, 179.
Douglas, folds of, 5.
    folds of, inflammatory shortening of, 181.
    fossa of, 5.
Drainage tube with transverse tube, 51.
Dropping tube, Schücking's, 51.
Dropsy of the Graafian follicle, 304.
    of the tube, 297.
Dysmenorrhœa, 19.
    in anteflexion, 182.
    in metritis, 156.
    in perimetritis, 279.
    in uterine myoma, 189, 231.
    membranous, 162.
Dysuria in anteflexion, 182.

Echinococcus of the kidney, 328.
    of the liver, 328.
    of the uterus, 264.
    pelvic, 295.
Ecraseur, 263.
    Kœberlé's wire, 248.
Ectopia of the bladder, 104.
Ectropium of the os uteri, 177.

Eczema of the vulva, 77.
Elephantiasis of the vulva, 82.
Elevation of the uterus, 225
Elytrorrhaphy, 213.
Emmet's operation for lacerated cervix, 178.
Endocervicitis, 168.
Endometritis, acute, 161.
        chronic, 161.
        fungosa, 162.
Endometrium, diseases of the, 161.
Endosalpingitis, 297.
Enterocele vaginalis, 206.
Enucleation of myomata, 244.
Episiorrhaphy, 218.
Epispadias, 73.
Epithelioma, 250.
Ergotin injections for myoma, 237.
Erosion of the cervix, Schroeder's method of operation for, 177.
Esthyomène, 84.
Evidement, 69.
Examination, combined, 27.
        combined, position of the hands during, 27.
        from the rectum, 28.
        with specula, 32.
        with the sound, 29.
Examinations, general remarks on, 25.
Excavation, vesico-uterine, 5.
Excision, conical, of the vaginal portion, 139.
External genitals, anatomy of, 1.
Extirpation, total, of the uterus, 258, 259.
Extrauterine pregnancy, 319.
Exudation, extraperitoneal, 265.
        intraperitoneal, anteposing the uterus, 198.
        large intraperitoneal, 282.

Fæces, importance of removal of, previous to operations, 53.
Fallopian tubes, danger of forcing fluids through, 69.
Fecundation and its consequences, 20.
Fibro-cystic tumors of the broad ligament, 295.
        tumors of the uterus, 229, 325.
Fibroma of the broad ligament, 295.
    of the ovary, 309.
    of the pelvic connective tissue, 295.
    of the round ligament, 296.
    of the tubes, 298.
    of the uterus, 226.
    of the vagina, 100.
    of the vulva, 83.
Fimbriæ of the oviduct, 6.
Fissure of the bladder, 104.
Fistula, due to carcinoma, 253.
    ileac, 103.
    intestinal 283.
    operation, after-treatment, 126.
    recto-vaginal, 102.

Fistula, recto-vesical, 127.
    uretero-vaginal, 116, 125, 126.
    urethro-vaginal, 116.
    vesical, in carcinoma, 116
    vesico-cervical, 116.
    vesico-vaginal, 115.
    vesico-vaginal, treatment of, 119.
    vesico-utero-vaginal, 116.
Follicles, Graafian, 7.
    of the vaginal portion, 172.
Follicular hypertrophy, 173.
Forceps, arterial, of Köberlé and Péan, 331.
    arterial, of Spencer Wells, 331.
    Billroth's, 333.
    Muzeux's, 36.
    Nélaton's, 332.
    Simon's, for operating on the female bladder, 114.
    spoon, 243.
    Wells's ovariotomy, 332.
Fossa navicularis, 1.
Frenulum, 2.
Freund's operation for laceration of the perinæum, 86.
    method of extirpation of the uterus, 259.
Fricke's episiorrhaphy, 218.
Fritsch's catheters, 48.
    leg-braces, 88.
    needle-holder, 124.
    operation for laceration of the perinæum, 87.
    pessary forceps, 190.
    trocar, 330.
    uterine dilators, 41.

Galvano-cautery, 66.
Gangrene of the vulva, 76.
Genitals, external, 1.
    internal, 3.
Gland, Bartholin's, anatomy of, 2.
    vulvo-vaginal or vulvo-vestibular, 2.
Globus hystericus, 342.
Graafian follicle, dropsy of the, 304.
    follicle, inflammation of the, 301.
    follicles, 7.
Growth of ovarian tumors, 311.
Gynatresiæ, operations for, 148.
    the, 130, 142.

Hæmatocele, 303.
    anteuterine, 290.
    extraperitoneal, 291.
    intraperitoneal, 289.
    pelvic, 289.
    retro-uterine, 289.
Hæmatocolpos, 143.
Hæmatoma, free, of the uterus, 264.
    of the ovary, 303.
    of the pelvic connective tissue, 291, 294.
    of the vulva, 77.

INDEX. 351

Hæmatometra, 143–145.
    differential diagnosis of, 236.
    in a rudimentary cornu, 143.
    of the tubes, 144.
    unilateral, 143.
Hæmatosalpinx, 144.
Hair in dermoid cysts of the ovary, 310.
Hegar's colporrhaphy, 220.
Hemorrhage in adenoma, 263.
    in carcinoma, 252, 254.
    in inversion of the uterus, 223.
    in myoma, 231.
    in retroflexion, 195.
    in sarcoma, 261.
    into ovarian cysts, 314.
    into the ovary, 303.
    into the tubes, 144, 297.
    of the bladder after fistula operation, 126.
    of the broad ligaments, 291.
Hermaphroditism, 73.
    bilateral, 74.
    lateral, 74.
    transverse, 74.
    unilateral, 74.
Hernia of the ovaries, 300.
    of the uterus, 225.
    of the vulva, 84.
Herpes of the vulva, 77.
Hiccough, hysterical, 342.
Hildebrandt's operation for laceration of the perinæum, 86.
Hodge pessaries, 193.
Homann, Katharina, external genitals of, 74.
Hook, double tenaculum, 36.
    sharp, 36.
Hydatid of Morgani, 299.
Hydrocele of the round ligament, 296.
Hydrometra, 143.
Hydronephrosis, 335.
    due to uterine carcinoma, 254.
    position of intestine in, 335.
Hydrops follicularis, 304.
    tubæ profluens, 297.
Hydrostatic disinfecting machine, 51.
Hymen, anatomy of, 2.
    atresia of the, 142.
Hyperæsthesia in hysteria, 339.
Hypersecretion of the uterus, 185.
Hypertrophy, follicular, 173.
    of the cervix, 209.
    of the clitoris, 82.
    of the ovaries, 301.
    of the uterus, 155.
    of the vulva, 82.
Hypoplasia of the uterus, 136.
Hypospadias, 73.
Hysteria, 338.
    castration for, 345.

Hysterical symptoms in retroflexion, 196.
Hystero-epilepsy, 343.
Hysterophores, 218.

Incarceration, symptoms of, in myoma, 230.
    symptoms of, in ovarian tumors, 316.
    symptoms of, in retroflexion of the gravid uterus, 197.
Injections, cold, for hemorrhages, 238.
    vaginal, 56, 57.
Instillation, 50.
Internal genitals, anatomy of, 3.
Inversion of the uterus, 222.
Involution, senile, 21.
Iodine, tincture of, in metritis, 164.
    tincture of, in myoma, 238.
    tincture of, in parametritis, 276.
    tincture of, in perimetritis, 287.
    tincture of, in vaginitis, 98.
Iron, tincture of sesquichloride of, for intrauterine application, 68, 167.
Irrigation, permanent, 50.
Irrigations, vaginal, 56, 57.
    vaginal, in parametritis, 275.

Knee-elbow position, 26.
Knife for incising the os uteri, 138.
    for scarification, 60.
Knives for fistula operation, Simon's, 122.
Köberlé's artery forceps, 331.
    wire écraseur, 248.

Labia majora, anatomy of, 1.
    minora, anatomy of, 1.
Laceration of the perinæum, 84.
Laminaria tents, 38, 39.
    tents, method of application, 40.
    tents, production of aseptic, 39.
Laparo-myomotomy, 239, 246.
Laparotomies, antisepsis during, 53.
Laparotomy in gynatresiæ, 149.
    in ovarian tumors, 329.
Laxatives, in general, 158.
Leeches, use of, 59, 60.
Leg-braces, Fritsch's, 88.
    for ano-dorsal position, 120.
    for elytrorrhaphy, 214.
Ligament, broad, anatomy of, 5.
    broad, cysts of the, 295.
    broad, hemorrhage into the, 290.
    broad, myoma of the, 295.
    broad, phlegmon of the, 265.
    ovarian, 7.
    round, anatomy of, 5.
    round, diseases of the, 295.
Ligamentum teres, 5.
Lipoma, abdominal, 327.

Lipoma of the omentum, 327.
of the tube, 298.
of the vulva, 84.
Liquor folliculi, 8.
Lister dressing after laparotomies, 54.
Lupus of the vulva, 84.

Machine, hydrostatic disinfecting, 51.
Malformations, diagrams illustrative of, 72.
of the vulva, 71.
Massage, 276.
Mastodynia, 340.
Mayer's rubber ring, 190.
Membrana granulosa, 8, 304.
Menopause, 19.
Menstruatio præcox, 19.
serotina, 19.
Menstruation, 15–20.
Metritis, acute, 152.
chronic, 154.
Mesosalpinx, 6.
Metamorphoses of ovarian tumors, 311.
Mons Veneris, anatomy of, 1.
Morgagni, hydatid of, 299.
Muzeux's forceps, 36.
Myoma, ergotin injections for, 237.
large cervical, 236.
large submucous, sessile on a broad base, 228.
of the anterior uterine wall, 228.
of the uterus, 226.
of the vagina, 100.
of the vulva, 83.
Myomata pediculated, 228.
uterine, as a cause of anteflexion, 182.
Myomotomy, 239.

Narcotics, to be used sparingly in hysteria, 346.
Needle-holder, 90.
Fritsch's, 124.
Nelaton's forceps, 332.
Nerves of the pelvis, 11.
of the uterus and ovary, 10.
Neuralgia in hysteria, 339.
Nitric acid as a caustic, 64.
Noma of the vulva, 77.

Obliteration, transverse, of the vagina for fistula, 125.
Oöphoritis, 301.
Operations, general considerations in, 55.
Ovarian fimbria, 6.
fluids obtained by tapping an aid in diagnosis, 334.
ligament, 7.
tumors, their relation to the neighborhood, 311.
Ovaries, accessory, 300.
diseases of the, 300.
dislocation of the, 301.
inflammation of the, 301.

Ovaries, new-formations of the, 304.
position of, 14.
supernumerary, 300.
Ovariotomy, 329.
after-treatment, 336.
treatment of pedicle in, 334.
Ovaritis, 301.
Ovary, adenoma of the, 305.
anatomy of, 7.
carcinoma of the, 305, 308.
cystomata of the, 305.
dermoids of the, 310.
fibroma of the, 309.
hæmatoma of the, 303.
hernia of the, 300.
papilloma of the, 305.
sarcoma of the, 309.
Oviduct, anatomy of, 6.
Ovula Nabothi, 172.
Ovulation, 15–20.
Ovum, anatomy of, 8.

Palpitation of the heart, hysterical, 342.
Papilloma of the ovary, 305.
of the vulva, 80.
Paquelin's thermo-cautery, 66.
Paralysis due to exudations, 270.
due to retroflexion of the gravid uterus, 196.
hysterical, 343.
Parametritis, 265.
Paravaginitis, 95.
Parovarium, anatomy of, 8.
Pedicle in ovariotomy, treatment of, 334.
of ovarian tumors, 311.
of ovarian tumors, torsion of 313.
Pelvic connective tissue, 267
connective tissue, hemorrhage into the, 291.
connective tissue, inflammation of the, 265.
connective tissue, new formations in the, 294.
organs, anatomy of, 3.
section, horizontal, schematic diagram of a, 267.
Perimetritis, 276.
Perinæum, laceration of the, 84.
laceration of, in prolapsus, 205.
laceration of, operations for, 85.
Peritoneum, anatomy of, 5, 277.
carcinosis of the, 250.
in prolapsus, 210.
Peritonitis, adhesive, 283.
pelvic, 276.
Pessaries, Hodge's, 193.
intra-uterine, in anteflexion, 186.
medicated, 59.
Schultze's figure-of-eight, 200.

Pessaries, transverse bar, 217.
Pessary, anteflexion, 188.
    forceps, Fritsch's, 190.
    Schultze's sleigh, 203.
    Thomas's, 200.
    Zwanck's, 217.
Phagedænic ulcers of the labia, 76.
Phlegmon of the broad ligament, 265.
Physiology of the female generative organs, 15-22.
Pincette, Sims' and Hegar's, for grasping folds of vaginal mucosa during excision, 214.
Plaques muqueuses of the vulva, 75.
Playfair's sound, 61.
Polypi, fibrous, 228, 232.
    mucous, 232.
    mucous, scissors for the abscission of, 178.
Polypus, fibrinous, of the uterus, 264.
    mucous, of the cervix uteri, 172.
    mucous, of the uterus, 262.
    scissors, 240.
Position for the perineal operation, 88.
    of patient for permanent irrigation, 52.
    of the Fallopian tubes, 14.
    of the ovaries, 14.
    of the patient for examination, 26.
    of the uterus, 12.
Præputium clitoridis, 2.
Priessnitz compresses in parametritis, 275.
Prolapse of the anterior vaginal wall: descensus of the uterus, 206.
    of the uterus by increased pressure from above, traction from below, or absence of the physiological supports, 208.
    primary, of the uterus after retroversion, 208.
    total, of the uterus and of both vaginal walls, 207.
Prolapsus of the uterus, 204.
    operation, Hegar's, 220.
Prophylaxis in gynecological surgery, 44.
Pruritus vulvæ, 75, 78.
Psychosis of hysteria, 343.
Pulse, necessity for watching, 50.
Pyometra, 143.

Questions for general diagnosis, 24.

Rachialgia, hysterical, 340.
Recto-vaginal fistulæ, 102.
Rectum, anatomy of, 9.
    examination from the, 28.
Remedies, the effect of, on the internal surface of the uterus, 67.
Reposition, bimanual of the uterus, 199.
Retroflexion, diagram of high degree of, 194.

Retroflexion, hysterical symptoms in, 196.
    of the uterus, 194.
    of the uterus with prolapse, 210.
Retroposition, retroversion with anteflexion, 181.
Retroversion, late puerperal, 191.
    of the uterus, 191.
Rheumatism, muscular, hysterical, 340.
Rods, uterine, 62.
Rosenmüller, organ of, 8.
Round ligament, diseases of the, 295.
Rupture of ovarian cysts, 311.
Ruptures, perineal, 84.

Salpinx, anatomy of, 6.
Sarcoma of the ovary, 309.
    of the uterus, 261.
    of the vagina, 100.
Scarification, 60.
    of the cervix in catarrh, 175.
Scarificator, 60.
Schroeder's division of the cervix for the elucidation of the anatomy of prolapsus, 209.
    extirpation of the uterus through the vagina, 260.
    method of dilatation, 42.
    method of operation for the removal of erosions, 177.
    supravaginal amputation, 259.
Schücking's catheter, 51.
    dropping tube, 51.
Schultze's dilator, 41.
    figure-of-eight pessaries, 200.
    sleigh pessary, 203.
    sound, 29.
    test tampon, 59.
Sciatica in hysteria, 341.
Scissors for the abscission of mucous polypi, 178.
    Siebold's polypus, 240.
Sea tangle tents, 38, 39.
Section, horizontal oblique, of the pelvis, 277.
Senile involution, 21.
Sensations, abnormal, in hysteria, 341.
Siebold's polypus scissors, 240.
Silver nitrate as a caustic, 63.
Simon and Hegar's operation for laceration of the perinæum, 86.
Simon's forceps for operating on the female bladder, 114.
    knives for fistula operation, 122.
    posterior colporrhaphy, 219.
    sharp spoon, 69, 70.
    speculum, 37.
Sims' and Hegar's pincette for grasping folds of vaginal mucosa during excision, 214.
    position, 34.

Sims' position, advantages and disadvantages of, 86.
speculum, 35.
Singultus, hysterical, 342.
Sitz-bath, saline, in parametritis, 274.
Sloughing in carcinoma, 252.
in fibroma, 233.
in ovarian cysts, 314.
of peritoneal exudations, cause of, 283.
Somnolence, hysterical, 343.
Sound, examination with the, 29.
Playfair's, 61.
uterine, 29.
Sounding, diagnostic points to be ascertained by, 30.
in dysmenorrhœa, 185.
of the tubes, 31.
perforation of the uterus by, 31.
Specula, cylindrical, 82.
examination with, 82.
Fergusson's, 32.
Speculum, Bozeman's, 89.
Byrne's, 66.
Cusco's, 83.
gigantic, 176.
opaque glass, 33.
Simon's, 37.
Sims's, 35.
Sphincter cunni, 3.
Spinal irritation in hysteria, 340.
Sponge tents, 38.
tents, method of application, 39.
tents, reasons for discarding, 38.
Spoon forceps, 243.
sharp, Simon's, 69, 70.
Spray, necessity of, 54.
Stem pessaries in anteflexion, 186.
pessaries, mode of inserting, 186.
Stenosis of the external os, 136.
of the external os, operation for, 137, 139.
of the internal os, 140.
Sterility in anteflexion, 182, 183.
narrowness of the os as a cause of, 183.
Suppositories, vaginal, 59.
Syphilis, 74, 129.

Tamponade, 57.
Tampon-carrier, 58.
Tapping, 322.
Temperature, necessity for watching, 50.
Tenaculum, Byrne's, 66.
hooks, single and double, 36.
Tents, dilatation of the uterus with, 38.
laminaria, 38, 39.
sea tangle, 38, 39.
sponge, 38.
tupelo, 38, 40.
Test tampon, 162.
tampon, Schultze's, 59.
Therapeutics, general, 55.

Thermo-cautery, Paquelin's, 66.
Thomas's pessary, 200.
Thompson's trocar, 323.
Toilet of the peritoneal space, 336.
Topography of the female genitals, 11.
Touch, the, examination by, 26.
Transmigration of the ova, 20.
Transverse obliteration of the vagina for fistula, 125.
Treatment, antiseptic, of gynecological wounds, 46.
of septic gynecological wounds, 47.
Trocar of Fritsch, 330.
of Spencer Wells, 330.
Thompson's, 323.
Tube, Fallopian, anatomy of, 6.
Tuberculosis of the bladder, 115.
of the vagina, 100.
of the uterus, 264.
Tubes, Fallopian, danger of forcing fluids through, 69.
Fallopian, diseases of the, 296.
Fallopian, inflammation of the, 297.
Fallopian, new-formations of the, 298.
Fallopian, position of, 14.
sounding of the, 31.
Tupelo tents, 38, 40.

Ureter, injury of, during fistula operation, 125.
Ureters, anatomy of, 9.
Urethra, anatomy of, 2, 8.
carunculæ of the, 128.
diseases of the, 127.
fissure of the, 127.
prolapse of the mucous membrane of the, 128.
syphilitic ulceration of the, 129.
varicose dilatation of the, 128.
Urethritis, gonorrhœal, 127.
traumatic, 127.
Uterine rods, 62.
Uterus, absence of, 134.
adenoma of the, 262.
anatomy of the, 4.
anteflexion of the, 179.
anteflexion of the, with prolapse, 210.
anteversion of the, 188.
arrests of development of the, 130.
at the end of the puerperium; perineal rupture, 205.
bicornis duplex separatus, 132.
bicornis duplex, vagina duplex, 131, 132.
bicornis seu arcuatus, 132.
bicornis unicollis, 132.
bicornuity of the, 132.
biforis, 132.
bilocularis duplex, 132.
bilocularity of the, 132.

Uterus, bipartitus, 134.
    catarrh of the internal surface of the, 161.
    carcinoma of the, 249.
    containing several myomata, 226.
    defect of the, 135.
    didelphys, 132.
    dilatation of the, forcible, 42.
    dilatation of the, mechanical, 41.
    dilatation of the, Schroeder's method, 42.
    dilatation of the, with tents, 38.
    dislocations of the, 179.
    echinococcus of the, 264.
    extirpation of the, 260.
    fibrinous polypus of the, 264.
    free hæmatoma of the, 264.
    hypoplasia of the, 136.
    incudiformis, 132.
    infantile, 136.
    inflammation of the, 151.
    inversion of the, 210, 222.
    malformations of the, 130.
    mucous polypus of the, 262.
    new-formations of the, 226.
    physiological mobility of the, 12.
    position of the, 12.
    prolapsus of the, 204.
    rarer displacements of the, 225.
    retroflexion of the, 194.
    retroflexion of the, with prolapse, 210.
    retroversion of the, 191.
    rudimentary solid, 134.
    sarcoma of the, 261.
    septus, 132.
    stenoses of the, 136.
    subseptus, 132.
    tuberculosis of the, 264.
    unicornis dexter, 133.
    unicornis sinister, 133.

Vagina, anatomy of, 3.
    carcinoma of the, 99.
    cysts of the, 99.
    diseases of the, 92.
    granulated, 93.
    inflammation of the, 92.
    myoma of the, 100.
    new-formations of the, 99.
    sarcoma of the, 100.
    tuberculosis of the, 100.
Vaginismus, 100, 341, 343.
Vaginitis, 92.
    adhæsiva, 93, 95.

Vaginitis, diphtheritic, 95.
    exfoliativa, 95.
    granulosa, 95.
    treatment of, 97.
Vesico-vaginal wall exposed in the speculum, 121.
Vesico-uterine excoriation, 5.
Vessels of the female generative organs, 9.
Vicarious menstruation, 19.
Vomiting, hysterical, 342.
Vulva, anatomy of the, 1.
    carcinoma of the, 81.
    cysts of the, 84.
    diphtheria of the, 77.
    diseases of the, 71.
    eczema of the, 77.
    elephantiasis of the, 82.
    gangrene of the, 76.
    herpes of the, 77.
    inflammations of, 74.
    lesions of the, 84.
    lipoma of the, 84.
    lupus of the, 84.
    malformations of the, 71.
    myoma of the, 83.
    new-formations of the, 80.
    noma of the, 77.
    papilloma of, 80.
    syphilitic affections of the, 75.
    thrombus of the, 77.
Vulvitis, 74.
    diabetic, 75.
    diagnosis of, 78.
    etiology and pathological anatomy of, 74.
    follicular, 77.
    gonorrhoic, 76.
    scrofulous, 77.
    symptoms of, 77.
    treatment of, 79.

Wedge-shaped excision, 139.
Wells's artery-forceps, 331.
    ovariotomy forceps, 332.
    trocar, 330.
Winckel's colporrhaphy, 220.
Wire écraseur, 247, 263.
Wounds, antiseptic treatment of gynecological, 46.
    treatment of septic gynecological, 46.

Xenomenia, 19.

Zwanck's pessary, 217.

www.ingramcontent.com/pod-product-compliance
Lightning Source LLC
Chambersburg PA
CBHW020235240426
43672CB00006B/533